AGRARIAN REFORM IN
CONTEMPORARY DEVELOPING COUNTRIES

Agrarian Reform in Contemporary Developing Countries

Edited by Ajit Kumar Ghose

A study prepared for the International Labour Office
within the framework of the World Employment Programme

CROOM HELM London & Canberra
ST. MARTIN'S PRESS New York

© 1983 International Labour Organisation
Croom Helm Ltd, Provident House, Burrell Row,
Beckenham, Kent BR3 1AT

British Library Cataloguing in Publication Data

Agrarian reform in contemporary developing countries
    1. Underdeveloped areas – Economic policy
    2. Underdeveloped areas – Rural conditions
    3. Rural development
    I. Ghose, A.K.
    330.9172'4 HC59.7

    ISBN 0-7099-1312-5

All rights reserved. For information, write:
St. Martin's Press, Inc., 175 Fifth Avenue, New York, NY 10010
Printed in Great Britain
First published in the United States of America in 1983

Library of Congress Cataloguing in Publication Data
Main entry under title:

Agrarian reform in contemporary developing countries.
    'A study prepared for the International Labour Office
within the framework of the World Employment Programme.'
    Includes index.
    1. Land reform – Developing countries – Case studies.
I. Ghose, Ajit Kumar, 1947-   . II. International
Labour Office. III. International Labour Organisation.
World Employment Programme.
HD1332.A37 1983   333.3'1   83-13703
ISBN 0-312-01445-7

Printed and bound in Great Britain

CONTENTS

ACKNOWLEDGEMENTS

The editor gratefully acknowledges the helpful suggestions and encouragement which he received from his colleagues — Dharam Ghai, Peter Peek and Samir Radwan — and the patience of the authors in responding to his queries and suggestions. To Sandra Deason and Lesley Brooks his warmest thanks for all the secretarial help.

PREFACE

This volume is a product of the ongoing work on rural poverty, employment and agrarian change. The earlier phase of this work focused on analysis and determinants of rural poverty and resulted in the publication of two volumes — *Poverty and Landlessness in Rural Asia* (ILO, Geneva, 1977) and *Rural Poverty and Agrarian Policies in Africa,* edited by Dharam Ghai and Samir Radwan (ILO, Geneva, 1983). Currently work is underway on analysis of rural poverty in a number of countries in Central America.

These studies led naturally to a consideration of agrarian systems as determinants of rural poverty, employment and growth. Two major results of research on this theme were *Agrarian Systems and Rural Development*, edited by D. Ghai, A.R. Khan, E. Lee and S. Radwan (Macmillan, London, 1979) and *Collective Agriculture and Rural Development in Soviet Central Asia* by A.R. Khan and D. Ghai (Macmillan, London, 1979). New areas currently under investigation include analysis of labour markets and of the change in the relative price of food with the emphasis in both cases being on rural poverty and employment.

The present volume represents continuation of the work on agrarian systems. Its main objective is to explore the political economy of agrarian reform including an assessment of the impact of reform on poverty, income distribution and growth. It seeks to achieve this objective through detailed case studies of significant experiments in agrarian reform in developing countries comprising a wide span of socio-economic conditions. Evidently, the circumstances under which reform is attempted have a critical bearing on the speed and scope of changes as well as on the nature of problems encountered in the process. Of the studies included here, the agrarian reforms in Ethiopia and Nicaragua were preceded by and associated with revolutionary changes in their political and economic systems. In the other cases studied here, even when the declared objectives were fairly radical, the reforms were attempted within the framework of the existing socio-economic systems. This fundamental difference is important in understanding the achievements and limitations of attempted reforms in different countries.

Another consideration which has a major impact on the outcome of reform is the extent to which the potential beneficiaries are organised to participate in the formulation and implementation of reforms. The active involvement of peasants and rural workers in the struggle to overthrow the Somoza regime in Nicaragua ensured a continuing role for them in agrarian reforms. In Ethiopia, the rural masses played little direct role in the overthrow of the imperial régime, though it was rural misery which ignited the spark. However, once the revolutionary government seized power, it embarked on a major campaign to associate actively the peasants, tenants and workers in the agrarian reform process, principally through the formation of peasant associations. In a different context, the government in West Bengal has encouraged the organisation of peasants, tenants and sharecroppers to accelerate the implementation of agrarian reforms. In contrast to these cases, there was relatively little effective participation of the deprived and the destitute in the reform process in other countries, although in Peru, at least, the official ideology put strong emphasis on participation and self-management. The extent and nature of participation by the rural poor in the reform process have an immediate impact on the content of the reform programme as well as on its longer-term evolution.

A related issue is the distribution of benefits from the attempted reforms. The case studies in the volume throw valuable light on this aspect. The large farmers and landlords lost out in all cases, but in the majority of cases it was the middle income groups — permanent workers in plantations and large farms in Chile and Peru, middle farmers in Chile, tenants and small farmers in Kerala and West Bengal — who were the major beneficiaries of reform. The poorest groups — the landless and agricultural workers in Kerala and West Bengal, and small peasants and seasonal workers in Peru and Chile, benefited marginally, if at all. It is only in Ethiopia and Nicaragua that the entire category of the rural poor appeared to have benefited from agrarian reforms.

In all cases, however, the immediate contribution of agrarian reform to alleviation of rural poverty has been a limited one. Rather, the key contribution of reform must lie in a removal of impediments to growth and the creation of an institutional framework and structure of incentives conducive to a broad-based and equitable process of sustained growth of rural production. Most of the reforms studied in this book have been too recent to allow for a definitive assessment in the light of the above criterion. The agrarian systems are still in a stage of transition and have not yet fully evolved into mature and established forms. Nevertheless, there are reasons to believe that the reforms have created conditions for long-term growth. Except in Nicaragua and

Ethiopia, however, this growth is likely to be accompanied by a process of differentiation of the peasantry. Problems of poverty and inequality, therefore, will remain important unless further reforms are undertaken in the near future.

Dharam Ghai
Chief
Rural Employment Policies Branch
Employment and Development Department
International Labour Office

PART ONE

INTRODUCTION

1 AGRARIAN REFORM IN DEVELOPING COUNTRIES: ISSUES OF THEORY AND PROBLEMS OF PRACTICE

Ajit Kumar Ghose

I Introduction

In poor agrarian economies, the pattern of landholding is a major correlate of political power structure, social hierarchy and economic relations. Possession of land confers on the possessor the mutually reinforcing attributes of political privilege and social prestige. The aggregate pattern of landownership, furthermore, determines the manner in which land and labour are combined for production purposes, with consequences for the quantum and distribution of the product. These, in turn, have implications for the relative and absolute material well-being of the population, particularly as food is the major product of land. Little wonder, therefore, that the 'land question' has so often inspired such passionate polemics and also been the subject of scientific controversy.

Current concern with the 'land question' derives from the accumulated experience of economic development over the last two decades. Past development efforts, contrary to expectations, have tended to produce what may be called a situation of rural crisis in the agrarian economies of the developing world. The following facts are by now well established. First, food and agricultural production *per capita* has been declining in a significant number of developing countries.[1] Second, rural poverty and malnutrition are problems of growing seriousness in large parts of the developing world.[2] Third, conditions of employment in rural areas appear to have worsened over the past two decades. In most countries, the rate of growth of employment in agriculture (resulting from a combination of expansion of acreage and increase in the intensity of cultivation) has not kept pace with that of the agricultural labour force, thereby implying a decline in employment per working person. Rural non-agricultural sectors have also tended to shrink in the face of competition from urban-industrial sectors. All this is usually reflected in declining trends in rural real wage rates and in the pattern of rural-urban migration.

Unfortunately, a careful and dispassionate analysis of the factors underlying these trends was thwarted for a considerable time by the social scientists' fascination with simple solutions. In the late 1960s and early 1970s, great concern was indeed expressed over the already observable tendency for the growth of food supply to fall behind that of population in some developing countries.[3] Famines and scarcities in some parts of the developing world appeared to confirm the worst fears. However, given the wide currency of the myth — that agricultural production in developing countries is carried out largely by undifferentiated subsistence-oriented peasants — such concern inevitably led to the idea that an acceleration in the rate of growth of food production was both essential and feasible through a reallocation of state revenues. This view helped to reinforce the seemingly strong argument for the technocratic panacea which was already being worked out in the form of the celebrated 'green revolution' strategy.

But in focusing on the crisis of food production in developing countries, this view did not offer a serious analysis of the underlying causes of this crisis. Nor did it take account of the fact that hunger and malnutrition had persisted (or increased) even in countries where *per capita* food production had shown positive growth. Instead it sought refuge in the simplistic assumption that production did not grow fast enough because the inputs necessary for such growth were not available to farmers. But the crisis could as easily be viewed as arising from demand constraints. In contemporary agrarian economies characterised by private property rights in land and large inequalities in its distribution, food production rises not in response to population growth but in response to the growth of effective demand. And it is obviously possible for *per capita* effective demand to decline while population increases.[4] Growing landlessness and near-landlessness, and declining levels of rural employment and real wages in developing countries suggest precisely such a process in rural areas.[5] Against this is, of course, the fact that the growth of industries and services in urban areas implies a growth in demand for the marketed surplus of food. Trends in the aggregate effective demand for food depend on the net effects of these two types of movements. A situation in which population is increasing, *per capita* effective demand for food in rural areas is declining and the demand for marketed surplus of food originating from urban-industrial sectors is rising, appears quite typical in developing countries. And in such a situation *per capita* effective demand for food in the economy as a whole may decline under certain quite plausible conditions.[6]

Indeed, the very growth of demand for marketed surplus of food could, under certain conditions, lead to a decline in the effective demand (in absolute and/or *per capita* terms) for food in rural areas.[7] The first effect of a rise in the demand for marketed surplus of food is a rise in its relative price. This in turn usually has the effect of depressing rural real wage rates, thus reducing the effective demand for food among rural landless labourers and small farmers. There may also be a rise in land rent. All these changes may have the effect of increasing the level and extent of indebtedness of poorer rural families, which allows the rural moneylenders to squeeze out an increased surplus by compressing the consumption of the indebted families. Thus, while the relative price of food and the share of marketed surplus in output rise, aggregate effective demand for food in the economy as a whole need not necessarily rise. Hence increased food production may not be warranted.

All this is not to argue that technical constraints on production do not exist but rather to emphasise that in general, a production (or supply) crisis has its obverse in a distribution (or demand) crisis (or, in other words, stagnant/declining food production *per capita* and the process of generation of rural poverty are inter-related phenomena). And both these crises have their roots in the existing institutional structure and its dynamic. To suppose that the production crisis can be tackled on its own is an error, since attempts to do so may actually deepen the distribution crisis. For the structural responses of the agrarian economy to the stimulus provided by the availability of modern inputs (usually at subsidised prices) often tend to be such as to further impoverish the already poor.[8] The result may be over-production[9] (i.e. production exceeding effective demand) which the state is then forced to sustain through price support schemes (for otherwise prices might slump, leading to a drastic fall in production in the next period). The state may thus be forced into a situation where it operates a subsidy programme (subsidised inputs, credit and prices) for the benefit of the richer strata of landowners. This, of course, may ensure adequate *per capita* food production in a purely statistical sense but does nothing to solve the real problems of rural poverty and unemployment.[10]

The above arguments, while representing a critique of past strategies of agrarian transformation, also serve to underline the relevance of agrarian reform for agricultural growth. When structural responses in agriculture to the stimuli provided either by price rises or by the availability of new inputs are such as to preempt the growth of aggregate effective demand for agricultural products (or, in other words, are concomitant with increasing poverty and unemployment), self-sustained growth of agricultural production is not likely. Poverty, unemployment and stagnation in production then need to be viewed as inter-related problems the solution to which can only be predicated upon a conscious restructuring of social relations of production in agriculture.

However, because land relations tend to be linked to political power relations, the process of agrarian reform is inherently a political process. In fact significant agrarian reform has tended historically to follow socio-political upheavals, and has rarely been implemented by stable governments in accordance with some preconceived plan. A recognition of this aspect has influenced the planning of the studies in this volume. The objective has not been to devise and recommend policy; but rather to understand the conditions in which specific agrarian reform programmes are formulated and implemented, the methods and instruments which are used and their consequences for growth, income distribution and poverty. This is why attention has been focused on those situations where significant agrarian reforms have been implemented in recent periods.

But the insights provided by the studies of specific instances of agrarian reform do suggest some general characteristics of the problematic of agrarian reform in contemporary developing countries. This essay will sketch these characteristics, and incorporate them into a theoretical perspective whose basic presupposition has already been discussed but may be worth restating. Problems of poverty, income distribution and growth cannot be considered in isolation from each other; they need to be viewed as different manifest characteristics of a particular economic system. Effecting substantive changes in these parameters may therefore involve a transition from one economic system to another. Agrarian reform has then to be viewed as an instrument for effecting such transition. A theoretical perspective on agrarian reform, therefore, requires characterisations of the existing agrarian system(s) as well as the alternative agrarian system(s) which the prevailing political processes may conceivably engender and which can provide suitable conditions for agricultural development.

The chapter accordingly is organised as follows. In section II, an attempt is made to isolate the essential features of the agrarian systems which characterise a large number of developing countries. Given the diversity of historical traditions, socio-political structures and economic institutions in the developing world, this is admittedly a difficult task, and generalisations usually embody some distortions. Nevertheless, relations of production in agriculture in these countries have certain basic features in common and these are what we seek to analyse in a schematic way. Certain immediate objectives of agrarian reform are then suggested. Section III attempts to identify, by abstraction from historical experience, those agrarian systems which can be regarded as providing the kind of conditions conducive to the attainment of the broad objectives outlined in the preceding section. In section IV, some important insights, thrown up by the studies, concerning the practice of agrarian reform are highlighted. An attempt is made also to evaluate the significance of the changes brought about by agrarian reforms in the rural economies of the regions studied.

II Agrarian Production Relations in Developing Countries: The Context of Agrarian Reform

Production relations in the agrarian sectors of developing countries appear bewilderingly complex at first sight. Hired-labour-based enterprises exist alongside feudalistic estates and traditional peasant farms; systems of labour hiring exist along with sharecropping relations and forms of serfdom; subsistence-oriented production exists together with market-oriented production. Traditional technology, meanwhile, displays great resilience, and production and productivity stagnate.

How does one conceptualise such a complex set of production relations? Very often, it has been described as 'semi-feudal' or 'pre-capitalist'. These, however, are essentially negative terms and their analytical content is far from clear. If the various types of production enterprises could be viewed as operating in isolation from each other, no serious conceptual difficulties would arise. But in reality they interact; both landowners and labourers often display multiple characteristics and are not easily placed in this or that standard category.

It is of course possible to classify countries according to landholding structure and certain other accompanying characteristics.[11] In some countries (mainly in Latin America and in some parts of Africa) the landholding structure is bimodal, while in others (mainly in Asia) it is unimodal. A bimodal structure tends to be associated with a high incidence of wage labour while a unimodal structure tends to be associated with a high incidence of tenancy. These broad groups could be further subdivided according to the level of technological development, the degree of commercialisation, and so on.

Such classifications are no doubt useful for certain purposes. However, they do not tell us much about the actual working of the production system, the internal laws of development and the linkages to national and international economies. Furthermore, it is not even clear that standard concepts such as landownership, wage labour, etc. can always be usefully employed in an analysis of these agrarian economies. A probing observer cannot but doubt, for example, that a wage labourer on a Latin American *hacienda* or in a South Asian village or on an African plantation shares many characteristics in common with his counterpart on a British or French farm. Similarly, landownership, in the usual sense of the term, does not exist in many parts of Africa; land rent in India is quite different from that in Britain; and so on.

In general, the use of the concepts and categories of conventional economic theory for the analysis of developing agrarian economies is difficult to justify. Empirical observations have demonstrated their inadequacy increasingly, and, at the level of practice, they seem to have inspired the wrong kind of policies. Indeed, the influence exerted by these concepts and categories on systems of date-collection and on the formulation of hypotheses has prompted a neglect of some rather obvious facts. There is a need to take a fresh look at the whole problem of constructing a conceptual framework to analyse these economies. In what follows, some preliminary formulations for such a framework will be proposed. These are based primarily on the studies included in this volume, but have drawn liberally from other studies as well.

Generally speaking, there are two clearly distinguishable agricultural classes in developing agrarian economies – one comprising those who work the land, and the other comprising those who do not work the land but receive a share in its produce. The first category includes peasants, sharecroppers, bonded labourers and various types of hired landless labourers, while the second category includes landlords, traders and usurers.[12] The first group can be further divided into two – one comprising those whom we may call independent peasants, and the other comprising marginal peasants, sharecroppers and landless labourers. Independent peasants are those who own enough land and working capital to be able to continue production and meet family consumption requirements without any need to lease in land or work for wages or borrow from moneylenders. We have, therefore, a three-fold classification of the agricultural population – the independent peasantry, the labouring class (i.e. the marginalised peasantry, share-croppers and landless labourers), and the non-labouring recipients of a share of the produce of land.

Historical tendencies suggest that the sources of dynamism (if any) of agrarian systems in developing countries must be sought in the sphere of relationships between the labouring class and the non-labouring recipients of a share of the produce of land, i.e. outside the domain of the independent peasantry. Historically, the independent peasantry has been steadily disintegrating: its members have slowly but inexorably been transformed into landlords (on rare occasions), marginal peasants, sharecroppers and landless labourers in these countries.[13] In some cases (mainly in Latin America and Asia), the process has already gone far enough to reduce the independent peasantry to an insignificant category. In others (mainly in Africa), it still constitutes an important category but displays unmistakable trends towards disintegration. On the whole, a consistent element in the general historical process has involved a steady shrinkage of the independent peasant sector.[14]

This process was, in general, set in motion by colonial intervention in these economies. The intervention took two basic forms: in some cases settlers from colonising countries were allotted land for purposes of commercial production on the basis of slave, serf, indentured or displaced peasants' labour (this was the case in much of Latin America, in the Caribbean countries and in Eastern Africa), and in others attempts were made to extract a surplus from agriculture without directly intervening in the production process (this was primarily the case in British India and West Africa). In the first situation, the peasants (sooner or later) found themselves confined to small areas with very little scope for expanding the land frontier; in the second situation, they found themselves subordinated to a newly created class of overlords — landlords, traders and moneylenders. A positive rate of population growth, coupled with periodic production crises caused by bad weather, and a growing external demand for agricultural products were then sufficient to start the process of disintegration of the peasantry.

With these brief observations, we may now attempt to characterise the relationships, as they exist today, between the labouring classes and the non-labouring recipients of a share of the produce of land. It should be noted that neither group, as defined here, is internally homogeneous. Nevertheless, there are good reasons for defining the groups in this manner. A marginal peasant may also be a sharecropper and a hired labourer just as a landlord may also be a moneylender and a trader. In general, these sub-groups are neither stable nor clearly distinguishable. What makes it worth while to distinguish two basic groups (and ignore the sub-groups) is the fact that an essential motivation can be defined quite unambiguously for each. It can be empirically observed that the first category is virtually identical with the group of rural poor. For this group, survival is the main problem and 'survival strategies' are dependent on employability and exchangeability of their labour. They receive incomes from a combination of sources such as self-cultivation of own land, sharecropping and wage labour. The members of the second group, on the other hand, seek to maximise their share of the produce of the land, which they receive in the form of profit,[15] rent, trading margin and usurious interest. It has to be remembered, however, that these categories of income often cannot be clearly differentiated. For example, the payment received by a landlord from his sharecropper often includes an element of profit, Ricardian differential rent and Marxian absolute rent. Furthermore, as noted above, a single individual may at the same time be landlord, trader and usurer. This is why the notion of a surplus − defined as the excess of net output over and above the share of direct producers (which essentially is remuneration for labour, whatever its form) − is a more appropriate concept in the context of developing agrarian societies. This is the surplus which is distributed as profit, rent, trading margin and usurious interest.

The observed production relations may then be viewed as the consequence of a simultaneous operation of two sets of mutually contradictory strategies − 'survival strategies' on the part of impoverished peasants and labourers, and 'surplus appropriation strategies' involving minimal direct intervention in production on the part of landlords-traders-moneylenders. This is the general context for the coexistence of a variety of production relations, and it provides the basis for two very general features of these agrarian systems, namely, personalised economic relations and a very low remuneration for labour. For individual members of the labouring class, the very employability and exchangeability of labour depends typically on the nature of their personal relationships with the landowning-trading-moneylending classes. A marginal peasant who is not creditworthy in the usual commercial sense but needs to borrow his working capital so as to be able to employ his own labour on his own farm can get credit only on the basis of his personal goodwill with particular moneylenders. A landless person seeking to lease land or to sell labour is competing with numerous others of his sort, and can acquire access to land and work only on the basis of his personal goodwill with particular landowners/employers. The situation is one where access to credit, land and work is more important than the terms and conditions of this access. Viewed in this manner, the differences between the diverse production relations would seem to be more apparent than real. Correspondingly, the differences between the diverse categories of surplus are also more apparent than real; surplus in all its forms is extracted essentially in the form of tributes from the direct producers rather than generated in the process of production itself. This is why the maximisation of surplus involves minimisation of remuneration for labour (i.e. the share of the direct producers in output), and not expansion of production and productivity.

It should be noted that in the above formulation, personalised economic relations arise from economic rather than socio-political compulsions. The basic sources of economic compulsions are the two inter-related phenomena: the poverty of labouring classes and a relative surplus of labour.[16] It should not be supposed, however, that personalised relationships emerged in the course and as a consequence of the emergence of a relative surplus of labour. Personalised relationships, sustained initially by state regulations and social sanctions, were in fact historically prior to a relative surplus of labour. But in the course of the historical process of disintegration of the independent peasantry, as the problem of relative surplus of labour became increasingly serious, the basis for personalised economic relations has tended to shift from the realm of the social and political to the realm of the economic. The role of state regulations and social sanctions as a result has tended to be increasingly irrelevant to their continued survival.

Given this characterisation of the non-peasant sector in developing economies, the essential features are: poverty and underemployment of labouring classes, personalised economic relations, and the extraction of tributes from direct producers as the basic mechanism of surplus appropriation. A few of the diverse implications of this formulation can be stated briefly. First, formal distinctions usually made between marginal peasants, sharecroppers and landless labourers are not very meaningful; they are all involved in qualitatively similar personal relationships with the surplus appropriators. Second, surplus appropriation in essence involves only a minimal direct intervention in the production process, and this is why the basic thrust of the system is towards lowering the remuneration for labour rather than towards stimulating productivity. Third, investment and technological change are largely absent from the scene precisely because the surplus appropriators are not directly involved in production. This does not mean that there is no accumulation but that it principally involves transfers of existing productive assets rather than creation of new ones. Fourth, it is fundamentally misleading to talk in terms of equilibrium values of such variables as the wage rate, the rate of usurious interest and the rental rate; because these are determined within a network of personalised relations, they can have multiple values even within small spatial limits.

It may perhaps be asked: if the apparently diverse production relations are indeed similar in essence (i.e. in terms of labour relations and surplus appropriation) then why should they exist at all in such diverse forms? Why do they not converge to a single form? The answer lies partly in the historicity of institutions and partly in the segmentation of the labour force. In most societies, a class of servile labourers (slaves or serfs) has existed historically. The origin of their existence can be traced to ancient tribal conquests in some cases and to colonial conquests in others. The conditions of their existence have changed over the years, but they were never able to become cultivators in their own right (as the peasants were) and have remained tied to the land to this day. They are the bonded labourers of today in the South Asian context and the permanent workers in Latin America *haciendas* and Caribbean and African plantations. Impoverished members of the peasantry, on the other hand, could not be reduced to this type of servile labour if only because impoverishment is a process and not a sudden event. The process through which peasants lose control of the means of production is slow, and on this account alone, they have to be absorbed as temporary (casual) workers or sharecroppers. The choice between these two modes of absorption probably depends upon the mechanism through which they lose control of the means of production and on the degree of their separation from the means of production. In order to be a sharecropper, the labourer needs to have at least some implements and draught animals.

The above formulations are, of course, in the nature of working hypotheses rather than established propositions. In this short essay, it is hardly possible to analyse in depth or to verify empirically all the concepts and relations suggested. But, as we shall see presently, they do fit the setting in which agrarian reforms were carried out in the cases studied in this volume. It should also be noted that it is in this setting that the 'green revolution' was promoted as an alternative to agrarian reform, rather too optimistically as it turned out. The consequences are well known, but their analytical significance remains inadequately understood. The really significant aspect of modern technologies is that they introduce the possibility of raising the productivity of labour. For this possibility to be realised in practice, however, investment and hence direct intervention in the production process are necessary. This is where modern technologies become inconsistent with existing production relations. The traditional surplus-appropriators face a difficult choice: either they continue with the old methods of surplus appropriation (which involve minimal direct intervention in the production process) so that new technologies do not prove attractive or they decide to adopt new technologies in which case personalised relationships become unnecessary, even obstructive. For in the new situation, improvement in labour productivity has to become the major concern and surplus has to be appropriated not in the form of tributes but in the form of capitalist profits. Which of these alternative choices is actually made depends on the structural specificities of an economy. In the Latin American context where large consolidated landlord farms exist, the second choice may be feasible (depending upon the buoyancy of the international markets, since most of these farms are export-oriented). In the South Asian context where petty landlordism is rampant and farms are highly fragmented, the first choice is the more likely.

An interesting situation arises in the case where an independent peasantry is still significant in terms of landholding and production. New technologies offer the independent peasants an opportunity to stall the historical process of pauperisation through the possibility of raising both land and labour productivities. They are likely to be both willing and able to adopt new technologies. In the process, they may well create expanded employment opportunities which in turn may make it possible for the already impoverished members of the peasantry to extricate themselves from the network of personalised relationships and attain some stability. It is the non-peasant sector that would then face a crisis and may even be forced to adopt new technologies. A possibility of genuinely capitalistic development may thus arise in such a context.[17]

The purpose of exploring these details is to point out that in situations where the independent peasantry has ceased to be significant, new technologies, by themselves, are unlikely to provide a solution to the problem of agricultural development. It is hardly surprising then that in many situations they are simply not adopted, or, if adopted, their full potential is not exploited. The 'green revolution' strategy was fundamentally misconceived in that it took no account of the vital interconnections between social and technical relations of production.

The profiles of the pre-reform situation drawn in the case studies included in this volume illustrate the relevance of some of the formulations presented above. The studies cover three regions from Latin America, one from Africa and three from Asia. Given the diversity in terms of ecological factors and political-administrative history, it is not surprising that these regions developed diverse institutional frameworks for agricultural production and distribution. In the Latin American regions (Chile, Nicaragua and Peru), rather impoverished peasant communities coexisted alongside *haciendas*, the dominant type of production unit. In Ethiopia, the dominant type of production unit was a peasant farm, but a class of *rentier* landlords was superimposed on the peasantry. In Iran, the dominant type of production unit was again a peasant farm, but here the state was in effect the sole landlord. In Kerala and West Bengal, the agrarian structure was more complex. Poor peasants constituted the vast majority of agriculturists and they were dependent on landlords and traders-moneylenders for access to land, work and credit. The landlords, who were quite numerous and not always large, resorted to both tenant-cultivation and self-cultivation with hired labour.

These apparent differences notwithstanding, it is possible to detect some essential similarities when one examines the nature of labour relations and the mechanisms of surplus appropriation. *Haciendas* typically used two types of labour – permanent and temporary. The permanent workers, who supplied the bulk of the labour employed on *haciendas*, were allotted small plots of land for the production of basic subsistence; they worked on the *haciendas* for minimal wages. The permanent workers were thus, in effect, paying a labour rent. The temporary workers were drawn from the ranks of impoverished peasants; they had little bargaining strength *vis-à-vis* the *hacienda*-owners and received an even lower remuneration for labour than the permanent workers. In Ethiopia and Iran the production process was controlled by the peasants, and hence the hired labour system was virtually non-existent. Here the agricultural surplus was appropriated by the ruling elite through a direct tribute system. In Kerala and West Bengal, hired labour played an important role in the process of production though peasants, a vast majority of whom were very small and economically non-viable, constituted the majority of the producers. However, with the exception of the traditional bonded labourers of Kerala, the poor peasants, tenants and labourers were neither stable nor clearly distinguishable groups. Moreover, well-developed labour markets did not exist and labour relations tended to be personalised. At the other end, the landlords, traders and moneylenders did not constitute stable or clearly distinguishable groups either. Thus, although, on the surface, a part of the agricultural surplus was appropriated in the process of production itself (in the case of farms operated with hired labour), it is at least arguable that even this concealed a form of extraction of tributes from the impoverished peasantry.[18] At any rate, there is little doubt that a major part of the surplus was extracted in the form of rent, trading margins and usurious interest. The class of independent peasantry was insignificant in all the cases.

An essential feature shared by all these sytems, therefore, was that surplus appropriation was largely divorced from the production process itself and this was related to two other common characteristics, namely the poverty of the direct producers and the virtual absence of productive investment.[19] Because the surplus appropriators were only minimally involved in the production process, their interests tended to be defined in terms of extracting the maximum from the direct producers rather than in terms of improving the technical conditions of production. If agrarian reform under these conditions was to stimulate productive investment and eradicate poverty, it had to negate these tendencies, and develop alternative modes of surplus appropriation which would have an inbuilt stimulus for productive investment.

One needs to bear in mind here the role that agricultural surplus played in the overall economy of these regions. Apart from being a principal source of food for the urban population, the agricultural surplus was often also the major source of export earnings and thus formed the basis for industrialisation. Established modes of surplus appropriation could not, therefore, be disrupted, even temporarily, without simultaneously disrupting the urban-industrial economy. On the other hand, if the established modes of surplus appropriation were left undisturbed, sustained growth was virtually impossible, since, with the stagnant or declining *per capita* agricultural production that they engendered, the volume of marketed agricultural surplus could not go on increasing indefinitely. Maintenance of the *status quo*, therefore, tended to preclude the possibility of growth and this perhaps explains the socio-political turmoil which accompanied agrarian reforms.

III Agrarian Transition in Developing Countries and the Role of Agrarian Reform

Historically, agricultural development has been associated with the emergence of three different agrarian systems — a capitalist system, a modernised peasant economy system, and a collective system. The capitalist system emerged in Western Europe and in North America. The modernised peasant economy system developed in Japan and, more recently, in South Korea. The collective system emerged mainly in the socialist countries, although some non-socialist developing countries have tried to experiment with it.

It is perhaps useful at this point to define the basic characteristics of the three agrarian systems. Although this may appear to be stating the obvious, the relevant terms have actually been widely misused and are frequently misunderstood, particularly because they carry ideological overtones in their non-rigorous commonplace usage. In a capitalist system, workers are separated from the means of production which are owned, controlled and utilised by a group of non-worker individuals who hire the workers on wages for purposes of production. The responsibility for the organisation of production, accumulation and investment thus rests with the non-worker-possessors of the means of production; their activities and motivations, consequently, are central to the working of the system. In a modernised peasant economy system, ownership and control of means of production are dispersed among basically family-consumption-oriented peasants who employ family labour for production purposes. Aggregate production is, therefore, dependent on decisions made by a large number of family farmers while the responsibility for accumulation usually rests with the state. In a collective system, the means of production are owned and controlled either by the state or jointly by a group of workers. Correspondingly, the responsibility for organising production and deciding on accumulation rests with either the state or a committee of workers' representatives.

Each of these systems is thus associated with a distinct pattern of ownership and control of the means of production, and this implies a specific mode of decision-making with regard to the use of resources (including labour), accumulation and investment. In other words, *ceteris paribus*, each of these systems is generally associated with a different profile related to the time-path of output and employment. Moreover, from the structural differences arise fundamental differences in terms of surplus appropriation and remuneration of labour. In a capitalist system, separation of workers from the means of production emphasises exchange between labour and other commodities, and both surplus appropriation and remuneration of labour depends on the nature of these exchange relations.[20] In a modernised peasant economy system, surplus appropriation generally involves extraction in some form by the state from the peasants and, consequently, remuneration of labour becomes a function of state policy. In a collective system, both the surplus and the remuneration of labour are in principle the outcome of decisions made by the workers themselves (though in practice such decisions are often made by agents of the state). Each of these systems is thus associated with a distinct pattern and trend in income distribution, both because productive assets are held very differently and because the mechanisms of determining the remuneration of labour are different. The details of this proposition need not detain us here, but it is perhaps pertinent to point out that historical experience has shown that income distribution tends to be very unequal in a capitalist system and relatively egalitarian in the other two systems.[21]

One can suggest, on the basis of an abstraction from historical experience, that the existing agrarian system in developing countries (which falls into no clear category) has to evolve into one of these systems if the problems of agricultural development are to be effectively resolved. Agrarian reform is an instrument available for achieving this transition. The actual choice between the alternative systems can of course only be made by the political processes in particular countries. At a theoretical level, the question that can be asked is: what are the general characteristics of the relevant transition processes?

That transitions must inevitably be processes stretched over time is not difficult to see. The task of agrarian reform can thus be viewed as the creation of an appropriate set of conditions for an eventual emergence of a particular system. In other words, agrarian reform, by itself, cannot immediately develop any particular system. The need for agrarian reform arises because the existing system, if left to itself, is likely to perpetuate itself indefinitely. And this means a perpetuation of rural poverty and agricultural stagnation.

Given this premiss, the following general proposition suggests itself: a revitalisation of the peasantry is a necessary first step towards evolving a new system irrespective of the nature of the new system: whether it is a capitalist or a modernised peasant economy or a collective system. In other words, the immediate task of agrarian reform is to re-establish the peasantry as the principal actor on the rural scene. This proposition obviously presumes that the peasantry has suffered serious disintegration, and this we argued earlier. It also implies that no direct transition (with or without state intervention) from the existing system to either a capitalistic or a collective system is feasible, and this requires a little elaboration. The point was made earlier that in a situation where neither plantations (or *haciendas*) nor independent peasants are significant, a capitalistic system is unlikely to develop. In principle, the possibility that a *hacienda* or a plantation may evolve into a capitalistic enterprise cannot be ruled out. Examples of a similar evolution can certainly be found in history.[22] However, conditions under which such a transition can take place are not easily created, in the short run at any rate. When, in addition, one recognises that such enterprises do not exist in large parts of the developing world, and that the existence of a few capitalistic enterprises does not imply a development of capitalism in the agrarian economy as a whole, the general proposition that a direct transition from the existing system to a capitalistic system is unlikely seems plausible enough.

As regards the case of a direct transition from the existing system to a collective system, this is difficult to visualise in a situation where large production units do not already exist. For, while it may be relatively easy to abolish private property rights in land, an integration of the existing work-force (an impoverished and differentiated peasantry) into a collective work-system is a much more difficult task. The fact that the development of a peasant agriculture was an intermediate step towards the development of a collective agriculture in a number of socialist countries was not merely accidental.[23] Once again, however, the possibility of transforming the existing large production units into collective farms cannot be ruled out, and indeed such transformations have been attempted in a number of Latin American countries. But the fate of these 'transformed' *haciendas* does not inspire much confidence; indeed they rather emphasise the point that a few collective farms in an otherwise private enterprise economy are unlikely to function properly.[24] And, once again, the existence of a few collective farms does not imply the development of a collective system in the agrarian economy as a whole.

If it is accepted that, in most situations, regeneration of the peasantry is the immediate task of agrarian reform, then attention has to be focused on two sets of questions:

(1) What are the mechanisms and immediate consequences of a regeneration of the peasantry?
(2) What are the basic follow-up requirements of a transition to either a modernised peasant economy or a capitalistic agriculture or a collective system?

Answers to the first set of questions can only be formulated with reference to specific situations. In view of our earlier discussion, however, it can be said that land transfers from the non-cultivating owners (landlords) to the actual tillers of the soil (and this usually implies an abolition of both wage-labour and tenancy) and, where relevant, an abolition of the stranglehold of the trading-moneylending classes over the peasantry (and this may involve a development of co-operative or state-

controlled marketing and credit institutions) must constitute the fundamental objectives. One immediate consequence of these measures, however, may be a decline in the volume of the marketed surplus available
to urban-industrial sectors, particularly in view of the prevailing poverty
of the direct producers. As has been argued, the poverty of the direct
producers is a consequence of the methods of surplus appropriation by
the landlords-traders-moneylenders (who control the marketed surplus).
The most important likely effect of abolishing these methods will be
(and no doubt should be) an increase in the consumption levels of the
direct producers and hence a decline in the marketed surplus (except in
the unlikely case of a dramatic rise in production). Such a decline may
involve either a shift in cropping pattern (from cash crops to food
crops) or a decline in the proportion of food output marketed or both.
In either case, considerable difficulty may have to be faced in the sustenance of the non-agricultural sectors of the economy. The problem
may be further complicated by disruptions in marketing and distribution networks.

Answers to the second set of questions have to be based on an understanding of the logic of peasant production. This has been explored in
terms of a schematic model, by A. V. Chayanov.[25] It will suffice here to
state some relevant conclusions which can be deduced easily from
Chayanov's model (though Chayanov did not deduce all of them). A
basic characteristic of a peasant economy system is that it is inherently
unstable in the sense that, left to itself, it tends to generate tendencies
which undermine its own basis. Except in situations where the land
frontier can be easily expanded, peasants tend to become socially differentiated over time, and, in modern conditions (i.e. when a capitalistic
industrial sector exists), the pace of this differentiation can be quite
rapid. This very characteristic, however, makes the transition to a capitalist system a relatively simple task. State policies are required to concentrate on two basic areas: industrialisation and the promotion of
modern technologies in agriculture. The first is about the expansion of
the market for agricultural products, and the second is about expansion
of production and productivity in agriculture as well as a quickening of
the pace of differentiation among the peasantry. The transition process
would involve a growing polarisation of the peasantry into capitalist
farmers and wage labourers, a growing concentration of landholdings
and other productive assets, a growing market-orientation of production, and a growing investment-intensity of production.

It is extremely doubtful whether a transition to a collective system
in agriculture is at all feasible so long as the other sectors of the

economy remain areas of private enterprise. The few experiments of this type seem to have run into considerable problems. The cases of Tanzania, Algeria and Mexico provide some relevant illustrations. Nevertheless, one can pose the question: what are the characteristics of such a transition process, assuming that the overall environment is favourable to such a transition? The experiences of China and Vietnam provide the basis for answering the question.[26] In broad terms, it is necessary to discourage private investment (which promotes social differentiation among the peasantry), to focus attention on the poorer strata of the peasantry, and to promote investment within a cooperative framework. This strategy thus calls for a high degree of involvement of the state at both political and economic levels.

The problems of modernising a peasant agriculture (while retaining its peasant character) arise from the contradictory requirements of ensuring a net inflow of resources into agriculture on the one hand (a peasant economy does not have an inbuilt mechanism for investment,[27] and if it acquires one then it is unlikely to remain a peasant economy for long) and of promoting industrialisation on the other. In the initial stages of development, industrialisation requires an outflow of resources from agriculture; modernisation of a peasant economy, on the other hand, requires a flow of resources into agriculture. In South Korea, the contradiction was resolved through reliance on a massive inflow of resources from external sources.[28]

These brief observations are intended to provide some broad criteria for evaluating the concrete achievements (and failures) of specific agrarian reform programmes. It is clear that an agrarian reform programme needs to have both short- and long-term objectives, and that the same short-term objectives can be consistent with a number of different long-term objectives. For short-term objectives concern a revitalisation of the peasantry, and a peasant economy system constitutes the point of departure for the eventual development of any of the three systems outlined earlier.

To this extent, one has to distinguish between the short- and long-term consequences of specific agrarian reform measures. These can be quite different. For example, agrarian reform may transform the existing system into a peasant economy and the latter may then evolve into a capitalist economy. In this case, although the immediate consequence of agrarian reform would be a reduction in inequality, as the process of capitalistic development gets under way, inequality is likely to increase. Similar observations could be made with regard to growth of production, shifts in employment patterns and changes in technology. But the point that needs to be emphasised is that it is not enough to consider only the short-term consequences of a particular agrarian reform programme. And in order to understand the likely long-term consequences, it is necessary to consider the probable dynamic of the particular structures brought about by agrarian reform.

IV Agrarian Reform in Practice: An Analysis of the Findings of the Case Studies

The socio-political context in which agrarian reform was implemented varied widely across the regions covered by the case studies included in this volume. In Nicaragua, they followed a revolutionary change of government; in Peru and Ethiopia, they followed military takeovers of state power; in Kerala and West Bengal, they followed the election to power of political parties supported by the impoverished sections of the peasantry; in Chile, they were carried out by popularly elected governments but were partially reversed by the army which subsequently seized power; in Iran, they were carried out by an absolute monarch. These different socio-political contexts naturally set limits to the scope of the reforms which, except in the case of Iran, essentially represented responses to crisis situations in agriculture (stagnation of production and technology, extreme poverty, severe unemployment and underemployment etc.).

The concrete forms which these responses assumed varied a great deal across the regions. In Kerala and West Bengal, land was redistributed in favour of the poor peasants, tenants and sharecroppers. In Peru, Chile and Nicaragua, *haciendas* were transformed into cooperatives. In Ethiopia, land was first redistributed in favour of the poor peasants and tenants, who were then integrated into co-operative networks, and efforts to develop collective production enterprises are now under way. In each of these cases, the beneficiaries were sections of the peasantry and the losers were sections of the traditional landed elite. This is one feature which the reform programmes in these regions shared in common: they were pro-peasant and anti-landlord in character. To this extent, they were designed to reverse historical trends and to undermine the basis of personalised relationships which characterised the non-peasant sector of the agrarian economy. Viewed in the light of our earlier discussions these can be considered as steps in the right direction.

Unfortunately, however, political constraints often rendered the responses rather half-hearted; most of the reform programmes lacked a well-defined long-term objective. In Kerala and West Bengal, the programmes went only some distance towards a genuine regeneration of the peasantry; in Peru and Chile, the transformation of *haciendas* into co-operatives was motivated more by expediency than by a genuine commitment to a collectivist model of development. In none of these cases, moreover, were structural changes attempted in the non-agricultural sectors of the economy. Subsequent weakening of the co-operative structure in Peru and reversals in Chile could be attributed, in part, to the sectoral anomalies thus created. Only in Ethiopia and Nicaragua, where old political structures were wholly replaced by new ones, do the reform programmes appear to have been formulated in the context of a wider objective; they aim to effect a transition to a collective agriculture and this aim is consistent with the changes which have been taking place in the non-agricultural sectors of the economy.

The case of Iran stands apart. Here the reforms did not constitute a response to the crisis in agriculture, though a crisis was undoubtedly developing. Rather, they formed part of a drive towards modernisation which was inspired by the availability of vast oil revenues. To the monarch and the technocrats around him, a traditional, technically backward peasant agriculture appeared to be an anachronism in the context of 'modern' Iran. The reforms, consequently, were based entirely on the presumptions of these 'modernisers' and were imposed 'from above'.

Given the characteristics of the reform programmes and given the fact that many of them are still in the process of implementation, it has often proved difficult to provide a definitive evaluation of their economic consequences. Apart from the usual identification problems, immediate results do not necessarily foreshadow long-term consequences and much depends on what follow-up policies are pursued by governments. For example, reforms may cause immediate disruptions in production while at the same time laying the basis of steady growth in the medium or long term; they may bring immediate benefits to sections of the rural poor but may, at the same time, set in motion processes which would tend to wipe out these benefits; and so on. The following observations should therefore be treated with caution and should be considered only as tentative.

The most clearly observable immediate effects of the reforms have been on the pattern of income distribution and on the levels of poverty. In all cases except Iran, the reforms did effect a significant redistribution of incomes and, in some cases, of assets in favour of groups of rural poor. This redistribution was achieved through three principal means:

through a redistribution of land, through a redistribution of the produce of land under tenancy, and through a rise in employment and wages. The first played the most important role in Kerala, West Bengal, Ethiopia and Nicaragua; the second played an important role only in Ethiopia where erstwhile tenants were freed from rental obligations; and the third played an important role in Peru and Chile where the governments, having transformed the *haciendas* into state farms and co-operatives, were able directly to effect increases in employment and real wages. In Kerala and West Bengal too, attempts were made to raise real wage rates, but such attempts were successful only when they were accompanied by either the emergence of strong rural trade unions or the implementation of significant rural works programmes. In these regions, however, labourers are employed mainly on a daily basis, and the effects of wage rises on the levels of living of the landless labourers are not clear since trends in employment remain unknown.

It has proved difficult to be precise about the extent of income redistribution which agrarian reforms have brought about in the regions concerned. What seems clear is that this has varied directly with the degree of pre-existing inequality. The redistributive impact of reforms has evidently been greater in Ethiopia, Nicaragua, Peru and Chile (1964-73) than in Kerala and West Bengal. It also seems clear that while the redistribution process in Ethiopia and Nicaragua favourably affected almost the entire category of rural poor, in Peru and Chile it benefited primarily the permanent workers on transformed *haciendas* (poor peasants and landless casual labourers remained unaffected), and in Kerala and West Bengal it benefited primarily the sharecroppers and poor peasants (landless labourers remained largely unaffected). These differences seem to be explained by three major factors. First, greater pre-reform inequalities in Ethiopia, Nicaragua, Peru and Chile principally reflected the sharp division of the rural population into a small but rich and powerful class of landlords and a large, impoverished class of peasants and workers. The degree of inequality within the class of peasants and workers, on the other hand, was low. Thus once the landlords were removed from the scene, the degree of income inequality dropped sharply. In contrast, the peasantry was highly differentiated in Kerala and West Bengal and the class of landlords-traders-moneylenders was quite large and not vastly superior to the peasantry in terms of income. Even a total disappearance of the class of landlords-traders-moneylenders would not have dramatically improved the income distribution in these regions. In the event, the reforms, which relied on ceiling laws for land redistribution and sought to improve the terms of tenancy, did weaken this class but failed to remove it from the rural scene. Second, while the direct producers in Peru and Chile were differentiated into classes – peasants, landless casual labourers and permanent workers, those in Ethiopia were almost exclusively peasants. Abolition of landlordism, therefore, benefited all direct producers in Ethiopia while the same measure benefited only a particular class of direct producers (permanent workers) in Peru and Chile. Nicaragua was closer to Peru and Chile than to Ethiopia in terms of structure, but here possibilities for a considerable expansion of acreage existed. As in Peru and Chile, peasants in Nicaragua did not benefit much from the abolition of landlordism, but they benefited from the easy availability of credit which made it possible for them to acquire land through reclamation. Third, while in Ethiopia and Nicaragua the reforms formed part of a much wider process of transition to a collectivist economic system, they had no clear long-term objectives in Peru and Chile; and in Kerala and West Bengal political constraints made it imperative for the governments to implement programmes which were less than adequate to achieve their stated objectives.

The case of Iran is exceptional in that here the reforms were in effect anti-peasant in character. The regime's ill-conceived attempts to transport Iran's agriculture overnight to the modern age through an active promotion of agribusinesses and farm corporations ignored the whole history and tradition of the existing rural society. They ended up destroying the fabric of the rural society, promoting production inefficiency and wasteful use of resources, accentuating income inequalities, and driving a section of the peasantry to urban slums. And yet, given Iran's vast oil resources, its economy could easily have withstood temporary disruptions in agricultural production; the state also had the capability to undertake huge investments in agriculture; conditions were thus most favourable for the development of a progressive and egalitarian peasant economy system.

The general distributional bias of the reforms in favour of sections of the rural poor would have been of little real benefit if production was unduly disrupted in the process. The evidence shows quite clearly, however, that the reforms did not cause any significant disruptions in production. Indeed, it is rather surprising that even in Ethiopia and Nicaragua, where the reform process was accompanied by political turbulence, disruption was minimal. On the other hand, it is not obvious that the reforms have had a positive impact on production. It is probably too early to judge this; reforms in the regions studied have been too recent to have any definitive impact on production trends, and there are reasons to believe (as will be discussed below) that they have created conditions for long-term growth. No direct evidence is available as to whether or not the reforms induced a change in the volume of marketed surplus. But since there was no dramatic change in the volume of production and the standards of living of sections of the rural poor appear to have improved, one must assume that there was in fact a decline in the volume of marketed surplus. Furthermore, there is little doubt that established marketing networks were disrupted to some extent and this could only have meant a greater decline in the marketed surplus.

The question of a continuation of the redistribution process and a sustenance of its beneficial effects on rural poverty and income distribution is intimately linked to the question of the probable structural dynamic of the agrarian economy, given the conditions created by the reforms. The question is difficult to answer given the unpredictability of the follow-up measures which the governments concerned might adopt for the future. The observations that follow relate to probable developments assuming that governments do not attempt to implement further reform measures. The discussion excludes the case of Iran because of the almost total lack of any reliable information on the developments there since the revolution.

In spite of their deficiencies, the reforms have reduced, and in some cases virtually eliminated, the importance of large landlords in the agrarian economy of the regions concerned. But in the majority of cases, even though the reforms have definitely brought benefits to some sections of the rural poor (poor peasants, sharecroppers, landless labourers, etc.), the rural poor still remain a significant category. On the other hand, land redistribution has enabled some poor peasants to join the ranks of independent peasantry, and this is the group which has emerged as the new dominant group in the rural scene. This trend is quite clear in the cases of Chile, Kerala and West Bengal, but probably also holds in the cases of Peru and Nicaragua.[29] With expanding irrigation and credit facilities, increasing availability of modern inputs and growing demand for an agrarian surplus for purposes of industrialisation, these medium landowners can conceivably emerge as dynamic entrepreneurs and can act as the agents of an incipient capitalism in agriculture. Such a development, while beneficial for growth, is likely to affect unfavourably the poorer beneficiaries of agrarian reform and lead to a renewed, albeit different, process of generation of inequalities. Although such a process need not necessarily generate poverty in all situations, in the context of the regions studied this seems a distinct possibility.

Notes

1. Cf. Food and Agriculture Organisation (FAO), *Fourth World Food Survey* (Rome, 1977); World Bank, *World Development Report* (Washington, DC, 1980).

2. ILO, *Poverty and Landlessness in Rural Asia* (Geneva, 1977); Asian Development Bank, *Rural Asia: Challenge and Opportunity* (Federal Publications, Singapore, 1978); Milton G. Esman *et al.*, *Landlessness and Near Landlessness in Developing Countries* (Cornell University, Ithaca, NY, 1980); C. A. Lassen, *Landlessness and Rural Poverty in Latin America* (Cornell University, Ithaca, NY, 1980); D. A. Rosenberg and J. G. Rosenberg, *Landless Peasants and Rural Poverty in Selected Asian Countries* (Cornell University, Ithaca, NY, 1978); D. Ghai and S. Radwan (eds.), *Agrarian Policies and Rural Poverty in Africa* (ILO, Geneva, 1982).

3. See, for example, R. Dumont and B. Rosier, *The Hungry Future* (André Deutsch, London, 1969); L. R. Brown and E. P. Eckholm, *By Bread Alone* (Oxford University Press, Oxford, 1974); A. N. Duckham *et al.* (eds.), *Food Production and Consumption* (North Holland Publishing Co., Amsterdam, 1976).

4. Some recent studies on famines have highlighted the complex network of relationships which govern food distribution in rural areas. See A. K. Sen, 'Starvation and Exchange Entitlements: A General Approach and its Application to the Great Bengal Famine', *Cambridge Journal of Economics*, vol. 1, no. 1 (1977); A. K. Ghose, 'Food Supply and Starvation: A Study of Famines with Reference to the Indian Subcontinent', *Oxford Economic Papers*, vol. 34, no. 2 (1982); M. Algamir, *Famine in South Asia: Political Economy of Mass Starvation* (Oelgeochlager, Gunn and Hain, Cambridge, Mass., 1980); and A. K. Sen, *Poverty and Famines* (Clarendon Press, Oxford, 1981).

5. The required conditions are: (a) incremental incomes (if any) accrue to high-income groups with a very low income elasticity of demand for food; and (b) there is a significant net transfer of incomes from low-income (high income elasticity) to high-income (low income elasticity) groups.

6. The required condition is that the growth of *per capita* effective demand for food in urban areas (a result of expansion in urban employment and incomes due to industrial growth) cannot fully cancel out the effect of the decline in *per capita* effective demand for food in rural areas. Given the typical urban-rural distribution of population and labour force in developing countries, this is highly plausible.

7. For an elaboration of the argument, see A. Ghose and K. Griffin, 'Rural Poverty and Development Alternatives in South and South-East Asia: Some Policy Issues', *Development and Change*, vol. 11 (1980).

8. Indeed, this is a basic lesson which can be elicited from the 'green revolution' experience on which there is by now an extensive literature. Among the most recent studies are A. Pearse, *Seeds of Plenty, Seeds of Change* (Clarendon Press, Oxford, 1980); B. H. Farmer (ed.), *Green Revolution? Technology and Change in Rice Growing Areas of Tamil Nadu and Sri Lanka* (Macmillan, London, 1977); International Rice Research Institute, *Economic Consequences of the New Rice Technology* (Los Banos, 1978); and K. B. Griffin, *The Political Economy of Agrarian Change* (Macmillan, London, 1974).

9. In this situation, marketed surplus tends to rise at a faster rate than production. If the urban-industrial economy is to absorb this increased marketed surplus, then the rate of industrial investment must be stepped up substantially. In a non-socialist economy, this requires a quick and adequate response from private capitalists which may or may not be forthcoming. Moreover, there may be real obstacles to a rapid expansion in industrial investment (e.g. shortage of foreign exchange, shortage of raw materials, shortage of skilled manpower, etc.).

10. To cite an example: around 1978-9, the government of India held huge stocks of foodgrains, acquired through procurement and price support schemes, while at the same time it officially admitted that about 50 per cent of Indians did not have enough to eat.

11. For some relevant data and typologies, see D. Ghai, E. Lee and S. Radwan, 'Rural Poverty in the Third World: Trends, Causes and Policy Reorientations' (ILO, Geneva, 1979; mimeographed World Employment Programme research working paper; restricted).

12. Capitalist farmers, who have only recently been appearing on the scene and who still constitute only a tiny minority among the landholders in most developing countries, can be left out of account, for the purpose of theoretical formulations, without introducing serious distortion.

13. In recent periods, with the introduction of modern technologies, some independent peasants have transformed themselves into capitalist farmers in some regions. Such a possibility will be considered in the course of subsequent discussions. Here we are concerned with a more remote past.

14. Several studies included in this volume highlight this trend. See also S. Barraclough and J. C. Collarte (eds.), *Agrarian Structure in Latin America* (D. C. Heath, Lexington, Massachusetts, 1973); K. Duncan, I. Rutledge and C. Harding, *Land and Labour in Latin America* (Cambridge University Press, Cambridge, 1977); C. Furtado, *Economic Development of Latin America: A Survey from Colonial Times to the Cuban Revolution* (Cambridge University Press, Cambridge, 1971); A. Pearse, *The Latin American Peasant* (Frank Cass, London, 1975); R. M. A. van Zwanenberg and Anne King, *An Economic History of Kenya and Uganda 1800-1970* (Macmillan, London, 1975); G. Ketching, *Class and Economic Change in Kenya, 1905-1970* (Yale, 1978); J. Iliff, *A Modern History of Tanganyika* (Cambridge University Press, Cambridge, 1979); A. G. Hopkins, *An Economic History of West Africa* (Longman, London, 1973); V. B. Singh (ed.), *Economic History of India 1857-1956* (Allied, Bombay, 1965); B. Moore, *Social Origins of Dictatorship and Democracy* (Penguin, Harmondsworth, 1969); G. Myrdal, *Asian Drama* (3 vols., Allen Lane, London, 1968); B. Sen, *Evolution of Agrarian Relations in India* (People's Publishing House, New Delhi, 1962); D. Thorner and A. Thorner, *Land and Labour in India* (Asia, London, 1962).

15. For lack of a better term, we are using the term 'profit' to denote the surplus accruing to a landowner when the land is operated with hired labour. It will be clear from subsequent discussions that in the present context this surplus is actually closer to feudal rent than to capitalist profit.

16. It should be emphasised here that this relative surplus of labour has nothing to do with land/man ratios. In many Latin American countries there is an abundance of land and yet the landless and near-landless face very similar situations to those in Asia. The term 'relative surplus of labour' only refers to the fact that given the existing social and technical relations of production the available labour is not fully utilised.

17. Such a possibility also obviously arises in a situation where a significant class of capitalist farmers already exists. We have excluded this situation from our account by assumption because it is not relevant for a majority of the developing countries. Certainly, none of the regions studied here had a significant class of capitalist farmers in the pre-reform situation.

18. See the study on West Bengal in this volume, Ch. 3.

19. An interesting difference between the Latin American and Asian situations is perhaps worth noting. The *haciendas* clearly differed from peasant farms in terms of cropping patterns and technical conditions of production. The landlord farms in Asia, however, differed little from peasant farms in these respects. On the

other hand, labour relations on *haciendas* were more clearly feudal than on land-lord farms in Asia.

20. This is the classical view of capitalism. The neo-classical view regards exchange as arising from specialisation and, consequently, is unable to provide a consistent explanation of the origin of capitalistic profits.

21. Thus rural income inequalities in Taiwan and South Korea, which have peasant economy systems, and in China and the Soviet Union, which have collective systems, are far lower than in the 'market economies', developed or underdeveloped. See S. Jain, *Size Distribution of Income: A Compilation of Data* (World Bank, Washington, DC, 1975); Ghose and Griffin, 'Rural Poverty and Development Alternatives'; K. Griffin and A. Saith, *The Pattern of Income Inequality in Rural China* (ILO-ARTEP, Bangkok, 1980); D. Ghai and A. R. Khan, *Collective Agriculture and Rural Development in Soviet Central Asia* (Macmillan, London, 1980).

22. This type of transformation has been described in the literature as the 'Prussian' or 'Junker' model of agrarian capitalism.

23. The experiences of China and Vietnam are particularly relevant here. Cf. J. Wong, *Land Reform in the People's Republic of China: Institutional Transformation in Agriculture* (Praeger, New York, 1974); E. E. Moise, 'Land Reform in China and North Viet Nam: Revolution at the Village Level', unpublished PhD dissertation, University of Michigan, 1977.

24. See the studies on Peru and Chile in this volume, Chs. 5 and 6.

25. A. V. Chayanov, *The Theory of Peasant Economy*, edited by D. Thorner, B. Keblay and R. E. F. Smith (Richard D. Irwin, Homewood, Illinois, 1966). For illuminating discussions of Chayanov's basic model, see T. Shanin, *The Awkward Class* (Clarendon Press, Oxford, 1972); M. Harrison, 'Chayanov and the Economics of Russian Peasantry', *Journal of Peasant Studies*, vol. 2, no. 4 (1975); and 'The Peasant Mode of Production in the Work of A. V. Chayanov', *Journal of Peasant Studies*, vol. 4, no. 4 (1977); A. K. Sen, 'Peasants and Dualism with or without Surplus Labour', *Journal of Political Economy*, vol. 76, no. 5 (1966); and Utsa Patnaik, 'Neo-populism and Marxism: The Chayanovian View of the Agrarian Question and its Fundamental Fallacy', *Journal of Peasant Studies*, vol. 6, no. 4 (July 1979).

26. Cf. Wong, *Land Reform in the People's Republic of China*; Moise, 'Land Reform in China'.

27. Chayanov had noted this feature in his study of the Russian peasantry. Recent anthropological works on the peasantry also support the proposition. See, for example, J. C. Scott, *The Moral Economy of the Peasant: Rebellion and Subsistence in South-East Asia* (Yale University Press, New Haven, Conn., 1976); S. L. Popkin, *Rational Peasant: The Political Economy of Rural Society in Vietnam* (University of California Press, Berkeley, 1979); E. Wolf, *Peasants* (Prentice-Hall, Englewood Cliffs, NJ, 1966); and *Peasant Wars of the Twentieth Century* (Harper and Row, New York, 1969).

28. S. H. Ban, P. Y. Moon and D. H. Perkins, *Rural Development: Studies in the Modernization of the Republic of Korea, 1945-1975* (Harvard University Press, Cambridge, Mass., 1980); Anne O. Krueger, *The Developmental Role of the Foreign Sector and Aid: Studies in the Modernization of the Republic of Korea* (Harvard University Press, Cambridge, Mass., 1980).

29. The exception is Ethiopia where the government has proclaimed the transition to a collectivist organisation of production as the basic goal of agrarian policy, and where further reforms are under way.

PART TWO

TOWARDS PEASANT ECONOMY SYSTEMS

2 AGRARIAN REFORM IN KERALA AND ITS IMPACT ON THE RURAL ECONOMY – A PRELIMINARY ASSESSMENT

K.N. Raj and Michael Tharakan*

I Introduction: A Historical Account of Land Relations in Kerala

Agrarian reform in Kerala over the last quarter of a century is generally believed to have been more far-reaching and effective than elsewhere in India, though carried out within the same administrative and political framework as in the rest of the country. There has, however, been no systematic analysis of the scope and content of the reform, and of what it has achieved and failed to achieve. This chapter is a preliminary effort in this direction. An attempt is made in the chapter to examine also, more specifically, the impact that agrarian reform has had on rural poverty and agricultural growth in the region, and to identify as far as possible the mechanisms and processes involved.

The measures of agrarian reform initiated after the formation of Kerala State in 1956 are rooted basically in the earlier history of land relations in this region, particularly during the period of British rule, and in the social structures and political movements and alignments they gave rise to. Since there were important intra-regional differences in this regard which are material to a clear appreciation of subsequent developments, we shall first outline briefly the main features in the evolution of land relations during this period and indicate the sources of these differences.[1]

For several centuries before the advent of the British rule, there was no common political authority in the region, only small independent principalities under the control of local chieftains and overlords often at war with each other. Nor was there at this time a concept of land-ownership which carried with it the rights later associated with private

*The material incorporated in section III was prepared by Michael Tharakan; the rest of the chapter (and the final draft) was done by K. N. Raj. The authors wish to acknowledge their debt to the Secretary, Land Board, Government of Kerala, for making available the latest data on the implementation of land reforms; to Mrs K. R. Gowrie, Minister for Agriculture, Government of Kerala, for clarifying certain issues connected with the interpretation of the data; to Mr E. Chandrasekharan Nair, Minister for Food, Civil Supplies and Housing, for securing the latest data concerning the working of primary credit co-operatives; and to Professor A. Vaidyanathan for his helpful comments on earlier drafts.

31

property in land under the British legal system (such as rights of sale, mortgage and hypothecation). As in the rest of India, there was only a hierarchy of privileges, rights and obligations relating to land distributed among different layers of society.

At the top of this hierarchy were Brahmins (known as Namboodiris in this region) who, in the absence of centralised political power, had managed to secure control over large tracts of land (either directly or through the temples they were associated with); local chieftains (known as 'naduvazhis', 'madampis', etc.); and their overlords (usually called Rajas) in each principality. Generally the Namboodiris were the more powerful in the northern parts of the region. Further south, the chieftains and the Rajas were stronger and the Brahmin influence correspondingly weaker.

The land was however only nominally in their hands, as almost all of it was conferred by them on favoured families belonging to the castes directly below them (such as Nambiars in the further north and Nayars elsewhere), in return for nominal payments of a customary nature and/or services in kind (such as provision of soldiers when required). These relatively privileged families, who thus secured land on superior tenures (such as 'kanom'), leased it out in turn in smaller parcels to others for cultivation through inferior tenures (like 'verumpattom'), particularly to members of the lower castes (like Ezhavas) and other non-Hindu communities (like Moplahs and Christians). While the burden of the rental and other customary obligations so imposed on cultivators was heavy, their right to cultivate the land was generally secure as long as these obligations were honoured. At the bottom of the hierarchy were communities at the periphery of the Hindu caste structure such as Pulayas, who generally supplied labour for the more arduous operations involved in cultivation (like ploughing, weeding and harvesting of land under paddy) and were in effect agrestic slaves.

This system was disturbed by several political developments towards the middle of the eighteenth century and by the subsequent establishment of British power over the entire region. The sequence of events and the directions of change they promoted were however different in the southern areas from those in the north. Areas in the north were invaded first by a Muslim ruler from neighbouring Mysore in alliance with the French, and later absorbed directly under British administration as part of the Malabar District (initially attached to the Presidency of Bombay and later to that of Madras). In the south, on the other hand, a large number of the principalities were over-run by one of the Rajas (of Venad) and unified under monarchical rule as part of a new

State of Travancore; similar unification on a smaller scale led to the creation of another State, Cochin, in the territory immediately adjacent to it
in the north. Though Travancore and Cochin acknowledged British suzer-
ainty they remained distinct political units headed by their respective
maharajas with their own administration (including a separate judiciary).

The main difference this made to the evolution of agrarian relations
was that in Travancore and Cochin the land under the control of chieftains, except of those who were identified with the new ruling families,
was annexed and formally brought under the ownership of the State (as
'sircar' land); they were presumed to have lost in the process of their
subjugation such rights on land as they had earlier. Nearly two-thirds of
the entire cultivated area thus came under direct State ownership in
Travancore by the first half of the nineteenth century. This in turn
made it possible for the tenants of such State-owned land to be conferred complete ownership rights through a royal proclamation in 1865,
and thereby for a new class of peasant proprietors to emerge in this
southern region. Cochin followed suit towards the end of the century.

In sharp contrast, in the Malabar District which came under direct
British governance, the Namboodiris and the chieftains were able not
only to retain their earlier control over land but also to strengthen their
hold over it. Their claim to have rights to land by birth ('janmom' rights
as they were called), supported by the contention that they were
similar to the rights of *dominium* in the soil recognised in Roman law,
found ready acceptance in the administration; this helped them to
acquire rights associated with private ownership under British law
which they did not possess under earlier customary law. Thus the
'janmies', who constituted no more than a microscopic proportion of
agrarian society, became (along with the temples they controlled) a
powerful oligarchy of landowners in the northern areas of the region.
The impartibility of the land held by temples (i.e. 'devaswom' land),
the traditional practice among Namboodiris permitting only the eldest
son of each family to marry, as well as the matriarchal joint family
system followed by Nayars and Nambiars, helped to preserve the concentration of power in this oligarchy.

More serious than the concentration of private ownership of land
among the 'janmies' in the north was the denial under British administration of the customary rights associated with 'kanom' and 'verum-
pattorn' tenures in the earlier traditional society. Security of tenure
came to be denied to even those in the upper strata of agrarian society
(such as the families who had earlier secured land on 'kanom' tenure),
and the bulk of the cultivators (who were previously left alone as long

as they carried out the traditional obligations) were reduced to mere tenants-at-will, subject to eviction at any time.

In Travancore also, Namboodiri families and upper-caste families associated with the monarchy, along with some temples, were able to acquire the status of 'janmies'; but the proportion of the land under their control was much smaller. Moreover, steps were taken by the State in the first half of the nineteenth century to protect the interests of tenants holding land under superior tenures such as 'kanom', in conformity with the traditional interpretation of their rights in land. The position in Cochin was closer to that of the Malabar District in this respect, as the 'janmies' were in possession of a larger proportion of the land and fewer constraints were placed on the enforcement of 'janmom' rights in conformity with British practice.

There were also other important differences in policy as between the northern and southern parts of the region. In Travancore, where waste lands were owned by the State, active encouragement was given from an early stage to the cultivation of such lands through tax exemption and later through conferment of ownership rights on the cultivators who reclaimed them. In the Malabar District waste lands were available in greater abundance, but they were in the hands of private owners and though some efforts were made by the administration to encourage their cultivation, the conditions of tenancy imposed by the owners and the judicial decisions which gave support to them proved to be serious deterrents. Only about 60 per cent of the total arable land was therefore under cultivation in this area even as late as the 1930s, while almost all the land available was under effective occupation in Cochin and Travancore.

As a result, ownership of land was much more diffused in the southern part of the region and the condition of tenants who secured land on even inferior tenures was somewhat better than in the north. When legislation enacted in the inter-war period enabled families belonging to the upper castes (particularly Nayars) to partition their joint family property — which they were hitherto not free to do — the process of diffusion of ownership was further accelerated. While this led to proliferation of small holdings, it helped other communities to acquire more land.

The oppressive conditions of tenancy in the north, and conflicts of interest on account of the changes that were taking place in the south, had considerable impact on the new political forces that were taking shape in the region as part of the wider national movement in the country. Tenancy reform became an important part of their political programmes (particularly in the Malabar area) and, even though only the relief of tenants belonging to the upper strata was highlighted initially, there developed also more radical forces interested in comprehensive changes within the agrarian structure. Despite legislation enacted by the British administration towards the end of the 1920s for extending relief to such tenants (including even those cultivating land under inferior tenures like 'verumpattom'), these movements continued to gather strength and resulted in the emergence of the Communist Party as a contender for political power in this region after the country became independent in 1947.

There were two important reasons for the strength and militancy of the agrarian movements that grew during the closing decade of British rule and in the decade thereafter. One was that, notwithstanding the legislation enacted in Malabar in 1929, the conditions of tenancy remained oppressive for those actually cultivating the land. The 'janmie' families could without much difficulty take possession of their land for 'self-cultivation', the 'fair rents' fixed by law were high, and the legal procedures involved in securing relief were too expensive, time-consuming and uncertain. As the pressure of population on land increased, there had been further sub-leasing at still higher rents, and the holders of land on such inferior tenures had no legal rights whatever. Even in Travancore and Cochin, which had earlier pioneered reform in respect of land held under superior tenures (particularly 'kanom'), there had been no attempt to give protection to the tenants lower down.

The other important reason was the rapid growth that was taking place in the number of labourers seeking wage employment in agriculture. Until the 1930s, agricultural labourers in Malabar, Cochin and Travancore together did not exceed 0.33 million in a total male working force of nearly 2 million; and more than two-thirds of the agricultural labourers were in Malabar. In the following two decades, the total number of agricultural labourers nearly doubled, and the growth was so much faster in the south that they formed about two-fifths of the working force in agriculture in Travancore by the end of the period (compared to less than one-sixth in 1931). This rapidly growing body of agricultural labourers, together with the lower strata of the tenants (particularly those from Malabar), offered a wide base for political organisation in the rural areas.

These features of the structure of agrarian society that had crystallised by the end of the British period, in the three political units that were subsequently merged, are reflected in Table 2.1, which presents data on the distribution of the economically active male population in agriculture according to the nature of their relationship to land in 1951. It will be observed that, while more than 80 per cent of all cultivators

Table 2.1: Composition of the Economically Active Male Population in Agriculture, 1951 (in thousands and per cent)

Category	Malabar District		Cochin State		Travancore State		Total	
			Territory Corresponding to					
Non-cultivating owners of land (i.e. *rentiers*)	15.8	(44.8)	7.0	(19.8)	12.5	(35.4)	35.3	(100.00)
Cultivators of land wholly or mainly owned	54.9	(9.6)	47.1	(8.2)	472.6	(82.2)	574.6	(100.00)
Cultivators of land wholly or mainly unowned (i.e. tenants)	184.4	(54.2)	69.4	(20.4)	86.7	(25.4)	340.5	(100.00)
Cultivating labourers	227.4	(31.3)	133.5	(18.4)	364.6	(50.3)	725.5	(100.00)
Total economically active population in agriculture	482.5	(28.8)	257.0	(15.3)	936.4	(55.9)	1 675.9	(100.00)

Note: 'Self-supporting persons' and 'earnings dependants', as defined in the population census of 1951, are taken here to constitute the economically active population.

Source: Census of India (1951).

Table 2.2: Decennial Rates of Growth of Population, 1901-51 (per cent)

	Districts	1901-11	1911-21	1921-31	1931-41	1941-51
Malabar	(Cannamore	+ 6.9	+ 2.8	+15.2	+12.6	+22.4
	(Kozhikode	+ 7.5	+ 3.5	+16.8	+12.2	+27.6
	(Molappuram	+ 9.6	+ 2.2	+14.4	+11.7	+17.7
	(Palghat	+ 7.3	+ 4.2	+10.2	+ 8.9	+18.5
Cochin	(Trichur	+12.7	+ 5.7	+22.1	+16.2	+21.5
	(Ernakulam	+14.0	+ 9.3	+26.6	+20.8	+19.9
Travancore	(Kottayam	+17.9	+16.1	+37.0	+21.8	+22.6
	(Alleppey	+14.8	+18.5	+24.3	+12.7	+19.4
	(Quilon	+14.7	+18.6	+27.9	+25.3	+29.0
	(Trivandrum	+17.5	+17.0	+28.6	+18.5	+30.8
Kerala State		+11.8	+ 9.2	+21.9	+16.0	+22.8
India		+ 5.8	−10.3	+11.0	+14.2	+13.3

Source: Census of India, various reports.

who had land of their own were in Travancore, about two-thirds of the *rentiers* (i.e. non-cultivating owners of land) and three-quarters of the tenants were in Malabar and Cochin; agricultural labourers were distributed more evenly, according to the distribution of the population.

II Intra-regional Differences in the Development of Agriculture before 1956

After the termination of British rule, Travancore and Cochin were first merged into one composite political unit in 1949 and later, with some territorial adjustments, integrated with the Malabar District to form the unified Kerala State in 1956. While this helped to bring together and consolidate agrarian organisations hitherto functioning separately and provide a broader political base for the Communist Party (which was able to come to power in the first State elections held in 1957), it also brought into focus some important differences in the pattern of agricultural development as between the northern and southern parts of the State. These differences affected the course of agrarian reform in a variety of ways.

A major reason for these intra-regional differences was the policy regarding land tenure followed in Travancore which fostered the growth of a substantial body of peasant proprietors from the last quarter of the nineteenth century (after the tenants of State-owned land were conferred complete ownership rights). Though the rates of land revenue payable were as high as in Malabar, these peasant proprietors, unlike the tenants in the north, were free from heavy rental obligations and had therefore greater incentive for raising productivity and more resources left for investment in agriculture. Along with peasant proprietorship there developed also an active market for land and, since some of the new owners (particularly among Christians in central and northern Travancore) were associated closely with financial and trading activity, it helped them to secure more land and establish links between agricultural development and the processes of commercial capitalism which had already taken root. Commercial banks advanced credit on the basis of land offered as security, and this was used to purchase more land, develop remunerative crops, and extend trading and processing of agricultural products.

To the incentives for agricultural development provided in these ways were added the pressures created by rapid population growth. It will be evident from Table 2.2 that the rate of population growth until 1931 was very much higher in the southern districts (which formed part of Cochin and Travancore) than in the northern.

As almost all the arable land in Cochin had been brought under cultivation by 1911 there was no scope for further extension, so the growth of population was reflected mainly in shifts from agricultural to non-agricultural occupations (though the latter already provided the means of livelihood for one-half of the people) and in a growing tendency to migrate. In Travancore, on the other hand, there was still land available and the pressure of population led to expansion of the area under cultivation; it was particularly rapid in the decade after the First World War, from less than 2.1 million acres in 1919/20 to over 2.5 million in 1928/9.

Significantly rice accounted for less than 10 per cent of this increase in area during the 1920s, even though it was the favoured foodgrain for consumption; tapioca absorbed about one-quarter of the additional area, and the rest went mainly into coconut, tea, rubber, pepper and the like. This pattern of crop preference was in part a reflection of the physical characteristics of the land that was being brought under cultivation, as it was not possible to grow tea or rubber on paddy fields (or any of them on land that would be just adequate for growing tapioca). Another reason was that rice could be freely imported, and about a third of the total rice consumption of Travancore was being met in this way.[2] But it also indicated that, while an inferior tuber such as tapioca was being used extensively as a source of calories in place of rice (evidently by those belonging to the lower-income groups), there was enough staying power and access to resources for a fairly large number to be able to bring new land under cultivation and develop perennial crops with relatively long gestation periods.

Perennial crops were of two kinds: those like coconut, jackfruit, mango and pepper which could be grown on garden land without any special form of organisation or investment for the purpose, and those such as tea, coffee and rubber which required more careful preparation of land, and management, and were usually organised as plantations on a larger scale. While broad-based commercialisation of agriculture was possible through the cultivation of the former in even small-sized holdings, the development of the latter had a more pronounced capitalist orientation. The relative pace and scale of expansion of area under the two types must therefore be considered separately.

Of the perennial crops grown on garden land no estimates are available on the area under jackfruit, mango, etc., but there is no reason to believe that there was any significant increase in this area. The area under pepper is likely to have grown but, even including other spices, the proportion of the land under these crops was not very significant except in particular localities. Coconut was therefore the only crop of importance in this category for the entire region, accounting for not less than a fifth of the area under cultivation at even the turn of the century.

The significance of coconut cultivation for the agrarian economy of Kerala cannot however be exaggerated. Traditionally it was valued perhaps mainly for its contribution to various household requirements such as nuts and oil for domestic consumption, husk for fuel, timber for house construction, and leaves for thatching of roofs. However, as the demand for coconut oil and coir products increased and other uses were found, not only did the price of coconut rise (relatively to even paddy) but coconut cultivation became the basis for a variety of processing industries as also for a wide range of trading activity.[3] Moreover, even on land leased in by tenants (as in Malabar), the security of tenure was greater and rents payable generally lower on garden land under coconut than on paddy land; and large coconut plantations under self-management were relatively rare. Thus, apart from being a major source of agricultural income, it helped to distribute such income to a greater extent among smallholders, and generate more income in other sectors of the economy.

The extent of improvement in the terms of trade of coconut relatively to paddy from the last quarter of the nineteenth century to just before the onset of the Great Depression can be judged from some data on prices relating to this period available for Travancore. In 1874 a coconut could fetch only about 0.25 kg. of paddy in exchange; by 1904 it could buy over 0.33 kg. and by 1928/9 about 0.75 kg. of paddy.[4] This not only raised the real income from coconut cultivation but provided considerable stimulus for expanding the area under coconut.

No information is available on the area under coconut in Malabar in the earlier decades of the twentieth century for assessing its response to this stimulus; all that can be said is that it was around 340,000 acres by the middle of the 1920s (about one-quarter of the net area cropped), and less than 400,000 acres even towards the close of the 1940s. In Travancore however the data available for the period after the First World War show that the area under coconut increased by nearly 45 per cent over less than a decade and a half, from about 460,000 acres in 1918/19 to 665,000 acres in 1931/2 (though it declined somewhat thereafter on account of the Depression). By the end of the 1930s the principal industries in Travancore were also those connected with the produce of the coconut palm, viz. the manufacture of coconut oil, the retting of coconut husk, the spinning of coir yarn and the weaving of coir mats and matting.[5]

Not only were direct linkages thus established between agriculture and industry through rapid expansion of coconut cultivation in Travancore and the process of commercialisation accelerated all round, but capitalist organisation developed within agriculture itself much more than in the northern districts through the emergence of plantations in tea, coffee and rubber on a fairly extensive scale. While the area under these crops in Malabar was only a little over 50,000 acres by the end of the 1940s, it was nearly four times as much in Travancore. Moreover, most of this was by then in the hands of local entrepreneurs, particularly Christians from Travancore.

Of the three main plantation crops, rubber was by far the most dynamic in terms of the rate of expansion and it was in it that the role of planters from Travancore was also the greatest. Though this expansion received a setback during the Depression the area under rubber in Travancore was well over 100,000 acres by 1950. As we shall see later, the expansion continued at an even more rapid rate thereafter; the area under rubber in Kerala was over 300,000 acres by 1960/1 and well over 0.5 million acres by 1978/9. The economic foundation on which such dynamism developed was however built mainly within the framework of the peasant proprietorship that took shape in Travancore between the two world wars.

Fortunately we have data on the distribution of landholdings in Travancore during this period, based on a sample survey conducted along with the 1931 population census.[6] Though they relate to individual ownership holdings (not household holdings taking into account the ownership holdings of all the individuals within each household), and estimates of area are available only for wet and dry land separately (not for holdings of both types of land put together), they help to give a fairly good idea of the position at that time.

It can be seen from Table 2.3 that, though about two-thirds of the total number of holdings were in the size-range of 0.4 to 5.0 acres (which could be categorised as small holdings), over 50 per cent of the area of both wet and dry land was in holdings above this size; less than 5 per cent of the holdings, 10 acres and above in size, accounted for well over a third of the total area. The largest-sized holdings belonging to the latter category are likely to have been owned, in the case of wet land, mainly by those who had still the status of 'janmies' (primarily Namboodiri families and upper-caste families associated with the monarchy). But the ones on dry land (including garden land and land used for plantation crops) would have been largely in the hands of those who had gained from the conferment of ownership rights on State-owned land in the latter half of the nineteenth century and were able to acquire a larger area subsequently.

An important factor affecting the distribution of ownership holdings

Table 2.3: *Distribution of Individual Ownership Holdings in Travancore, 1931*

Size of Ownership Holdings (in acres)	Wet Land				Dry Land				All Land	
	Number of Holdings (000)	(%)	Area (000 acres)	(%)	Number of Holdings	(%)	Area (000 acres)	(%)	Number of Holdings	(%)
Below 0.40	90.0	(32.7)	0.02	(3.7)	122.4	(19.7)	0.03	(1.6)		15.6
0.40 to 0.99	81.9	(29.8)	0.06	(9.8)	152.2	(24.5)	0.08	(5.6)		22.5
1.00 to 1.99	49.8	(18.1)	0.09	(14.2)	146.6	(23.6)	0.21	(13.2)		24.3
2.00 to 4.99	35.7	(13.0)	0.13	(21.3)	135.4	(21.8)	0.41	(25.6)		24.8
5.00 to 9.99	10.5	(3.8)	0.08	(13.5)	41.0	(6.6)	0.26	(16.5)		8.3
10.00 and above	7.2	(2.6)	0.23	(37.5)	23.6	(3.8)	0.60	(37.5)		4.4
Total	275.1	(100.0)	0.60	(100.0)	621.3	(100.0)	1.60	(100.0)		100.0

Note: Wet land denotes paddy land; dry land covers garden land as well as land under plantation crops held by individuals. There may be slight discrepancies due to rounding.
Source: Census of India (1931), vol. XXVIII, *Travancore*, Part I, Report, Appendix IV on 'The Economic Condition of the People'.

in the inter-war period was the partitioning of Nayar (and Ezhava) joint families allowed by special legislation (to which reference has been made earlier), which resulted in about 0.4 million acres of land (i.e. around one-fifth of the area under cultivation) being partitioned between 1926 and 1930.[7] Nearly two-fifths of this area went into individual shares of less than 1 acre in size, and over two-fifths into shares between 1 and 5 acres in size; less than a fifth of the area remained therefore in holdings of more than 5 acres, and only about one-tenth in holdings of 10 acres and above in size.

By the early 1930s, most of the area in the largest-sized holdings on dry land is therefore likely to have been owned by relatively affluent Christians having close association with trade (and often industry as well). There is evidence that the land transactions in the following decade also moved in their favour.[8] Even in the case of wet land a high proportion of the area in the largest-sized holdings in Travancore was probably land reclaimed from backwaters (in Kultanad) and leased in by Christian entrepreneurs from Namboodiri landlords for a highly capitalistic form of paddy cultivation.[9] Moreover, unlike other land under paddy cultivated mainly in small holdings (in Travancore as well as in Cochin and Malabar), very little of the produce from this highly fertile land (on which double cropping was possible) was for self-consumption; almost all of it was marketed.

It was therefore this class, representing the most progressive and powerful elements emerging from among the peasant proprietors of Travancore, that the popular movements for agrarian reform had to contend with after unification of Kerala in 1956. The much higher concentration in the ownership of land and the predominance of small tenant holdings in Malabar (and Cochin) made it relatively easy to isolate the landlords ('janmies') politically; nor had they any deep involvement in agriculture itself. On the other hand, those with relatively large holdings in Travancore were not only actively associated with commercial agriculture but could make common cause with the smaller peasant proprietors.[10] Consequently, though the small holdings in Travancore had got steadily smaller under the pressure of population growth and partitioning, and were providing recruits to the growing body of agricultural (and other) labourers seeking wage employment, it was more difficult to bring about basic changes in the ownership structure.

III The process of Land Reforms

III.1 Phase 1, 1947-57

The intra-regional differences in agrarian structure and in the pattern of agricultural development noted above were becoming reflected in the attitude of the Congress Party to land reform even in the decade 1947-56, i.e. between the end of British rule and the integration of Kerala as a single political unit. In Malabar, which continued to be part of Madras State during this period, the local organisation of the Party was under pressure to ensure fixity of tenure to those holding land under both superior and inferior tenures (which had already been granted in principle by legislation in the 1930s), and lower rents on land under paddy as well as garden crops; but this was about as far as its interest in land reforms went. In Travancore, where Christians formed the hard core of the Congress Party (following the pursuit of policies hostile to them by the earlier regime, from about the middle of the 1930s), not only was the scale of tenancy very much less but many of those who leased out land were themselves peasant proprietors who had only moderate-sized holdings. The Party had therefore very little interest even in tenancy reform, and actually withdrew its support in 1954 to a government in the composite State of Travancore-Cochin (led by a minority party) which sought to impose ceilings on landholdings along with such reform.

At the national level also, the position of the Congress Party on land reform remained unclear and ambivalent. Some elements in the Party favoured a radical programme (as outlined in the *Report* of its Agrarian Reforms Committee towards the middle of 1949), and pressed for comprehensive tenancy reform, imposition of ceilings on ownership holdings, distribution of waste and surplus land to marginal farmers and the landless, and formation of agricultural co-operatives. But there was strong resistance to such proposals from landed interests entrenched within the Party in various States. An *ad hoc* solution therefore was to let the Planning Commission formulate a broad approach to land reform (as part of the First and Second Five Year Plans), and let the actual programme be settled (including important issues of policy such as on imposition of ceilings) in the light of the perceptions and pressures at the political level of each State.

The position of the Communist Party was clearer and sharper. It was committed on the national plane to the abolition of 'landlordism', which was interpreted to mean taking over land from owners who did not engage themselves in manual work related to cultivation, nonpayment of compensation for land so acquired, and conferment of ownership on actual tillers (defined as only those who contributed their own or their family members' labour in cultivation). The Party proposed fixation of ceilings on landownership at levels that would ensure large enough surpluses of land becoming available, and distribution of such surplus land) along with waste land under government and private ownership) among poor peasants and the landless. It visualised also a variety of measures for consolidation of holdings, land management, and development of co-operative and village panchayats, to be implemented on the basis of a broad consensus among the mass of the peasantry.

However, it was only when the Party was voted to power in Kerala that it found itself compelled to concretise this approach and formulate a precise programme for implementation. The manner in which it did so is of interest, not only because it determined the direction and character of the changes in agrarian structure that were to follow in Kerala but even more for the reason that it was the first major exercise in agrarian reform in India attempted by a political party with revolutionary objectives functioning within the framework of a liberal parliamentary democracy. The Party's programme for land reform was set out in the Kerala Agrarian Relations Bill, which was introduced in the State Legislative Assembly in December 1957 and passed in June 1959.

It was certainly clear from the outset that there could be no escape from payment of compensation to landowners. The Constitution of India guaranteed the right to property as a 'fundamental right', and all State governments were bound by it. But the compensation payable could be reduced by specifying it as a multiple of the annual rent and by fixing a 'fair rent' that was itself a relatively low proportion of the produce from land. This was how the problem posed by compensation was sought to be resolved. The fair rent proposed for paddy land was as low as one-sixth to one-twelfth of the gross produce, and tenants were to pay only 16 times this fair rent to acquire ownership rights. (In contrast, the Malabar Tenancy Act of 1929 had stipulated the fair rent on paddy land as the difference between two-thirds of the average annual gross produce of the three previous years and 2.5 times the seed required for an agricultural year, which together was likely to have worked out at over one-half of the gross produce from land.)

Who should be excluded from, and who would qualify for, conferment of ownership rights proved to be a more difficult issue. Elsewhere in India the Congress Party had already accepted the position (in the context of abolition of *zamindari*) that the actual performance of all or some of the manual tasks of cultivation was neither necessary nor indispensable for qualifying as a cultivator. Persons belonging to some high castes, it had been argued, were traditionally regarded in India as cultivators even though they were prohibited from such manual work on account of their special status. Hence 'a person who hires agricultural labour either permanently or casually for the performance of all or some manual tasks must still be regarded as a "tiller of the soil" provided that he finances and supervises agricultural production and takes the risks involved'.[11]

In its broad policy formulation the Communist Party was clear that only those who contributed their own or their family members' labour in cultivation could qualify as actual tillers and that ownership rights were to be conferred only on them. But the question remained whether this criterion would be politically acceptable and enforceable in Kerala, where many 'cultivators' (owners as well as tenants) depended on others for manual labour. Significantly, the proposals formulated in 1957 were silent on this question. In providing for fixity of tenure and right of purchase of ownership to tenants, no distinction was made between those who 'only personally supervise cultivation' and those who, by contributing their own or their family members' labour in cultivation, were truly tillers. This naturally meant that the proposed benefits of land reform could accrue also to the upper strata of tenants operating their holdings with hired labour.

There was similar (and no less significant) concession in regard to the proposed ceiling on landholdings. Though the ceiling itself was to be not on individual but on family holdings (unlike many other States of India where it came to be introduced later), and was fixed at 15 acres of double crop paddy land or its equivalent in the case of a family of five members (extendable up to 25 acres in the case of larger families at the rate of 1 acre per additional member), it was not to be applicable to plantations and cashew estates, private forests, and land owned by religious and charitable organisations. Plantations and cashew estates so exempted were themselves to be identified not with reference to any specifications regarding area or form of organisation but wholly with reference to the actual crops grown. Thus not only land under tea and coffee plantations in the hands of companies was outside the purview of the ceiling but so also was all land under tea, coffee, rubber, cardamom and cashew even when it was in holdings operated on a family basis or through partnerships.

These exemptions from the proposed ceilings had important implications for landless agricultural labourers, as their prospects of acquiring land depended primarily on the extent of the surplus found available for distribution. The only other provision in the Bill from which they could gain was limited to those among them who were in possession of small allotments of land for erection of hutments (permitted by owners as a source of supply of cheap labour); these hutment dwellers ('kudikidappukars') could not be extended the status of tenants, as they had been leased no land for cultivation, and so they could only be freed from threats of eviction from the hutment land and given fixity of tenure on it.

The proposed reform of agrarian relations was therefore in effect a programme for reform of tenancy with a view to its abolition. The rest of the agrarian structure was to be left largely untouched, particularly in the segment of plantation crop agriculture where capitalistic forms had been rapidly developing in close association with trading interests. Even in the measures contemplated for abolition of tenancy there were some compromises, though mostly of a less fundamental nature. For instance, since Travancore had already had a long period of peasant proprietorship and landowners with relatively small holdings were numerically an important segment of agrarian society, it was prudent not to alienate them. Tenants who happened to have leased-in land from owners of small holdings (defined as those owning 5 acres or less of paddy land) were therefore denied the opportunity of acquiring ownership rights.

Nevertheless, there was still enough in the Agrarian Relations Bill to stir up considerable political opposition. The provision for granting fixity of tenure and ownership rights to tenants covered not only tenants-at-will ('verumpattomdars') but sharecroppers ('varomdars'), licensees and other still inferior holders of land under tenancy. To ensure effective implementation, all the rights of landlords and intermediaries on leased-out land were to vest in the State government (as from a particular date to be notified) for eventual transfer to the tenants. This meant that the responsibility for transfer of ownership was not to be left to the initiative of tenants, some of whom could have been perhaps 'persuaded' out of it by landlords. All of them were to become *de facto* owners as from the date on which the rights were vested in the government; and, since the three stages of the transfer of ownership (namely granting fixity of tenure, fixing 'fair rents', and conferment of ownership) were to be gone through simultaneously, the possibilities of delays and resultant retrogression were practically foreclosed.

Moreover, there was an important provision crucial to the entire programme: the constitution of a Land Board at the State level, and of a multitude of local Land Tribunals working under it, entrusted with the tasks of implementation including settlement of disputed issues. There was to be also political representation on them, through a member elected by the State Legislative Board for the Land Board, and members elected by local bodies for the Tribunals. In the prevailing political situation, most such representatives would have been from among those strongly sympathetic to the provisions of the Bill, and were likely not only to promote its speedy and effective implementation but also to act as a check on possible biases introduced at the level of the bureaucracy. However, precisely for that reason, this provision was viewed with suspicion and hostility by the propertied interests in land as well as by those who were on more general grounds opposed to the party in power.

III.2 Phase II, 1959-69

The opposition to the Kerala Agrarian Relations Bill (hereafter referred to as KARB) took concrete shape within days of its passing by the State Legislative Assembly in July 1959. It took the form of a 'liberation struggle' ('vimochana samaram'), organised by all the political parties in opposition, for creating conditions in which the government would not be able to function. Though support for the movement was mobilised on a variety of issues going well beyond those raised by the Bill, its provisions were pointedly characterised as a threat to ownership of land in general; and this became a rallying point (particularly in the old Travancore region) for unifying a very wide segment of all owners of land irrespective of the size of their holdings. The resulting political upheaval led ultimately to the removal of the Communist regime, through the central government applying special provisions of the Constitution for the purpose.

However, even though the party which sponsored the Bill was displaced from power, the proposed land reform programme could not itself be jettisoned or drastically modified as several of its provisions (particularly those concerning abolition of tenancy) had wide political appeal. A new coalition government, headed by the Congress Party, made some changes but the Kerala Agrarian Relations Act, passed in 1960 (KARA, 1960), under its guidance left intact the main body of the earlier Bill.

The changes made were however significant, beginning a process of slow erosion. The new Act redefined 'small owners' to include those with rights on areas up to 10 acres of double crop paddy land if the area in their actual possession was less than 5 acres of such land. The earlier provisions for elected representation on Land Tribunals and the Land Board, and for restoration of rights to tenants evicted after 1956 (in anticipation of reform), were deleted. Moreover, the Act excluded not only plantations proper from the ceiling provisions but all the contiguous and interspersed agricultural land within the boundaries of plantations, and specifically permitted eviction of labourers from small allotments of land given for hutments ('kudikidappu') within plantation areas. These provisions were intended in the main to protect a wide range of interests with considerable influence in the region that previously constituted Travancore.

Even with these changes, however, the new Act (KARA, 1960) could not be implemented, this time on account of legal obstacles linked with the rights to property, equality and freedom guaranteed in the Constitution. Though laws restricting property rights had been given protection earlier by an amendment of the Constitution, the Supreme Court ruled that the inclusion of holdings of land not strictly conforming to the definition of an 'estate' as given in the amending Article was open to challenge. Since extensive areas in Malabar (which was regarded as a 'ryotwari' tract for fixation of land revenue under the British administration) were declared as outside the purview of this definition, as also some further areas in the Travancore region, major provisions of the Act became inoperative in effect over most parts of the State. The Act was declared *ultra vires* of the Constitution in 1963.

Constitutional questions were also being raised at this time, elsewhere in the country, about the 'fundamental' nature of private property rights and the violation of the right to equality on account of discriminatory rates of compensation fixed in land reform legislation. It was therefore only after another amendment of the Constitution was passed by the Indian Parliament, protecting all legislation included in its Ninth Schedule from legal scrutiny, that the threads could be picked up again. The defunct Act (KARA, 1960) was then replaced by a fresh Kerala Land Reforms Act (KLRA, 1963), after its inclusion in the Ninth Schedule.

More changes were however introduced at this stage diluting further some of the earlier provisions. The ceiling was raised to 36 acres for a family of five, and the exemptions were enlarged to include cashew estates of 10 acres and over, pure pepper and coconut gardens of more than 5 acres, and paddy holdings on land reclaimed from backwaters (i.e. 'kayal' land in the Kuttanad area of Travancore). 'Small owners' were redefined as those having interests in land up to 24 acres, and tenants could claim ownership rights on the land operated by them only if they agreed to the owners resuming cultivation of half the area. The earlier provision giving representation to members of the State Legislature in the Land Board, and to members of local bodies in the Land Tribunals, was deleted in this Act also.

Though the Act (KLRA, 1963) came into force soon afterwards and was free from legal challenge, its implementation turned out to be very slow. A Land Reforms Survey conducted towards the end of 1966 showed that, even three years after the enactment, the Land Tribunals had been approached for fixation of 'fair rents' on less than 4 per cent of the total area under tenancy in the State, no purchase at all of ownership rights under the provisions of the Act had taken place, and enforcement of ceilings had not begun. One reason was the complex administrative and legal procedures involved; but no less important was the absence of an elected government in the State during most of this period, on account of the political instability which set in soon after the Act came into force, the implementation of the legislation being therefore left entirely to the bureaucracy without clear and firm political direction.

Meanwhile land transfers were taking place on a fairly extensive scale, both by transfer of ownership (by sale, partition and gift) and transfer of possession (by means of mortgage, sub-leasing, surrender of leased-in land to landlords, etc.), covering more than 0.4 million acres over the decade 1957-66. Since the protection given to tenants against eviction in 1957 (immediately after the Communist Party came into power) had lapsed (along with KARB, 1959), and the 1963 Act gave no such protection with retrospective effect, they could be evicted without any legal check till the new Act came into force in April 1964. Though the area involved in these land transfers was only about 10 per cent of the total area operated in the State it was a more substantial proportion of the area that was intended to be transferred under the provisions of the Agrarian Relations Bill.

This phase came to an end only when another swing of the political pendulum brought to power in 1967 a coalition government led by the left wing of the (by then) divided Communist Party. The new government immediately introduced legislation to restore tenants evicted after April 1964 as well as to prevent future evictions, followed soon after by a comprehensive amendment to KLRA, 1963, restoring to a large extent the original provisions of the 1959 Bill. Moreover, it gave the landless hutment dwellers ('kudikidappukars') not only fixity of tenure but the right to purchase up to one-tenth of an acre at a nominal price.

The amended Act (KLRAA, 1969) is the land reform law that has been in force since then.

The 1969 Act vested in the government, as from the beginning of 1970, the ownership rights on *all* land leased out to tenants, thereby depriving the owners and the tenants of the option to continue their tenurial arrangements in any form. The creation of new tenancies was also banned, with retrospective effect from April 1964.

III.3 Phase III, 1970-80

With ownership rights vested in the government, tenants became *de facto* owners from 1970. What was left to be done was only conferment of *de jure* ownership on them through the due processes. Tenants had the option to pay the compensation in instalments to the government (with the government independently compensating the owners in such cases) or settle the matter directly with the owners on a mutually acceptable basis. The issue of purchase certificates by the Land Board to tenants signifying *de jure* ownership appears to have been separated however from the payment of compensation.

According to data available from the Land Board, a little over 3.6 million applications were received from tenants up to October 1980 for transfer of ownership; about two-thirds of them had been allowed, and 2.4 million purchase certificates issued on that basis to the applicants signifying their *de jure* ownership. There are however some vital gaps in the information available with the Land Board which make it difficult to judge how far the provisions of KLRAA, 1969, in regard to abolition of tenancy have been adhered to in the course of its implementation. Each application has been administratively treated by the Board as a 'case'; and no effort has been made so far to tabulate the relevant data on the area involved in each such application for transfer of ownership and the number of applications made by a tenant for all the different plots that could constitute the total area leased in. From the total number of applications allowed so far, it is not possible therefore to infer either the number of tenants who have gained from this measure or the total area of land on which ownership rights have been so transferred.

A somewhat glaring inconsistency that must be noted here concerns the total compensation payable by tenants for the land so transferred. The purchase price fixed was 16 times the fair rent on the leased-in land, with option to the tenant to save one-quarter of the total if the entire amount due was paid in one lump sum. The expectation was that about Rs 2,500 million could be collected in this way. Though the transfer of ownership rights to tenants under the Act is now believed to be nearly complete, the total amount realised so far is only a little over Rs 200 million; the balance to be collected in further instalments seems unlikely to exceed Rs 100 million.

One possible explanation for the wide discrepancy is the provision in the Act which permitted settlements to be made on the basis of mutual consent between the owners and the tenants involved. Such cases of settlement numbered a little over 0.25 million by October 1980; if the average area involved in these cases was very much higher than in those settled through the Land Tribunals, a substantial part of the estimated compensation would have been paid directly to the owners without being channelled through the government.

At the same time, if settlements were reached in this manner for transfer of ownership rights in respect of a large part of the area that was proposed to be covered by tenancy reform, one could also entertain doubts as to how far the terms of such settlement were favourable to tenants. It is not inconceivable that some tenants were persuaded to refrain from claiming ownership rights over all the land to which they were entitled, in return for the rights being transferred on part of such land without any payment. The Land Reforms Survey of 1966/7 had found that ignorance of eligibility and desire to maintain good relations with the owners (on whom possibly the tenants were dependent in other ways) were responsible for tenants not claiming ownership rights (under KLRA, 1963) on more than 1 million acres of leased-in land; such factors could have worked in the same direction in the implementation of the 1969 Act. They remain open questions however and cannot be settled decisively without systematic tabulation of the relevant data by the Land Board (which has not been done so far).

Still there is little doubt that the enforcement of the provisions for tenancy abolition have been far more effective than those relating to ceilings. The amended Act of 1969 had scaled down the ceiling limit to 20 acres for a family of five and confined the exemptions to rubber, tea and coffee plantations, private forests and other such non-agricultural land, and land belonging to religious, charitable and educational institutions of a public nature. It has also laid down that the area to be declared surplus was to be assessed as on a particular date, stipulated in the Act, namely 1 April 1964, and that all the gifts, partition, etc. allowed for should have been completed before that date. Nevertheless, the total surplus land identified by the Land Board and available for takeover by October 1980 was only a little over 0.11 million acres, i.e. less than 2.5 per cent of the total area of land operated in the State.

An important reason was that, even though the ceiling was fixed on the total area a family of five could hold, there was not only additional provision for other members in a household but an unmarried adult could claim up to one-half of the ceiling fixed for a family. Since households with large holdings of land had also generally several adult members the effective ceiling on holdings was correspondingly higher.

In implementing the provision that surplus land was to be assessed as on a stipulated date, and that all the gifts, partition, etc. allowed for should have been completed before that date, the law relating to the joint family system practised by Hindus posed further problems of a legal and political nature. According to this law, all members of a joint family had inherent rights to an equal share in the family property irrespective of their age, without such rights having to be sanctioned by the head of the family through a documented gift or partition. Insistence on gifts and partitions being completed before the stipulated date by Christian and Muslim families alone (which did not have this legal protection) was therefore held to be discriminatory. Consequently, a further amendment had to be passed in 1972 validating transfers of surplus land through gifts and partitions to children and grandchildren that had been effected after the date stipulated in the Act. When this amendment was later struck out by the judiciary, on the ground that all potential surplus land had already been vested in the government and transactions of this nature by the owners concerned could not therefore be legalised retrospectively, still another amendment had to be passed in 1979 validating *all* such transfers from 1964.

It needs also to be noted that one of the important provisions of the 1959 Bill requiring elected representation in the Land Tribunals and the Land Board — which was later deleted — did not find a place again in the 1969 Act. It was only through an amendment in 1972 that 'popular committees', consisting of non-official members to be nominated by the recognised political parties, were formed to advise the official agencies entrusted with the enforcement of ceilings. A Review Committee (with non-official members) had also been constituted at the State level. It is doubtful however whether these committees were in a position to scrutinise individual cases and exercise their judgement in the manner earlier expected of elected representatives functioning as an integral part of the official agencies of implementation.

Some efforts were made through political agitation over the period 1970-80 to create awareness of the failures in enforcement of ceilings and compel corrective action. They were not however successful in preventing the evasions that were taking place openly through bogus claims and transfers (such as by making surplus land part of tourist hotels, film studios and other business enterprises which had been specifically exempted from the purview of the ceiling legislation) or even in identifying specifically the available surplus land on a large enough scale. This was in contrast to the collective action taken by agricultural labourers in the early 1970s for directly staking their claims to the small plots of land given to them for hutments ('kudikidappu') and which played an important part in the effectiveness of that part of the land reform legislation.

IV Economic Consequences of Land Reforms

IV.1 Changes in the Structure of Landholding

The implementation of land reforms after the formation of the Kerala State in 1957 thus proved to be a tortuous and time-consuming process, undoubtedly affecting the scope and content of the programme itself. Nevertheless, a substantial part of whatever survived from the original proposals had been implemented by 1980, particularly the provisions relating to abolition of tenancy and conferment of ownership rights on landless 'hutment dwellers'. Though acquisition of surplus land by imposition of ceilings and its distribution were not yet complete, it was already clear that there was not much more that could be secured in this way. It is therefore not premature now to consider what effect all these measures taken together have had on the pattern of distribution of landholdings and thereby on the rural economy of Kerala.

It would have been ideal if some year in the second half of the 1950s could have been taken as a benchmark and the distribution of landholdings then compared with that in a year in the second half of the 1970s. Unfortunately, no data are available on landholdings in the State as a whole in the 1950s; and even the estimates available for subsequent years are not strictly comparable with each other for a variety of reasons. Some inferences can, however, be drawn from careful analysis of the available data, covering the period 1961/2 to 1976/7. This is what we shall attempt to do, noting the precise coverage of the data in each case and the qualifications that need to be borne in mind.

Though some estimates are available from a survey of landholdings in Kerala conducted by the National Sample Survey in 1960/1, they cover only operational holdings and offer no idea of the pattern of ownership of land at that time. A fresh survey in 1961/2, organised as part of a more comprehensive effort by the National Sample Survey covering the entire country, removed this lacuna. While this survey was confined to households in the rural sector (which is a serious limitation in Kerala since the distinction between rural and urban areas is very much less clear-cut than elsewhere in India), the sample for the State consisted of a little over 2,000 households in this sector (drawn from 136 villlages), and the selection procedure was such as to ensure that households with landholdings of 1 to 10 acres would receive much higher weightage than those with holdings below 1 acre, and holdings of 10 acres and above would get still higher weightage. One should, therefore, expect the data to reveal reasonably well not only the pattern of distribution of operational holdings on all land in the rural sector 'used wholly or partly for agricultural production' but also the distribution of ownership over that proportion of such land as was owned by households within the sector.

Table 2.4 presents the data separately for the ownership and operational holdings of rural households in 1961/2. The following inferences can be drawn from them:

(a) Since the estimate of the total area under operational holdings (3.3 million acres) is about 30 per cent lower than the estimate of net cropped area available independently (from the Board of Revenue of the State government) for 1961/2 and the difference is much larger than can be accounted for by the exclusion of the urban sector, one must presume that the survey underestimated the area. It is probable that this was because the number of villages and households in the sample from the upland areas of the State was not enough to capture adequately the large-sized holdings under plantation crops; to that extent, the distribution of both ownership and operational holdings is also likely to have been more skewed than indicated by the estimates in Table 2.4.

(b) Of the total number of rural households in 1961/2 (about 2.5 million), nearly a third (about ¾ million) owned no land at all; even after leasing in from others households possessing no land numbered over 0.4 million accounting for one-sixth of all rural households.

(c) The proportion of total owned area leased out (8.5 per cent) does not reflect adequately the scale of tenancy in the sector at that time, as it covers only area leased out by rural households; a closer estimate would have been possible if data had been collected on the total area leased in by rural households, but this had not been done.

Table 2.4: Ownership and Operational Holdings of Rural Households in Kerala, 1961-2

Size-class of Holding (in acres)	Ownership Holdings Number of Households (000)		Total Area (000 acres)		Area Leased-out (000 acres) [% of area owned]	Operational Holdings Number of Households (000)		Total Area (000 acres)	
0.00	770		Nil		Nil	417		—	
0.01–0.49	695	(40.4)	143	(4.6)	6.0 [4.2]	930	(44.8)	136	(4.1)
0.50–0.99	339	(19.7)	241	(7.7)	4.1 [1.7]	384	(18.5)	275	(8.3)
1.00–2.49	368	(21.4)	569	(18.1)	22.8 [4.0]	392	(18.9)	619	(18.4)
2.50–4.99	184	(10.7)	634	(20.2)	13.9 [2.2]	218	(10.5)	767	(23.1)
5.00–9.99	96	(5.6)	634	(20.2)	29.8 [4.7]	106	(5.1)	708	(21.4)
10.00–14.99	19	(1.1)	230	(7.3)	15.0 [6.5]	27	(1.3)	333	(10.0)
15.00–19.99	10	(0.6)	168	(5.4)	24.0 [14.3]	8	(0.4)	132	(4.0)
20.00–24.99	5	(0.3)	96	(3.1)	6.0 [6.3]	5	(0.2)	99	(3.0)
25.00 and above	6	(0.3)	423	(13.5)	145.9 [34.5]	5	(0.2)	254	(7.7)
0.01–25.00 and above	1 722	(100.0)	3 138	(100.0)	267.5 [8.5]	2 075	(100.0)	3 314	(100.0)
Total (including 0.00 size-class of holdings)	2 492		3 138		267.5 [8.5]	2 492		3 314	

Source: *Tables with Notes in Some Aspects of Landholdings in Rural Areas* (State and All-India) (estimates), The National Sample Survey, no. 144, Seventeenth Round, September 1961-July 1962, Tables 3.6, 4, 5.6, 7.6 and 8.

Table 2.5: Ownership and Operational Holdings of Households in Kerala, 1966-7 (rural and urban)

Size-class of holding (in acres)	Ownership Holdings				Operational Holdings			
	Number of Households (000) (%)		Total Area (000 acres) (%)	Area Leased Out (000 acres) [% of area owned]	Number of Households (000) (%)		Total Area (000 acres) (%)	% of Area Leased-in to Area Operated
0.00	2 027		Nil	Nil	1 051		Nil	Nil
0.01–0.99	903	(60.0)	348 (10.2)	2.6 [0.8]	1 481	(59.7)	561 (12.4)	46.2
1.00–2.49	337	(22.0)	510 (15.0)	11.2 [2.2]	548	(22.1)	843 (18.7)	46.0
2.50–4.99	135	(9.0)	481 (14.2)	21.1 [4.4]	250	(10.1)	887 (19.6)	47.7
5.00–9.99	88	(5.8)	602 (17.7)	30.2 [5.0]	139	(5.6)	958 (21.2)	35.7
10.00–14.99	23	(1.5)	272 (8.0)	33.9 [12.5]	37	(1.5)	447 (9.9)	42.8
15.00–19.99	5	(0.3)	86 (2.5)	29.1 [33.7]	9	(0.4)	151 (3.3)	41.6
20.00–24.99	3	(0.2)	76 (2.2)	21.4 [28.2]	5	(0.2)	110 (2.4)	51.3
25.00 and above	10	(0.7)	1 021 (30.1)	651.6 [63.8]	11	(0.4)	560 (12.4)	35.7
0.01–25.00 and above	1 504	(100.0)	3 396 (100.0)	801.1 [23.6]	2 479	(100.0)	4 516 (100.0)	42.6
Total (including 0.00 size-class of holding)	3 531		3 396	801.1	3 531		4 516	42.6

Note: There may be slight discrepancies due to rounding.
Source: Land Reforms Survey in Kerala, 1966-7: Report (Bureau of Economics and Statistics, Government of Kerala, 1968), Tables 19 and 17.

(d) A little over 1 million rural households had ownership holdings less than 1 acre in size, but the total land owned by them was less than 0.4 million acres; moreover, though the number of households with operational holdings in this size-range was over 1.3 million, the area in their possession was only marginally higher, indicating that the net area leased in from households with larger holdings within the rural sector was negligible.

(e) More than four-fifths of the total area leased out by rural households was from those who owned holdings of 25 acres and more in size.

A more comprehensive survey of landholdings was conducted towards the end of 1966 by the Bureau of Economics and Statistics of the Government of Kerala, specially focusing on the issues raised by land reform. It covered both rural and urban areas and had nearly 3,500 households in the sample (drawn from 144 villages and 18 urban wards). Moreover, the villages from which the sample households were drawn were allocated between the different natural regions of the State according to the proportion of the population in the lowland, midland and highland regions of each district (which ensured that the areas in which plantations were located were adequately covered); and, in each village, equal representation was given (wherever possible) to five categories of households distinguished according to the nature of their interest in land, two of which related to households leasing in land and one to those leasing out land (thus ensuring that households involved in such leasing received adequate weightage).

Table 2.5 presents the estimates from this Land Reforms Survey of 1966/7. The total number of households was about 1 million more than the estimate for 1961/2, reflecting the inclusion of urban households (which were probably about 0.7 million) and growth in the number of rural households in the intervening period (due to growth of population and possibly some partitioning). The estimated total area under operational holdings was also about a third higher, amounting to over 4.5 million acres (which was seven-eighths of the net cropped area as reported by the Board of Revenue). The coverage of the 1966/7 survey was, therefore, clearly much wider, taking in the holdings of urban households and of rural households in the uplands of the State which were probably under-represented in the sample for the earlier survey. Since temples, families of chieftains and royal households, and other religious and charitable institutions were not covered by the survey, the land owned by them could not be included in the estimates of the area under ownership holdings but, in so far as such land was leased out to rural or urban households, it would have been counted in as part of the total area under operational holdings; indeed the discrepancy between the estimate of the total area in ownership holdings (3.4 million acres) and of the total area in operational holdings (4.5 million acres) is explained by the area leased in by rural and urban households from these sources.

An important difference made by the wider coverage was that ownership holdings in the size-range of 25 acres and above were captured more adequately. Though only 0.66 per cent of the total number of households owning land (and 0.33 per cent of all households) they held over 1 million acres (as can be seen from Table 2.5), accounting for more than 30 per cent of the total area under ownership holdings. Considered along with the land owned by temples, families of chieftains and royal households, and other religious and charitable institutions (which could not have been less than 1.1 million acres, since this was the area of land leased in by other households, from them), it is ample evidence of the extent of concentration in the ownership of land in Kerala even as recently as a decade and a half ago.

This did not mean, however, that a large surplus was available from these holdings from the imposition of ceilings as proposed. The report on the survey had the following observations on this question:

> A notable feature of the size distribution of holdings is that there is a sharp fall in the number of holdings to an insignificantly low level in the two size-groups just above 15 acres. This, in all likelihood, is the result of purposeful subdivision and alienation in order to circumvent the provision of the K.L.R. Act fixing a ceiling on land-holdings. Holdings above 25 acres in size are mostly plantations which are exempted from the above provision.
>
> Above 83 per cent of these households [with operational holdings of 15 acres and above] . . . do not have any excess land under the provisions of the law. The main reason for this is that the average number of family units (i.e. the number of married couples and adult unmarried persons) per household is about three and consequently, the households are able to retain a much larger area than the ceiling.

An estimate of surplus land made in the report (on the basis of the then operative provisions of KLRA, 1963) indicated that it was not likely to be more than about 0.11 million acres, almost wholly from the northern areas of the State. This corresponds closely to the area identified so far as surplus (in terms of the provisions of KLRA, 1969).

The proposed reform could, therefore, affect significantly the concentration in ownership of land only through its provisions for transfer of ownership of land leased in by tenants. This could, however, be considerable. It will be seen from Table 2.5 that as much as 0.65 million acres of the total 0.80 million acres leased out by households was from ownership holdings in the size range of 25 acres and above. To this should be added 1.1 million acres of land leased in by households from temples, families of chieftains and royal households, etc., since ownership of such land was also to be transferred to tenants. In a State that had only about 5 million acres of land under cultivation, one would expect significant changes in agrarian relations and in the distribution of wealth and income from transfer of as much as 1.75 million acres from the upper strata of society (including the institutions hitherto under their control) to others lower down.

An important question to be examined, however, is who exactly would have gained from such transfer and to what extent. For reasons indicated earlier, it is clear that landless households in the rural sector (estimated at 0.4 million in 1961/2) could have gained only to the extent that they were agricultural labourers in possesion of hutment dwellings. According to the 1966/7 survey, the total number of 'kudikidappukar' households was about 0.35 million, but the total area they were entitled to acquire (at the rate of one-tenth of an acre at the maximum per household) could not be more than 0.04 million acres. We need, therefore, to consider how far tenant households with relatively small holdings could have gained.

Table 2.6 presents the estimates available from the 1966/7 survey on the distribution of land owned and leased in by tenant households according to the size of their operational holdings distinguishing between households which had already some land of their own and those which did not. Table 2.7 shows the intra-regional variations in the distribution of leased-in land as between these two categories of tenant households. From Tables 2.5, 2.6 and 2.7 it is possible to draw some inferences as to who among the tenant households stood to gain from transfer of ownership on leased-in land, to what extent, where, and at whose cost. We shall briefly indicate them below:

(a) Of all the households with operational holdings (about 2.5 million, just over one-half had leased-in land and were tenant households; more than three-quarters of these tenant households

Table 2.6: *Distribution of Land Leased in by Tenant Households, According to Size of Operational Holdings, 1966/7*

Size of Operational Holdings (acres)	Tenant Households Depending Entirely on Leased-in Land		Tenant Households with Leased-in Land Supplementing Own Land		All Tenant Households		
	No. (000)(%)	Area Leased In (000 acres)	No. (000)(%)	Area Leased In	No. (000)(%)	Area Owned (000 acres)(%)	Area Leased In (000 acres)(%)
Below 1	655.6 (67.1)	240.9 (21.5)	63.5 (21.8)	17.9 (2.2)	719.1 (56.7)	18.8 (3.1)	258.8 (13.5)
1.00–4.99	285.4 (29.2)	576.0 (51.5)	155.4 (53.4)	234.1 (29.1)	440.8 (34.8)	192.1 (32.1)	810.1 (42.2)
5.00–14.99	33.6 (3.4)	260.9 (23.3)	59.7 (20.5)	272.5 (33.9)	93.3 (7.4)	257.6 (43.1)	533.4 (27.8)
15.00 and above	1.9 (0.2)	40.0 (3.6)	12.3 (4.2)	279.3 (34.8)	14.2 (1.1)	129.6 (21.7)	319.3 (16.6)
	976.5 (100.0)	1 117.9 (100.0)	291.1 (100.0)	803.6 (100.0)	1 267.6 (100.0)	598.0 (100.0)	1 921.5 (100.0)

Note: There may be slight discrepancies due to rounding.
Source: Land Reforms Survey in Kerala, 1966-7: Report.

Table 2.7: Intra-regional Variation in the Scale of Tenancy and in Type of Tenant Households, 1966/7

District	Tenant Households Depending Entirely on Leased-in Land		Area Leased In (000 acres) (%)		Tenant Households with Leased-in Land Supplementing Own Land		Area Leased In (000 acres) (%)	
	Number (000)	(%)			Number	(%)		
Malabar	723.7	(74.1)	861.8	(77.1)	89.6	(30.8)	312.5	(38.9)
Cochin	204.2	(20.9)	218.4	(19.5)	92.1	(31.6)	255.2	(31.8)
Travancore	48.6	(5.0)	37.6	(3.3)	109.3	(37.5)	236.2	(29.3)
Kerala State	976.5	(100.0)	1 117.9	(100.0)	291.1	(100.0)	803.6	(100.0)

Note: There may be slight discrepancies due to rounding.
Source: Land Reforms Survey in Kerala, 1966-7: Report.

had no land of their own. Approximately 1.25 million house-holds stood to gain, therefore, directly from transfer of owner-ship on leased-in land, nearly 1 million of whom could newly acquire ownership status through the proposed transfer.

(b) Of the tenant households that could newly acquire ownership status more than two-thirds (0.66 million) had, however, hold-ings below one acre in size, and the total area of leased-in land over which they could acquire ownership was less than 0.25 million acres; they had only very small parcels of land (about one-third of an acre on average) and their gains could be, there-fore, only marginal. More than 70 per cent of the tenant house-holds with holdings in this size range were from districts falling within the earlier territory of Malabar.

(c) About 0.8 million acres of leased-in land were in the hands of tenants (0.44 million in number) with holdings of 1 to 5 acres in size who could acquire ownership over them. More than two-thirds of this area was also in Malabar.

(d) Less than a tenth of all tenant households (a little over 0.1 million) had operational holdings above 5 acres in size; they could gain substantially through transfer of ownership on the land leased in by them (about 0.85 million acres). Of them, approximately 14,000 households with relatively large-sized holdings above 15 acres in size were in a position to increase the area owned by them from 0.13 million to 0.45 million acres in this way.

(e) Of the total area of land leased in (over 1.9 million acres), more than 86 per cent was in the territory of earlier Malabar and Cochin. More than 90 per cent of the area leased out from ownership holdings of 25 acres and above in size was also in Malabar and Cochin. The proposed tenancy reform implied therefore primarily redistribution of ownership in land in the northern districts of the State.

(f) Of the land so redistributed, more than two-fifths could be claimed by tenants with holdings of 5 acres and above, a signifi-cant proportion of it by those with holdings of relatively large size (15 acres and above); most of these holdings were also located in the northern districts.

(g) Since the total area leased out by households in Travancore was but a very small proportion (about 3 per cent) of the total area leased out by all households in the state (0.8 million acres), the possible loss to them on account of transfer of ownership rights to tenants was negligible.

We need now to compare the pattern of distribution of land in the 1960s (before the land reform was implemented) with the position thereafter. The data problems are more serious here. Two surveys on landholdings conducted by the Bureau of Economics and Statistics of the Government of Kerala in 1970/1 and 1976/7 as part of an agricultural census were concerned only with operational holdings (though they covered the urban and the rural sectors and included plantations as well). Another survey designed and carried out by the Bureau in 1978/9 was concerned with both ownership and operational holdings, and was organised specifically for the purpose of studying (among other things) the changes in the distribution of holdings after 1966/7 following the land reform, but the findings from this survey are not yet available and cannot be used. The only source of data available on ownership holdings is, therefore, a survey conducted by the National Sample Survey in 1971/2, as part of a larger effort covering the entire country. Unlike the one for 1961/2 (referred to in detail earlier), this survey covered both the rural and urban sectors and furnished data separately for each sector. But it could not have captured fully the changes in distribution brought about by land reform as neither the legal transfer of ownership to tenants (through issue of purchase certificates) nor acquisition of surplus land from large holders of land was near completion at that time.

The most serious of all the limitations of data concerning landholdings in the 1970s is that the implementation of land reform had itself the effect of distorting the information supplied in response to surveys, resulting in considerable under-reporting of the area in the relatively larger-sized holdings. This affected even surveys of operational holdings, like the one conducted in 1970/1 in which under-reporting of area was attributed by the Bureau to some special reasons (among others) which are of relevance here: (i) the legislation imposing ceilings on landholdings had led to large-scale partitions in families and, though the land belonging to a family might be operated as a single unit for all practical purposes, only the land legally belonging to the head of the household would be reported as the holding area; (ii) if a member of a household operated land on behalf of others who were residing away from the household, the details of the land belonging to such persons might not be reported by the household member; (iii) the liability on account of procurement of foodgrains by the State through a progressively graded producers' levy had also created a tendency to under-report the area under paddy. 'The extent of under-reporting is seen largest in the holdings of more than four hectares. This shows that the Land Reform Act has heavily influenced the cultivators while reporting the area of their operational holdings.[12]

Since it was the area in the larger holdings that was under-reported most, and such holdings accounted for a high proportion of the total area under both ownership and operational holdings, the estimates of total area arrived at through these sample surveys were also thereby affected (as it reduced to that extent the average size of holdings in the selected samples). Even when the upland areas of the State (where the size of holdings tends to be larger) have been given adequate weightage in the samples, the estimates made of the total area under operational holdings have been generally about a third lower than the estimates of net cropped area in the State available from the Board of Revenue. When the total area is seriously under-estimated on account of greater under-reporting of area in the relatively large-sized holdings, one has obviously to be very careful in drawing inferences about the size-distribution of holdings from the estimated percentage shares of different-sized holdings in the total. There is no reason to assume that the Land Reforms Survey conducted by the Bureau in 1978/9 will be free from this problem.

We shall, nevertheless, attempt an assessment of the impact of land reform on the distribution of holdings on the basis of such estimates as are available, keeping in mind the nature and direction of the biases introduced by the factors indicated above. We shall consider data relating to both ownership and operational holdings (particularly since after the abolition of tenancy there should be no significant difference in distribution between the two), selecting in each case estimates for the 1960s and 1970s that are most comparable with each other.

In regard to ownership holdings our choice is limited for the rural sector to the data from the NSS of 1961/2 and 1971/2, and for the rural and urban sectors together to the data from the Land Reforms Survey of 1966/7 and the NSS survey of 1971/2. Though the latter two are perhaps less comparable due to differences in sampling design than the former, we shall examine them all and see what inferences can be reasonably drawn. Table 2.8 summarises the estimates in a convenient form.

For operational holdings we shall consider households in the rural and urban sectors taken together and compare the estimates of the land held by them in two different sets of years, first 1966/7 and 1971/2 and then 1970/1 and 1976/7. The estimates for 1976/7 are still provisional, and moreover, some of the size-classes for which data have been furnished so far (particularly at the lower end of the distribution) are

Table 2.8: Household Ownership Holdings in Kerala, 1961/2, 1966/7 and 1971/2

Size class of Ownership Holdings (in acres)	Rural Sector								Rural and Urban Sectors							
	Number of Households (000)				Area (000 million)				Number of Households (000)				Area (000 million)			
	1961/2	(%)	1971/2	(%)	1961/2	(%)	1971/2	(%)	1966/7	(%)	1971/2	(%)	1966/7	(%)	1971/2	(%)
0.00	770	–	400	–	–	–	–	–	2 027	–	730	–	–	–	–	–
0.01–0.99	1 034	(60.1)	1 437	(60.0)	384	(12.3)	449	(16.8)	903	(60.0)	1 805	(70.4)	348	(10.2)	514	(17.7)
1.00–4.99	552	(32.1)	605	(28.2)	1 203	(38.3)	1 292	(48.4)	472	(31.4)	646	(25.2)	991	(29.2)	1 382	(47.6)
5.00–24.99	130	(7.6)	99	(4.5)	1 128	(36.0)	851	(31.9)	119	(7.8)	111	(4.3)	1 036	(32.4)	931	(32.1)
25.00 and above	6	(0.3)	2	(0.1)	423	(13.5)	79	(3.0)	10	(0.7)	2	(0.1)	1 021	(30.1)	79	(2.7)
0.01 and above	1 722	(100.0)	2 144	(100.0)	3 138	(100.0)	2 670	(100.0)	1 504	(100.0)	2 564	(100.0)	3 396	(100.0)	2 904	(100.0)
Total (0.00 and above)	2 492		2 544		3 138		2 670		3 531		3 294		3 396		2 904	
Total net cropped area as estimated by Board of Revenue	–		–		NA		NA		–		–		4 724		5 407	
Total estimated area under ownership holdings as % of total net cropped area	–		–		–		–		–		–		71.9		53.7	

Note: There may be slight discrepancies due to rounding.

Source: Land Reforms Survey in Kerala, 1966/7: Report; NSS Survey 1971/2.

Table 2.9: Household Operational Holdings in Kerala, 1966/7 to 1976/7

(1) Rural Sector

Size-class of Operational Holdings (in acres)	Number of Households (000)				Area under Holdings (000 acres)			
	1966/7	%	1971/2	%	1966/7	%	1971/2	%
0.00	1 051		697		–		–	
0.01–0.99	1 481	(59.7)	1 822	(70.2)	561	(12.4)	554	(18.1)
1.00–4.99	798	(32.2)	653	(25.1)	1 730	(38.3)	1 430	(46.8)
5.00–24.99	190	(7.7)	119	(4.6)	1 666	(36.8)	991	(32.4)
25.00 and above	11	(0.4)	2	(0.1)	560	(12.4)	81	(2.6)
0.01 and above	2 479	(100.0)	2 597	(100.0)	4 516	(100.0)	3 056	(100.0)
Total (0.0 and above)	3 531		3 294		4 516		3 056	
Area under holdings as % of net cropped area (as reported by Board of Revenue)			–		87.4		56.5	

(2) Urban Sector

Size-class of Operational Holdings (in hectares)	(in acres)	Number of Households (000)				Area under Holdings (000 acres)			
		1970/1	%	1976/7	%	1970/1	%	1976/7	%
Below 0.5	Below 1.23	2 030	(71.9)	2 611	(75.4)	728	(18.3)	942	(23.7)
0.5–1.0	1.23–2.46	368	(13.0)	426	(12.3)	637	(16.0)	739	(18.6)
1.0–2.0	2.46–4.92	268	(9.5)	277	(8.0)	902	(22.7)	939	(23.6)
Below 2.0	Below 4.92 (approx. below 5 acres)	2 666	(94.4)	3 313	(95.7)	2 267	(57.0)	2 620	(65.8)
2.0–10	4.92–24.71	153	(5.4)	145	(4.2)	1 206	(30.4)	1 197	(30.1)
10 and above	Above 24.71 (approx. 25 acres and above)	4	(0.1)	4	(0.1)	495	(12.5)	165	(4.1)
Total		2 823	(100.0)	3 462	(100.0)	3 969	(100.0)	3 982	(100.0)
Area under holdings as % of net cropped area (as reported by Board of Revenue)		—		—		74.0		73.2	

Note: There may be slight discrepancies due to rounding.

not quite comparable with those for 1966/7 and 1971/2; they are, however, the only estimates at hand on distribution of landholdings in the State in the second half of the 1970s, and are wholly comparable with those for 1970/1 as they are both from surveys undertaken by the Bureau of Economics and Statistics of the Government of Kerala as part of an agricultural census for these two years. The data are presented in summary form in Table 2.9 for selected size classes.

It will be seen from Table 2.8 that the number of households with ownership holdings was significantly higher in 1971/2 than earlier, about 0.4 million more when only the rural sector is taken into account (and the comparison is with 1961/2), and about 1 million more when both the rural and urban sectors are taken together (and the comparison is with the estimate for 1966/7). The larger numbers are predominant in the holdings below 1 acre in size (accounting for well over four-fifths of the additional households with ownership holdings), but part of the increase is to be found also among holdings of 1 to 5 acres in size. The number of households with holdings above this size was lower in 1971/2 than before.

The area in ownership holdings was also higher in 1971/2 than earlier in the two lower size-classes (i.e. in holdings below 1 acre, and 1 to 5 acres); the increase was over 0.5 million acres in the rural and urban sectors together compared to 1966/7, though it was only about one-third as much in the rural sector alone compared to 1961/2. The area in the higher size-classes was, however, lower than before, very much lower in the holdings of 25 acres and over. Consequently, the total estimated area in ownership holdings in 1971/2 was only about 85 per cent as high as in the earlier years.

In view of the differences in sampling design of the two surveys from which these estimates are derived, and the probable sampling and non-sampling errors, it would be illegitimate to draw any conclusions about the precise extent of the changes in the number and area of ownership holdings belonging to the different size-classes as between 1966/7 and 1971/2. The directions of change and the broad orders of magnitude indicated by the estimates do, however, conform to some of the inferences drawn earlier (from Tables 2.5, 2.6 and 2.7) on the probable impact of land reform on the distribution of ownership holdings. One can thus affirm that (a) the households that gained were in number mostly in the size-class of holdings below 1 acre but the additional area gained by them was relatively small; and (b) though the number of ownership holdings 1 to 5 acres in size increased much less, their gain in terms of area was probably very much more (particularly when households in the urban sector are also counted in).

There was, however, no reason why the total area in ownership hold-ings or the area in holdings of 5 to 25 acres in size should have been lower in 1971/2 than in 1966/7 (as appears from Table 2.8). Since the proposed reform visualised transfer of ownership households that had leased-in land owned by temples, families of chieftains, etc., the total area in household ownership holdings should have been considerably higher in 1971/2 than in 1966/7. As a significant part of the area leased in by households in 1966/7 was into holdings in the size-range of 5 to 25 acres, one would expect the area owned by such households to be also larger in 1971/2. All one can say, therefore, is that the lower esti-mates of area in ownership holdings as a whole and in the intermediate size-range of 5 to 25 acres could have been due to one or more of the following reasons: (i) the larger ownership holdings were probably less adequately covered than in 1966/7, particularly the ones in the upland areas, due to the sampling design adopted by the NSS being different from that used earlier for the Land Reforms Survey; (ii) since the imp-lementation of land reform was incomplete at that time, households with relatively large holdings could have been unsure how much of their land they would be able to retain; and (iii) the area under owner-ship holdings could have been generally under-reported, more so the area in the larger holdings, because only the land belonging to the head of the household was reported and the shares of the others in the family were left out (either from abundant caution or with a view to concealment).

Whatever the reason, the 1971/2 survey did fail to capture a large part of the area that it was intended to cover. The extent to which it failed will be evident from Table 2.9, which presents in summary form the estimates in respect of operational holdings. Not only was the estimated area in 1971/2 lower than in 1966/7 for all size-classes but the total area under operational holdings was less than 57 per cent of the net cropped area in the State that year (as reported by the Board of Revenue), whereas it was more than 87 per cent in 1966/7. No defini-tive inferences can, therefore, be drawn about the changes in the inter-vening period, except perhaps the two concerning relatively small-sized holdings indicated earlier.

It is possible, however, to carry the analysis further and reach some-what firmer conclusions on the basis of the estimates available from the two surveys conducted by the Bureau of Economics and Statistics in 1970/1 and 1976/7. Though they relate to operational holdings only, it would be safe to assume that, while not all the area owned might be reported as part of such holdings (particularly if there had been evasion or avoidance of land reform) only area actually owned by households will have been generally reported as operational holdings after the abolition of tenancy; the pattern of distribution of operational holdings can therefore be taken to reflect also the pattern of distribution of ownership holdings, with the qualification that some of the area owned (particularly in the relatively large-sized holdings) might not have been reported at all.

It will be seen from Table 2.9 that the total estimated area under operational holdings was about the same in 1976/7 as in 1970/1, accounting in both cases for nearly three-quarters of the net cropped area in the State (as reported by the Board of Revenue). This is also much closer to the proportion of net cropped area covered by the survey conducted in 1966/7. It is, therefore, possible to compare the changes in distribution between these years with greater confidence.

In 1966/7, the estimated number of households with ownership holdings below 5 acres in size was about 1.33 million and the total area owned by them was also about 1.33 million acres; by 1976/7, the number of households with operational holdings in this size-range (which we might assume was in their ownership as well) was 3.33 million and the area under them was nearly 2.66 million. The increase both in the number of households and in area, it will be noticed, conform fairly closely to the estimates inferred earlier from Tables 2.5, 2.6 and 2.7 on the probable impact of tenancy abolition on holdings below 5 acres in size.

The lower number of households and the lower area in operational holdings of 25 acres and above in size in 1976/7 is also in conformity with what one would expect from our earlier analysis of the estimates for 1966/7; since 0.85 million acres of the total 1 million acres in the ownership holdings of this size had been leased out earlier, and this area would have been transferred to the tenant households, they would have been left with only the balance (which is what one observes in Table 2.9).

But what is clearly not in conformity with earlier estimates, and the inferences we have drawn therefrom, is the area in holdings of between 5 and 25 acres in size. Though there were only about 120,000 households with ownership holdings in this size-range in 1966/7, with total owned area of a little over 1 million acres, leasing in of land had raised the number of operational holdings in this size range to 190,000 and the total area to 1.66 million acres, and tenancy abolition should have led to their acquiring ownership over this area. The estimates for 1976/7 show that there were only 145,000 households with holdings of this size and the area within them was less than 1.25 million acres.

Though this could have been partly due to the holdings of some households falling into the lower size-class through partition of land, it seems highly probable that there has been considerable under-reporting of area among households in this size-range. In fact, much of the evasion and avoidance of land reform is likely to have been among this small group of households accounting for less than 5 per cent of all households. Together with the still smaller proportion of households with holdings of 25 acres and above, it is possible that they have still in their possession around 2 million acres of land (since holdings below 5 acres in size account for only 2.66 million acres of the total 5 million acres of net cropped area in 1976/7). While there is clear evidence therefore of reduction in concentration in landownership, particularly at the extremities (i.e. in holdings below 1 acre and above 25 acres), about 150,000 medium owner households could have generally strengthened their position in the course of this period.

IV.2 Changes in Rural Economic Relations

As will have been evident from our brief account of the earlier history of agrarian relations in the region, the southern part of the State (which formed part of Travancore) had been freed more than a century ago from the oppressive system of tenancy dominated by 'janmies' which persisted in the north (particularly in Malabar) till very recently. This had helped to promote a highly commercialised pattern of agricultural development based on peasant proprietorship and on an upper stratum of entrepreneurs with strong capitalist orientation. The land reform implemented over the last two decades after the formation of Kerala was not designed to change this pattern; their main thrust was in effect to extend to the northern districts what had already been accomplished earlier in the south.

Nevertheless, there were important differences in the political and social milieu within which the recent reform of land tenures was undertaken and therefore both in the details of the changes made and in the structure and content of the larger programme of agrarian reorganisation of which it was a part. Unlike the case in Travancore earlier, tenancy was now being abolished totally and there was no scope for further leasing out of land on any terms; since land revenue was also in effect abolished (and replaced by a basic land tax of Rs 2 per acre, retained only to ensure maintenance of proper records of landholdings), the entire income from land could accrue to the owner. Active assistance was also being given by State agencies, through public investment in irrigation, development of co-operative credit, and a variety of other measures for increasing agricultural production. At the same time, on the political plane, agricultural labourers were being organised in trade unions for collective bargaining so that some of the gains from agrarian reform and development could be secured by them. Taking all these together, one might expect the recent agrarian reform in Kerala to promote a higher rate of development of agriculture and a more equitable distribution of the income from it than before.

Whether enough time has passed by for effects of this nature to show up is, however, debatable, since the actual implementation of land reform has taken place only over the last decade. On the other hand, it is possible to argue that the *de facto* position in the agrarian sphere had already changed some years earlier on account of the radically transformed political environment in the State; that tenants had not only enough security of tenure but even the payment of rents by them had become irregular and nominal; and that collective bargaining by agricultural labourers had taken root in the 1960s.

We shall therefore only note some of the trends in evidence in the course of the last two decades within the rural sector and speculate on the impact that agrarian reform could have already had or is likely to have on these trends. We shall confine ourselves to those aspects of the rural economy on which one could expect land reform and other related measures to have some effect, e.g. the market for agricultural labour, organisation of co-operative credit, agricultural production and rural poverty. Needless to say, this should be regarded as only a very tentative exploration of issues which require more systematic investigation and analysis.

At the turn of the twentieth century, and even around 1911, the total number of agricultural labourers in Kerala was no more than about 0.75 million, of whom one-half were females. Nearly two-thirds of them were in the Malabar District in the north; the number in Travancore was less than 150,000. The agricultural labourers at this time were drawn chiefly from the lowest castes who were previously agrestic slaves (e.g. Pulayas, Pariahs, Cherumars), and this appears to have been the case as much in Malabar as in Travancore and Cochin. They were employed mainly in paddy cultivation and, though in some cases attached to particular estates, were generally paid by the day.[13]

The total number of agricultural labourers does not seem to have increased very much up to 1931, and such increases that took place were confined largely to Travancore. However, as will be evident from

Table 2.10, there has been phenomenal growth in their number since then, particularly after 1951, and it rose to nearly 2 million by 1971.

Table 2.10: Agricultural Labourers in Kerala, 1901-71 (in millions)

	Male		Female	Total
1901	0.33	(52.4)a	0.17	0.50
1911	0.35	(53.0)	0.38	0.73
1921	0.35	(53.0)	0.33	0.68
1931	0.38	(53.5)	0.30	0.68
1941	—			—
1951	0.71	(88.8)	0.41	1.12
1961	0.79	(87.8)	0.55	1.35b
1971	1.20	(115.4)	0.71	1.91

Note: a. Figures in parentheses denote the percentage of male agricultural labourers to male cultivators, as recorded in the decennial censuses.
b. The estimates for 1961 given here are the adjusted figures worked out by the Bureau of Economics and Statistics of the Government of Kerala allowing for the obvious under-estimation in the figures recorded in the population census of that year. See *Working Force Estimates, 1951 and 1961 Kerala* (State Income Unit, Bureau of Economics and Statistics, Government of Kerala, 1968), p. 38. There may be slight discrepancies due to rounding in the 1961 figures.
Source: *Census Report*, 1901-71.

This is not a full measure of the scale of proletarianisation in the rural sector, as one would then need to take account also of a more nondescript category of 'general labourers' (as recorded in the earlier censuses), some of whom have also been on occasions seekers of wage employment in agriculture.[14] A more adequate picture can be had, particularly of the trends in the recent period, from Table 2.11 based on data collected through the Rural Labour Enquiry (conducted by the Labour Bureau, Government of India) in 1964/5 and 1974/5.

It will be seen that, by the middle of the 1970s, more than two-fifths of all rural households were primarily suppliers of wage labour and that about two-thirds of such households were agricultural labour households. The total number of rural wage earners was over 2.75 million, of whom again two-thirds were agricultural labourers (though, unlike earlier, only a third of them were from scheduled caste and scheduled tribe households). It can also be observed that, while the number of male agricultural labourers increased at only about the same rate as the total number of rural households, the number of female labourers grew at twice that rate.

With such rapid growth in the supply of rural labour one would

Table 2.11: Rural Labour in Kerala, 1964/5 to 1974/5

	1964/5	1974/5	% Change
	(in millions)		
Total rural households	2.48	3.23	+ 30.2
Total rural labour households	1.04	1.36	+ 30.8
of which agricultural labour households	0.70	0.89	+ 27.1
Total wage earners in rural labour households			
Men	1.24	1.67	+ 34.7
Women	0.77	1.10	+ 42.9
Children	0.04	0.04	−
	2.05	2.82	+ 37.6
Total labourers in agricultural labour households			
− agricultural labourers	1.29	1.75	+ 35.7
of whom Men	0.79	1.02	+ 29.1
Women	0.45	0.72	+ 60.0
Children	0.02	0.01	− 50.0
− non-agricultural labourers	0.12	0.17	+ 41.7
	1.41	1.92	+ 36.2
Total labourers in non-agricultural labour households			
− agricultural labourers	0.09	0.11	+ 22.2
− non-agricultural labourers	0.53	0.78	+ 47.2
	0.62	0.89	+ 43.5
Total agricultural labourers in rural labour households	1.38	1.86	+ 34.8
of whom those belonging to scheduled castes and tribes	0.50	0.60	+ 16.7

Note: There may be slight discrepancies due to rounding.
Source: *Rural Labour Enquiry, 1974-5, Final Report on Wages and Earnings of Rural Labour Households* (Labour Bureau, Ministry of Labour, Government of India, June 1979).

expect considerable downward pressure on wages unless there was corresponding increase in the demand for such labour for agriculture and related activities. The movement in rural wage rates in Kerala since the middle of the 1950s does not, however, indicate any such pressure; unlike in most other States of India there has been some increase even in real terms. The real wage rate of male field labourers is estimated to have risen by more than 50 per cent between 1956/7 and 1971/2; there has been some further rise since then.[15]

Two factors that have been mentioned as possible causes for this rise in the real wage rates of agricultural labourers in Kerala are: (a) rise in productivity of land during this period, and (b) the emergence and growth of trade union organisation among agricultural labourers, and the support given to it by the State government as part of its policies for agrarian reform. The correlation between productivity increase and rise in the real wage rates of agricultural labour in different States during this period has not been found to be particularly strong, though it is true that the output per worker engaged in agriculture was higher in Kerala than in any other State (including Punjab) and could, there-fore, sustain a higher wage rate than elsewhere. The available evidence on the impact of collective bargaining is somewhat stronger, as two of the three districts which recorded the highest increase in real wage rates between 1956/7 and 1971/2 were those in which agricultural labour unions had been active all through the 1960s.[16]

It is probable, however, that a more important factor than either of these was the acquisition of ownership over small holdings of land by a much larger proportion of households supplying wage labour than before. Even prior to the implementation of land reforms most tenant households (including those that supplied wage labour) had begun to enjoy security of tenure and often made only nominal payments of rent to the owners. By the early 1970s, as indicated earlier, rural households with full ownership rights over small-sized holdings (of less than 1 acre) were 0.4 million more in number than in 1961/2; a high proportion of them are likely to have been agricultural labour households. The extent of the increase in the number of agricultural labour households with land in the course of two decades will be evident from Table 2.12. Though the holdings acquired were generally small, one must expect their acquisition by agricultural labour households to have raised at least to some extent the reserve price of the labour forthcoming from these households.

Another factor that could have worked in the same direction is the much lower rate of growth in the number of agricultural labourers belonging to scheduled castes and tribes in recent years. It will be seen from Table 2.11 that, while the total number of wage earners in rural labour households increased by nearly 38 per cent between 1964/5 and 1974/5, the number of agricultural labourers belonging to this category increased by less than 17 per cent during this period. The reasons for this are not clear — how far it is due to a lower rate of growth of

Table 2.12: Agricultural Labour Households with Land

	1956/7	1964/5	1974/5
Total rural households (million)	2.20	2.48	3.23
Total agricultural labour households (million)	0.50	0.70	0.89
Agricultural labour households with land (million)	0.26	0.49	0.77
as % of total rural households	11.8	19.8	23.8
as % of total agricultural labour households	51.6	70.0	86.5

Sources: *Agricultural Labour in India, Report on the Second Enquiry, 1956-7* (1960), and *Rural Labour Enquiry, 1974-5.*

population among scheduled castes and tribes, and how far due to their withdrawal from agricultural labour — and need to be investigated further. There is little doubt, however, that since the bulk of the labour for paddy cultivation (particularly for certain operations such as land preparation) has traditionally come from the lower castes, the lower rate of growth of labourers from among them is likely to have had some effect on the supply of field labour and thereby on the wage rates for such labour.

To the extent that land reform resulted in decrease in the relative share of large-sized holdings in the total operated area one would also expect some reduction to have taken place in the demand for wage labour. However, this may not have been very significant as (a) the dependence on wage labour is not inconsiderable in Kerala even among households with holdings of 0.5 to 2 hectares in size (see Table 2.13),

Table 2.13: Percentage of Landholding Households Dependent on Wage Labour, According to Size of Holdings, 1970/1

Size-class of Holdings (in hectares)	Households Dependent for Agricultural Work Mainly (%)		
	Wage Labour	Family and Wage Labour	Family Labour
0.04–0.25	14.2	15.5	70.3
0.25–0.50	22.1	26.6	51.4
0.50–1.00	29.9	32.1	38.0
1.00–2.00	38.7	33.2	28.1
2.00–3.00	44.0	35.6	20.4
3.00–4.00	54.3	28.7	17.1
4.00–5.00	63.3	25.6	11.1
5.00–10.00	67.7	24.0	8.3
Over 10.00	79.7	14.4	5.9
Holdings of all sizes	23.6	23.3	53.1

Source: *The Third World Census of Agriculture, 1970-71, Report for Kerala State*, vol. II, Table XXI.

(b) the scale of land transfer of holdings below this size has been limited, and (c) the share of holdings above this size in the total operated area remains large even after land reform.

Nevertheless, it would appear from a more disaggregated analysis that, within Kerala, the districts where relatively large operational holdings (of more than 2 hectares, i.e. 5 hectares in size) accounted for a higher proportion of the total operated area, and the number of agricultural labourers available in relation to the area under such holdings was comparatively lower, have been able to maintain relatively high wage rates over the last quarter of a century. The supporting evidence, though by no means conclusive, is presented in Table 2.14.

It will be seen from Table 2.14 that in 1965/6 the highest wage rates were in Cannanore, Trichur and Ernakulam where the share of the relatively large operational holdings in total operated area was much higher, and the number of agricultural labourers in relation to the area under such holdings comparatively much less, than in the State as a whole. One would have expected the wage rate in Palghat to be higher (in view of the high share of the relatively large-sized holdings in the total operated area), but it is well known that seasonal immigration of labour into this district from the adjoining areas of Tamil Nadu tends to keep the wage rate low. On the other hand, the wage-rate prevailing in Alleppey appears to be higher than one might expect (in view of the share of the relatively large operational holdings being so much lower, and the number of agricultural labourers in relation to the area under such holdings being so much higher, than in the State as a whole); evidently this was the effect of collective bargaining, which started here before anywhere else and has been generally more effective than in the rest of the State.[17]

It will also be noticed that, while Cannanore, Trichur, Kozhikode and Quilon have improved their relative position in regard to the prevailing wage rates, there has been deterioration over a period in the relative position of Palghat, Ernakulam , Kottayam and Alleppey. For an adequate explanation of these movements in wage differentials over time one would have to take into account several other factors such as extension of area under cultivation, changes in cropping intensity and in the cropping pattern (such as those associated with shift from paddy to coconut cultivation in many districts), demand for labour for non-agricultural purposes (such as from traditional industries like coir and cashew, trade and transport, and construction activity), and possibly also some of the indirect effects of emigration (which have perhaps been significant in some areas since 1975/6). The decline in the relative

Table 2.14: Average Daily Wage Rates of Male Field Labour and Some Factors Governing its Demand and Supply

District	Average Daily Wage rate (in Rs)[a]					Percentage of total Operated area in holdings of 5 acres and Above[b]	Average No. of Acres in holdings of 5 acres and above per Agricul. Labourer[c]
	1958/9	1965/6	1970/1	1975/6	1979/80		
Cannanore	2.01 (124.8)	3.70 (115.6)	6.23 (122.4)	11.42 (140.8)	13.25 (138.3)	61.6	8.5
Kozhikode	1.85 (114.9)	3.18 (99.4)	4.39 (86.2)	9.13 (112.6)	10.17 (106.2)	24.4	1.6
Palghat	1.27 (78.9)	2.71 (84.7)	4.05 (79.6)	6.00 (74.0)	7.15 (74.6)	53.2	3.0
Trichur	1.53 (95.0)	3.21 (100.3)	5.62 (110.4)	8.50 (104.8)	10.09 (105.3)	61.6	6.3
Ernakulam	—	4.01 (125.3)	5.83 (114.5)	9.00 (111.0)	10.25 (107.0)	50.4	4.8
Kottayam	1.47 (91.3)	2.93 (91.6)	5.04 (99.0)	7.38 (87.3)	7.40 (77.2)	46.7	5.1
Alleppey	1.68 (104.3)	3.14 (98.1)	5.43 (106.7)	7.08 (87.3)	8.53 (89.0)	27.1	1.3
Quilon	1.60 (99.4)	2.92 (91.3)	4.47 (87.8)	7.50 (92.3)	10.04 (104.8)	37.5	2.9
Trivandrum	1.46 (90.7)	2.96 (92.5)	4.75 (93.3)	7.58 (93.5)	8.90 (92.9)	31.8	1.5
Kerala State	1.61 (100.0)	3.20 (100.0)	5.09 (100.0)	8.11 (100.0)	9.58 (100.0)	49.3	3.6

Notes: a. Figures in brackets indicate the percentage level of the wage rate in each district relative to the State average.
    b. The estimates are based on data from the *Land Reforms Survey in Kerala* (1966/7).
    c. The estimate for each district includes only male agricultural labourers and plantation labour, and is taken from the 1961 population census.
Sources: Various sources.

level of the wage rates for field labour in Alleppey and Palghat (where agricultural labour unions have been more active than elsewhere) provides, however, enough suggestive evidence to indicate that, unless the population dependent on agricultural labour for livelihood declines or the demand for such labour grows considerably from the larger-sized holdings (say, due to more intensive use of land for agricultural production), it might prove very difficult to maintain for long the relatively high wage rates that have been prevailing in many parts of Kerala.

It needs to be emphasised in this context that, even while daily wage rates have been rising, available data suggest that the average number of days of employment secured by agricultural labourers over a year has been declining, particularly since the middle of the 1960s (as will be evident from Table 2.15). One cannot, therefore, be sure how far their annual real income has increased over this period, though it is possible that there has been some increase in this income, partly on account of the rise in real wage rates and partly from self-employment on the small holdings acquired in the process of land reform.

Table 2.15: Average Annual Days of Wage Employment of Male Agricultural Labourers Belonging to Agricultural Labour Households

	Agricultural Employment	Non-agricultural Employment	Total Employment
1950/1	170	25	195
1956/7	156	14	170
1964/5	173	14	187
1974/5	138	12	150

Sources: The sources of the data are the Agricultural Labour Enquiries of 1950/1 and 1956/7 and the Rural Labour Enquiries of 1964/5 and 1974/5; the estimates for 1974/5, which were communicated privately to us by the Labour Bureau, Government of India, in June 1979, were provisional and subject to revision.

In the sphere of rural credit, professional moneylenders as a class have been generally less prominent in Kerala than elsewhere in India, and various forms of indigenous credit based on mobilisation of savings on group or community basis had developed at an early stage to meet requirements for both productive and non-productive purposes. Along with the growth of an active market for land in Travancore in the early decades of the present century there had also developed a number of small unit banks in the region which, despite some prominent failures, laid the foundation for extensive commercial banking in rural areas.[18]

Nevertheless, rates of interest remained high (though disguised in various ways) and credit was not easily accessible to those who could not offer adequate security and to small holders in general even when they had the security of their produce to offer.[19] The need to develop co-operative credit was therefore recognised all along as a necessary and integral aspect of agrarian reform.

The factors in favour of the development of institutional credit in Kerala, particularly of co-operative credit, were also noted by the All India Rural Credit Review Committee appointed by the Reserve Bank of India in July 1966.

> With hardly any area susceptible to famine, it has a variety of crops, including the remunerative plantation crops and fruit orchards, which made the gross value of agricultural output per acre in the state the highest in the country in 1966-7. In regard to per acre consumption of fertilizers also, Kerala ranked first among the states. Two other factors congenial to the growth of cooperatives are the predominance of cultivators with small holdings, who have relatively more to gain from a system of cooperative credit, and the high rate of literacy in which Kerala leads all the other states. The density of population, which is the highest for any state, may also be considered as a favourable circumstance in that it should enable a credit society to achieve viability even within a relatively small area of operation. The traditional familiarity of the rural population in Kerala with banking habits is another advantage.[20]

Nevertheless, until then the record of co-operative agricultural credit was 'a somewhat mixed one', and not only was the coverage of co-operative membership limited but both the proportion of borrowing members and the level of advances per member were low and deposit mobilisation by the co-operative credit societies was unsatisfactory.[21]

It was only after 1969/70 that the picture changed. As will be evident from Table 2.16, there has been not only weeding out of dormant societies and considerable expansion in membership and in the number of borrowing members, but phenomenal growth in the deposits of primary agricultural credit societies in the course of the subsequent decade. Along with a nearly fivefold increase in the loans advanced between 1969/70 and 1978/9 there has also been a decline in the percentage of overdues to outstanding loans.

Though credit made available by commercial banks for agriculture has also been rapidly expanding over the last decade, the share of

Table 2.16: *Progress of Primary Agricultural Credit Societies, 1956/7 to 1978/9*

	1956/7	1960/1	1965/6	1969/70	1974/5	1978/9
Total number of societies	2 098	2 336	2 440	2 160	1 787	1 616
Total number of members (000)	456	806	1 283	1 536	1 962	3 004
Total number of borrowing members (000)	NA	386	499	594	947	1 348
Owned funds (in Rs crores)	1.6	2.8	6.1	10.7	18.8	32.6
Deposits (in Rs crores)	0.4	1.2	4.7	8.8	17.0	85.3
Borrowing (in Rs crores)	NA	2.8	10.4	22.3	37.9	59.5
Total loans issued during the year (in Rs crores)	1.5	5.1	12.4	25.8	44.5	122.0
Total loans outstanding at the end of the year (in Rs crores)	1.6	5.1	14.6	29.8	52.7	141.9
Per cent of overdues to outstanding loans	19.8	16.6	24.9	24.0	35.9	18.7

Sources: *Report of the All India Rural Credit Review Committee* (1969), p. 240; *Review of Agricultural Development and Cooperative Credit in Kerala* (All India Debt and Investment Survey, 1971-2) (1977), pp. 96-7; 'Progress of Cooperative Credit Societies, 1977-78 and 1978-79' (mimeo.). All the above publications are of the Reserve Bank of India.

primary agricultural credit societies in the total insitutional credit extended to the rural sector was over one-half in 1974/5;[22] with the rapid increase in the loans advanced by these societies in the subsequent period, their share is unlikely to have fallen since then. However, it would appear from the data available on the distribution of the loans extended by the credit societies that, while even agricultural labourers without land of their own have been among the beneficiaries, the percentage share of owners of small holdings (i.e. below 1 hectare) in the loans advanced has not come up to their percentage share in the total operated area. This can be seen from Table 2.17. Further investigation

Table 2.17: Percentage Share of Holdings of Different Size-groups in Total Area Operated and in the Loans and Advances of Primary Agricultural Credit Societies

Category of Borrower	Area under Operational Holdings, 1976/7	Loans and Advances from Primary Agricultural Credit Societies, 1977/8			
		Number of Borrowers	Amount Advanced	Amount Out-standing	Amount Overdue
Holders of owner-ship holdings					
Below 1 hectare	42.3	37.9	33.1	30.6	30.9
1 to 2 hectares	23.6	21.7	24.5	22.4	23.8
2 hectares and above	34.1	26.7	34.1	39.3	40.1
Tenant cultivators	NA	2.2	1.1	1.2	1.6
Agricultural labourers	NA	5.8	4.1	3.6	1.6
Others	NA	5.8	3.1	3.0	2.4
Total	100.0	100.0	100.0	100.0	100.0

Sources: For data on distribution of landholdings, Table 2.9 of this chapter; for data on loans and advances, *Statistical Statement relating to Co-operative Movement in India, Part I, Credit Societies*, Table 29.

is needed to find out the precise purposes for which credit has been extended by co-operative societies and how far it has helped to raise agricultural production in the relatively small-sized holdings.

Since co-operative marketing has developed very much less than co-operative credit (the total value of agricultural produce sold by primary marketing societies being less than Rs 16 crores in 1977/8), and there are large margins appropriated by traders, it is indeed doubtful whether small holders of land have adequate incentive for raising agricultural production. Very little attention has been given so far to this aspect of agrarian reform, despite numerous recommendations made by various committees for improvements in agricultural marketing.[23] The glaring failures in this sphere of public policy reflect perhaps the considerable power and influence still wielded by commercial interests in the State over even political parties committed to radical agrarian reform.

IV.3 Agricultural Production, Incomes and Poverty

The most striking and potentially the most serious of all the trends in the rural economy of Kerala in the period following the implementation of land reform has been in agricultural production itself. Until the 1970s the area under cultivation as well as the output of crops were growing at a reasonably rapid rate; consequently, there was an increase of nearly 25 per cent in the gross income from land (estimated at constant prices) between 1960/1 and 1970/1.[24] Thereafter there has been no increase in either net or gross sown area. Production continued to rise (though at a slower rate) till 1974/5, but there has been a significant decline in output since then,[25] with the result that the income from agriculture towards the end of the decade was lower (when estimated at constant prices) than at the beginning.

This decline in agricultural output and income is attributable to a large extent to a sharp fall in the production of coconuts (from nearly 4,000 million nuts in 1970/1 to around 3,200 million nuts in 1979/80) caused by disease affecting this crop in some of the districts. However, there has been a decline in production in other crops also: for instance, the output of rice has fallen from 1.38 million tons in 1972/3 to 1.28 million tons in 1979/80, and the output of tapioca from 5.4 million tons in 1975/6 to 4.2 million tons in 1979/80. There have been perceptible increases in production only in the case of rubber, coffee, cardamom and bananas (including plantains).

The impact of the overall decline in agricultural output in the course of the 1970s has been perhaps less in the northern districts (particularly in Cannanore and Kozhikode) than in the southern (such as Alleppey and Quilon), because (a) the extension of net sown area in the course of the 1960s was much greater in the northern districts; and (b) not only were the southern districts more affected by coconut disease but the relative share of coconut in the total cropped area was also greater in many of these districts (see Table 2.18). This could have been among the factors that kept the wage rates higher generally in these northern districts (as observed earlier from Table 2.14).[26] At the same time it is evident that in all districts south of Trichur, where the percentage increase in net area sown was much lower, increase in cropping intensity (as reflected in a higher rate of increase of gross cropped area) was

Table 2.18: *Net Sown Area and Gross Cropped Area, District-wise, 1957/8 to 1973/4, and Gross Area Under Rice and Coconut, 1963/4 and 1973/4 (in 000 hectares)*a

District b	Net Area Sown			Gross Cropped Area			Area under Rice		Area under Coconut	
	1957/8	1963/4	1973/4	1957/8	1963/4	1973/4	1963/4	1973/4	1963/4	1973/4
Cannanore	214.2 (100.0)	275.4 (128.5)	316.7 (147.9)	251.4 (100.0)	312.1 (124.1)	350.0 (139.2)	95.7 (30.7)	98.1 (28.0)	67.2 (21.5)	91.2 (26.1)
Kozhikode and Palghat (including Malappuram)	523.1 (100.0)	583.0 (111.5)	668.7 (127.8)	637.5 (100.0)	693.1 (108.7)	879.5 (131.5)	305.9 (44.1)	339.4 (38.6)	134.8 (19.4)	190.9 (21.7)
Trichur	129.1 (100.0)	134.6 (104.3)	138.6 (107.4)	187.2 (100.0)	202.5 (108.2)	247.8 (132.4)	108.5 (53.6)	109.9 (44.4)	35.5 (17.5)	56.9 (23.0)
Ernakulam and Kottayam (including Idikki)	461.9 (100.0)	501.0 (108.5)	533.3 (115.5)	504.9 (100.0)	560.8 (111.1)	666.5 (132.0)	124.2 (22.1)	133.4 (20.0)	111.1 (19.8)	142.1 (21.3)
Alleppey	157.9 (100.0)	160.0 (101.3)	162.9 (103.2)	197.1 (100.0)	220.2 (111.7)	240.0 (121.8)	82.3 (37.4)	92.0 (38.3)	69.1 (31.4)	79.9 (33.3)
Quilon	205.8 (100.0)	217.9 (105.9)	229.6 (111.6)	237.8 (100.0)	277.0 (116.5)	371.4 (156.2)	49.6 (17.9)	51.2 (13.8)	70.4 (25.4)	106.8 (28.8)
Trivandrum	147.3 (100.0)	150.1 (101.9)	152.2 (103.3)	195.0 (100.0)	196.1 (100.6)	244.3 (125.2)	38.8 (19.8)	39.8 (16.3)	56.9 (29.0)	77.0 (31.5)
Kerala State	1 839.0 (100.0)	2 022.0 (110.0)	2 202.0 (119.7)	2 211.0 (100.0)	2 462.0 (111.4)	3 000.0 (135.7)	805.1 (32.7)	874.7 (29.2)	545.0 (22.1)	744.8 (24.8)

Notes: a. The District of Malappuram was carved out of Kozhikode and Palghat District in 1969; likewise, the District of Idikki was carved out of Ernakulam and Kottayam District in 1972.

b. The figures shown in brackets in the case of net area sown and total cropped area are index numbers with 1957/8 as the base year; the figures shown in brackets in the case of area under rice and area under coconut are the percentages they form of total cropped area in the respective years.

Source: Bureau of Economics and Statistics, Government of Kerala.

considerably greater.

It is probable, therefore, that the decline in the rate of growth of output in agriculture after 1970/1 was due to a combination of circumstances, the non-availability of land for further extension of cultivation even in the northern districts (after the rapid increase in net sown area which took place there between 1957/8 and 1973/4) and the difficulties everwhere in raising the yield per unit of cropped area on land that was either too heavily dependent on rainfall alone or could not be easily reached by reservoir-based surface irrigation. Neither of these can be traced to land reforms as such, though greater attention paid to the location and type of irrigation projects undertaken during this period as well as to other related aspects of planning in the agricultural sector could have arrested the decline.[27]

Some other factors contributing to the declining trend in agricultural output have also been mentioned, particularly in explanation of the fall in rice output after 1972/3. These are the sharp decline in the price of paddy in Kerala (following substantial increase in the imports from elsewhere in the country), rise in the price of nutrients, and rise in the wages of field labour.[28] It needs to be emphasised, however, that though the substitution of higher-value crops like coconut for rice could result in contraction of employment for agricultural labourers, it could signify more efficient use of the available land, more so if such substitution takes place on paddy land that cannot be double-cropped (which is probably what has been happening). From this point of view what matters is not whether there is a decline in area under any particular crop but the direction and extent of change in the value added per unit of land as a consequence of the changes in the cropping pattern with which the decline is associated.

It needs to be mentioned also in this context that, though agricultural output as a whole (and the income from it at constant prices) has been rising at a much slower rate since 1970/1 (and actually declining since 1974/5), the real income from agriculture in Kerala could have been still significantly higher at the end of the decade than at the beginning on account of the substantial increases that have taken place in the prices of its export products. It can be seen from Table 2.19, which gives the percentage increase in the wholesale prices of selected agricultural products since 1960, that the prices of products used extensively for export to the rest of the world (including other parts of India) have risen very much more than of those primarily consumed within the State (such as rice and tapioca).

This is not to under-rate the seriousness of the crisis in the agricultural

Table 2.19: Percentage Increases in Wholesale Prices of Selected Agricultural Products

	April 1960-April 1970	April 1970-April 1979
Rice	+ 78.9	+ 52.6
Tapioca	+ 74.5	+123.6
Coconut	+140.4	+189.2
Rubber	+ 30.6	+128.9
Cashew nut	+130.8	+152.9
Pepper	+ 1.3	+146.7
Tea	+ 5.5	+139.0

Source: Bureau of Economics and Statistics, Government of Kerala.

sector but rather to emphasise the fragile basis on which the economy of Kerala has been resting over the last few years. Table 2.20 indicates very well the impact of the trends in this sector on *per capita* income in the State as a whole.

Table 2.20: Net State Domestic Product at Factor Cost by Industry of Origin (in Rs million)

	Primary Sector	Secondary Sector	Tertiary Sector	SDP (net)	Per Capita Income (in Rs)
At constant 1960/1 prices					
1960/1	2 420 (100.0)	659 (100.0)	1 244 (100.0)	4 322 (100.0)	259 (100.0)
1969/70	3 001 (124.0)	1 072 (167.2)	1 924 (154.7)	5 997 (138.7)	292 (113.0)
At constant 1970/1 prices					
1970/1	6 471 (100.0)	1 946 (100.0)	3 414 (100.0)	11 834 (100.0)	564 (100.0)
1974/5	6 827 (105.5)	2 241 (115.2)	4 012 (117.5)	13 080 (110.5)	569 (100.9)
1978/9	6 273 (96.9)	2 631 (135.2)	4 432 (129.8)	13 335 (112.7)	538 (95.4)

Source: Bureau of Economics and Statistics, Government of Kerala.

Evidently, it is the improvement in the terms of trade (referred to above) and the large-scale inflow of remittances from migrants abroad (which do not by and large enter the estimates of SDP) that have maintained some semblance of prosperity in the State in recent years. This cannot last very long, unless the opportunity is used for strengthening the technological basis for more rapid agricultural growth in the future: the land reforms that have been implemented provide a sound institutional framework for such strengthening but cannot obviously be a substitute for it.

It should be evident from the analysis in the preceding sections that nothing definitive can be said about the impact of agrarian reform on poverty in Kerala. The measurement of poverty itself presents a variety of problems that are not easy to sort out.[29] In addition there have been significant changes over the last two decades both in the distribution of income derived from agriculture and in the rate of growth of income in this sector. Some of the factors involved in these changes are of such bewildering complexity, varying considerably in their importance from region to region within the State, that any attempt to sum up their over-all impact on poverty is likely to be self-defeating. All that can be said perhaps, impressionistically, is that the scale of poverty is likely to have gone down significantly between 1956/7 and 1972/3, both on account of the redistributive effects of land reform and the relatively high rate of growth of agricultural output during this period; and that, if the gains of this period have been subsequently maintained, it could only have been due to other extraneous factors such as improvement in the terms of trade of Kerala and the considerable inflow of remittances from migrants.

Notes

1. For a more detailed account, see T. C. Varghese, *Agrarian Change and Economic Consequences: Land Tenures in Kerala, 1850-1960* (Allied Publishers Private, 1970).

2. There is reason to believe that, where access to imported foodgrain was more difficult (as in many regions of India, particularly in the rural areas), the foodgrain requirements of a growing agricultural population after the First World War led to retardation of the process of commercialisation of agriculture. See Ashoka Mody, 'Population Growth and Commercialisation of Agriculture: India – 1890-1940' (Working Paper no. 117, Centre for Development Studies, Oct. 1980).

3. 'For the people of Kerala the use of the coconut is manifold. It has both food and commercial value. Its luscious kernel is an indispensable ingredient in the diet of the Keralites. The trees are also extensively tapped, and they give sweet toddy which when fermented forms an intoxicating beverage, the "beer" of the working classes of Kerala. The distilled toddy which is known by the name "cocobrandy" has won for itself a place in the liquor trade of India. The coconut oil is an excellent media for cooking and has a variety of industrial uses in the manufacture of soaps, cosmetics and medicinal preparations. The oil-cake is also valuable as a good cattle-feed and a rich plant food. The coconut husk which is a by-product of the coconut forms the raw material for the coir industry . . . The coconut shell lends itself to the making of exquisite works of art in the hands of expert craftsmen. The shell is also used for producing activated carbon which is a gas absorbent and a decolourising agent in the manufacture of vegetable and other oils. The plaited leaves of the coconut tree are widely used as thatch to give shelter to the common man. The coconut stem which generally serves as timber for house construction in Kerala . . .[is] indispensable for building operations.

Thus the uses of the coconut palm are legion, but its commercial value is of vital importance to the Kerala economy.' V. R. Pillai and P. G. K. Panikar, *Land Reclamation in Kerala* (Asia Publishing House, 1965), pp. 34-5.

4. Data for 1874 and 1904 are from Nagam Aiya, *Travancore State Manual*, vol. III (1906).

5. See T. K. Velu Pillai, *The Travancore State Manual*, vol. III (1940).

6. See *Census of India* (1931), vol. XXVIII, *Travancore*, Part I, Report Appendix IV on 'The Economic Condition of the People'.

7. *Census of India, Travancore*, p. 477. The share of Ezhavas in the total area partitioned during this period was only 0.06 million acres; the rest belonged to Nayars.

8. See A. N. Thampy, *Report of the Enquiry into the Sub-division and Fragmentation of Agricultural Holdings in Travancore* (1941); also Varghese, *Agrarian Change and Economic Consequences*.

9. 'In the old reclamation areas [of Kuttanad] tenancy is the rule rather than the exception. Much of the land is owned by the landlords who lease it to a class of tenants who are often themselves owners of large areas under their cultivation' (Pillai and Panikar, *Land Reclamation*, p. 45). 'The most significant point to be noted . . . is the fact that many of the tenant-cultivators of Kuttanad pocket operate large-sized holdings' (Varghese, *Agrarian Change and Economic Consequences*, p. 169).

10. The following observations about the cultivators of paddy on reclaimed land in Kuttanad should also be noted. 'All of them own coconut orchards, and some have rubber, cardamom and tea estate also. Even their cultivation of rice is not limited to capitalistic farming in the lake areas, but extends to *Karappadams* where cultivation is carried on in the conventional style. Nor are the activities of these farmers confined to agriculture. Several of them have different lines of business; one of them, for example, conducts banking, another has a boat-building business, a third carries on trade in spices and hill products, and so on' (Pillai and Panikar, *Land Reclamation*, pp. 80-1).

11. See *Report of the U. P. Zamindari Abolition Committee* (1948), vol. I, p. 365. As Professor Daniel Thorner was to observe later, 'what this formulation provides is a mode of considering the non-cultivating owner as contributing to production and accordingly entitled to a share of it even when, in practice, he leaves to others the functions of planning, of direction, of conducting, and of physically carrying out the entire round of agricultural operations'. Cf. *The Agrarian Prospect in India* (Delhi School of Economics, 1956).

12. *The Third Decennial World Census of Agriculture, 1970-71, Report for Kerala State, Volume I, General Report* (Bureau of Economics and Statistics, Government of Kerala, Nov. 1973), Ch. VIII. The report offers the further observation: 'In 1966 when the Land Reforms Survey was conducted, the situation was different. Though the Land Reforms Act had been passed in 1963 strict measures in implementing the provisions of the Act were not taken till 1966. As a result the cultivators were practically enjoying a rather settled condition in respect of their properties and so they were not motivated to under-report the areas of the holdings held by them' (ibid., pp. 267-8).

13. 'The agricultural labourers are chiefly the Pulayas in Central and North Travancore and the Pariahs and Shanars in South Travancore. They are a skilled class of people and indispensable for the cultivation of wet lands all over the country . . . The Pariahs in South Travancore and the Pulayas in North Travancore may be found even at present attaching themselves as soil-serfs to their old estates. They do not wish to go off from them, and are therefore content with whatever wages their old masters are pleased to give them' (V. Nagam Aiya, *The Travancore State Manual*, vol. III (1906), p. 172). 'The landless labourers consist of two classes, farm servants and small labourers. In this country where the hold-

ings are small and where labour is available without much difficulty and is comparatively cheap, farm servants are rarely engaged except in large estates. We have therefore but 7,534 workers with 3,083 dependants under this category, while the labourers paid by the day aggregate 89,345 excluding 54,834 dependants' (*Census of India, Volume XXVI, Travancore, The Report on the Census*, by N. Subramhanya Aiyar (1903), p. 396). 'The bulk of the agricultural labourers are emancipated serfs, who are still paid in kind and at the same old rates. Their condition as a class cannot be said to have improved to any appreciable extent, though individuals among them have, by obtaining employment in plantations and gardens, begun to earn better wages in recent years' (C. Achutha Menon, *The Cochin State Manual* (1911), p. 246). 'They [Cherumars] work very hard for the pittance they receive; in fact nearly all the rice land cultivation used to be in former days carried on by them. The influx of European planters, who offer good wages, has had a marked effect on releasing this class from some of their bonds, and the hold which their masters had over them has been proportionately relaxed. It is said that the difficulty of providing for their women is the chief obstacle to their complete release from their shackles. The women must have dwellings of some sort somewhere, and the masters provide the women with huts and allow the men to work on plantations on condition that they return in good time for rice cultivation and hand over a considerable portion of their earnings' (William Logan, *Manual of the Malabar District* (1906), vol. I, p. 150).

14. There is reason to believe that agricultural labourers have been classified in the census returns as 'general labourers' not only in 1961 but even earlier, particularly in 1931. One has to be careful, however, lumping the two categories except for particular purposes; workers classified under 'general labour' were numerically a small proportion of 'agricultural labourers' in Malabar, while they were the dominant category in Travancore, all through the period 1901-31, and it is possible that this reflects an important difference in the nature of the work for which agricultural labourers were primarily employed in the two areas.

15. For the period 1956/7 to 1971/2, see A. J. Jose, 'Trends in Real Wage Rates of Agricultural Labourers', *Economic and Political Weekly*, vol. IX, no. 13, Review of Agriculture (30 Mar. 1974). Between 1971/2 and 1978/9 money wage rates rose further by about 80 per cent, while the consumer price index for agricultural labourers in the State rose only by a little over 50 per cent.

16. See *Poverty, Unemployment and Development Policy: A Case Study of Selected Issues with Reference to Kerala* (Centre for Development Studies, 1975), Ch. VII, pp. 92-4; also *Distribution of National Income by States, 1960-61* (National Council of Applied Economic Research, 1963), Table 5.

17. For an account of some of the historical factors affecting trade unionism among agricultural labourers in this region, see A. V. Jose, 'Trade Union Movement among Agricultural Labourers in Kerala: The Case of Kuttanad Region' (Working Paper no. 93, Centre for Development Studies, July 1979).

18. 'Owing to commercialisation and consequent monetisation, there also arose in this region indigenous credit institutions known as *kuries* and *chitties*. The growth of these institutions paved the way for the rapid development of commercial banking, and together had a very profound effect on the further growth of the economy of Kerala. The *kuries* and *chitties* were the initial mobilisers of savings and purveyors of credit . . . However, when commercial banking developed in the region, and began to cater to credit requirements for productive purposes, one of the main activities of many banks in the region continued to be running of *kuries* and *chitties* for the purpose of mobilising savings and for attracting depositors' (Varghese, *Agrarian Change and Economic Consequences*, Ch. 6). See also C.P.S. Nayar, *Chit Finance* (Vora, 1973).

19. See T. V. Narayana Kurup, 'Price of Rural Credit: An Empirical Analysis of Kerala', *Economic and Political Weekly*, vol. XI, no. 27 (3 July 1976).

90   *Agrarian Reform in Kerala*

20. *Report of the All India Rural Credit Review Committee* (Reserve Bank of India, Dec. 1969), pp. 239-41.

21. Ibid., p. 241.

22. *Review of Agricultural Development and Co-operative Credit in Kerala* (1977), pp. 24-6.

23. Given below, for instance, are some extracts from the *Report of the Committee on Commodity Taxation* (Government of Kerala, May 1976): 'The small grower is legion in the field of coconut production. The age-old practice of the itinerary merchants advancing loans to the farms and thereby virtually pre-empting their output is still very widespread. All studies known to the Committee, including the Legislative Committee's Report on Regulated Markets and the study conducted by the Coconut Directorate, have shown that the coconut trade is controlled by the middlemen who generally subject the farmer to a number of unwarranted deductions from the ruling market price ... The Legislative Committee on Regulated Markets has recommended to the Government an eminently practical scheme on the basis of the working of regulated markets in several States ... To take care of the financial requirements of small farmers, it would be necessary to bring them within the fold of co-operative marketing societies ... It is extremely important to replace the merchant traders completely if the small growers are to be duly protected ... this extended network of co-operative marketing societies appears to the Committee as a very essential adjunct to the regulated market scheme' (pp. 10-12). The Committe on Commodity Taxation made similar recommendations in respect of other products also. No action has however been taken so far to set up regulated markets; nor has there been much progress in extending the network of co-operative marketing.

24. See 'A Quarter Century of Agricultural Development in Kerala – An Overview', Background Paper prepared for the formulation of the Five-Year Plan, 1978-83 (State Planning Board, Government of Kerala, Oct. 1977).

25. See *Economic Review, 1980*; also *Statistics for Planning, 1980* (State Planning Board, Government of Kerala, 1981).

26. It would appear from the data on wage rates available for 1950-1 that they were at that time higher in Travancore than in Cochin and Malabar (see *Report on Intensive Survey of Agricultural Labour, Vol. IV, South India*, Agricultural Labour Enquiry, 1955, pp. 25-6 and p. 126). If this is correct, the higher wages in evidence in the northern districts are a subsequent development.

27. 'Some Notes on Possibilities of Decentralised Development in Kerala' (Centre for Development Studies, Feb. 1980).

28. Paddy Price, Nutrients Price and Field Labour Wages, 1972-9

| | Farm Price of Paddy | Imports of Rice (000 tonnes) | Paddy (in kg.) required to Buy | | | |
			1 kg. of N	1 kg. of P_2O_5	1 kg. of K_2O	1 Man-day of Field Labour
1972/3	1.19	511	1.75	2.44	0.77	4.86
1974/5	2.38	509	1.83	2.24	0.86	3.38
1978/9	1.20	1 596	2.73	2.59	1.12	7.49

Source: P. G. K. Panikar, 'Recent Trends in the Area under [Rice] and Production of Rice in Kerala' (Working Paper no. 116, Centre for Development Studies, 1980).

29. For a fuller discussion, see *Poverty, Unemployment and Development Policy: Case Study of Selected Issues with Reference to Kerala* (Centre for Development Studies, 1975).

3 AGRARIAN REFORM IN WEST BENGAL: OBJECTIVES, ACHIEVEMENTS AND LIMITATIONS

Ajit Kumar Ghose

I Introduction

On the surface, West Bengal is one of the richer States of India; its *per capita* income for 1979/80 was estimated to be Rs 723 (in 1970/1 prices) as compared to Rs 677 for India as a whole.[1] Its rural economy, however, displays extremely disquieting characteristics. To begin with, rural-urban gaps in living standards are very considerable. Although reliable estimates of levels of income in rural and urban areas are difficult to obtain, consumer expenditure surveys generally show that the average level of consumption expenditure in urban areas tends to be about twice that in rural areas.[2] In addition, of course, the inequality in the distribution of consumption expenditure in rural areas is significant; the Gini coefficient was 0.3 in 1973/4, according to a recent estimate.[3] A combination of a low average income and its unequal distribution implies a high incidence of rural poverty. As a matter of fact, 66 per cent of the rural population were found to be living in abject poverty in 1973/4.[4]

As regards agricultural production, the rural sector has consistently failed to meet the food requirements of the State.[5] Yield rates of crops, particularly food crops, are low, and the potential for increasing them remains large.[6] Less than 25 per cent of the net cropped area is irrigated, and production technology is still largely traditional. In 1978/9, 39 per cent of the gross area under cereals was planted with high-yielding varieties, and chemical fertilisers used per hectare of gross area under crops (excluding tea) amounted to 34.8 kg.[7] Data from the 26th round of the National Sample Survey, pertaining to the year 1970/1, showed that there were 0.01 power pumps, 0.03 mechanical threshers, and no tractors per hectare of operated area in West Bengal.[8]

The structural parameters of the agrarian economy[9] portray a dismal picture. Although landlessness is not very significant, near-landlessness is. According to the 26th round of the National Sample Survey, 9.78 per cent of the rural households did not own any land, and 7.85 per cent

91

neither owned nor operated any land. But 67.84 per cent owned and 42.22 per cent operated less than 2.5 acres. There is a high degree of inequality in the distribution of land. On the basis of the data from the 26th round of the National Sample Survey, the concentration ratio works out to be 0.6176 for household ownership holdings and 0.4819 for household operational holdings.[10] About 18 per cent of the total operated area is cultivated by tenants.[11] Sharecropping is by far the predominant form of tenancy; about 92 per cent of the total leased-in area is under sharecropping. The degree of fragmentation of holdings is absurdly high; a 5-acre holding may be scattered in 7 to 8 plots. The use of hired labour is widespread; according to the 1971 population census, hired labourers constituted about 44 per cent of the agricultural work-force.[12] Institutional credit network is underdeveloped, and a large proportion of the small farmers and sharecroppers are subject to exploitation by rural moneylenders.

The above description is intended to provide a brief profile of the rural economy of West Bengal. The picture that emerges is clearly disquieting. But even more disquieting have been the time trends. Over the past twenty years, there was a significant rise in both the incidence of poverty and income inequality in rural West Bengal.[13] This would not seem surprising in view of the trends in production and structural changes over approximately the same period. These will be discussed in detail in a later section of this chapter, and here we shall limit ourselves to a few brief observations. Growth of foodgrains output, as also of total agricultural output, generally failed to keep pace with that of population. As we shall see, this is mainly attributable to the extremely slow pace of technological change. The proportion of gross cropped area irrigated remained virtually constant; the spread of high-yielding varieties (HYVs), once they arrived on the scene, was slow; the level of use of chemical fertilisers, though increasing, remained low; and there was little trend towards mechanisation. Parallel changes in structural parameters were, however, significant. The structure of landholding became progressively disaggregated, the relative importance of very small farms increased, and there was a consequent rise in the degree of fragmentation of holdings. The proportion of agricultural labourers in the agricultural work-force consistently rose, while the proportion of operated area under sharecropping declined slightly. The pattern of structural change was thus somewhat curious; increasing smallness and fragmentation of holdings were combined with a development of a seemingly capitalistic labour process.

This then is the context in which the present State government of

West Bengal is attempting to implement an agrarian reform programme. And it is with this context in mind that we have to formulate criteria for judging the relevance and possible consequences of the programme. It is evident that any reform programme, to be worth while, must aim simultaneously to reduce poverty and inequality, and to promote growth and technological change. Growth that generates poverty and inequality, if achievable at all in the given context, will defeat its own objective. On the other hand, an attempt to reduce poverty and inequality through redistributive measures that stifle growth will be self-defeating in the end.

The formulation of an agrarian programme that can simultaneously reduce poverty and inequality, and promote growth and technological change requires, first and foremost, an identification of the basic causes of stagnation (or slow growth) on the one hand and of increasing poverty and inequality on the other.[14] This demands a proper appreciation and a satisfactory resolution of several outstanding issues. Are there structural constraints on growth and technological change? If so, what are these? What are the possible alternative ways of removing them? What are the structural determinants of poverty and inequality? Would a removal of structural constraints on growth and technological change automatically remove the structural causes of poverty and inequality? If not, what is the structural framework that, if evolved, would ensure growth and technological change on the one hand, and reduction of poverty and inequality on the other? How and to what extent is it possible to steer the agrarian economy, given its present character and past history, towards such a framework?

Whether or not the present State government of West Bengal consciously sought answers to these questions in the course of formulating its programme is irrelevant; such answers are implicit in the programme itself. Before discussing these, however, it would be useful for us to seek our own answers, so that we can develop a perspective against which the programme can be judged. This chapter is therefore organised as follows. Section II presents an analysis of the historical trends in production and technology as also of the process of structural evolution of the agrarian economy. On this basis, we seek to identify in section III the causes of relative stagnation and the basic elements of the process of generation of poverty and inequality. As a logical follow-up of this analysis, we attempt to draw the outlines of a structural framework which can effectively handle the problems of growth, technological change, poverty and inequality. In section IV, we examine the elements of the reform programme and the administrative machinery being relied upon for its implementation. We also try to evaluate the relevance of the programme, its possible consequences for growth, technological change, poverty and inequality, and discuss the possible appropriate follow-up programmes. Section V draws some implications for the theory of agrarian reform.

II   Production, Technology and Agrarian Structure:
A Historical Perspective

The currently extant features of West Bengal's agrarian economy are the outcome of historical processes over a long period, and have to be viewed in this context if their essential characteristics are to be identified. It is this presupposition that underlies our attempt to analyse, in this section, the historical trends in West Bengal's agriculture with respect to production, technology and some selected structural variables. The presupposition also defines the limits of our analysis. It is not our intention to attempt to resolve outstanding issues of history. Rather, we shall choose to focus on some well-established historical facts, study their mutual coherence, and draw the implications in so far as they relate to the principal characteristics of agriculture in West Bengal today.

The historical period to which we shall confine attention begins around the turn of this century and extends up to the current period. The choice of the period has been dictated by two principal considerations. First, this is roughly the period over which major structural changes in West Bengal's agriculture took place. Second, statistical materials relating to earlier periods are scarce.

II.1 Production

There are reasons to believe that around 1880, the then Bengal Presidency, comprising present-day West Bengal, Bihar, Orissa and Bangladesh, had a solvent agricultural sector. According to one estimate, it was producing a net annual surplus of 1.2 million tons of foodgrains.[15] Since then, however, the agricultural sector experienced an almost steady impoverishment until 1947. After 1947, the performance of agriculture improved. Growth of production, however, still failed to keep pace with that of population so that *per capita* output continued to decline. The tendency was particularly pronounced in the case of foodgrains. It is through this long process of relative stagnation that the solvent agriculture of the 1880s moved towards the state of acute insolvency in which it is found today.

Some relevant data are presented in Table 3.1. The major features revealed by these data may be described as follows. During the last 56 years of British colonial rule (1891-1947), according to Blyn's data, agricultural output was declining at a rather rapid rate. This was wholly due to the fact that foodgrains output was declining at a sharp rate; non-foodgrains output was actually increasing. The decline in food output, in turn, was caused by a significant decline in yields per unit of land. Blyn's analysis, however, suggests that decline in yield rates was sharp in Bihar and Orissa, and not so much in Bengal proper.[16] Islam's data for the period 1920-46 tend to confirm this. They show that the decline in yield was small and was counteracted by relatively larger growth in acreage, so that growth in food output in Bengal proper was positive. But it was less than the growth of population, implying a steady decline in *per capita* food output. Growth of non-foodgrains output was, however, significant. Thus total agricultural output kept just ahead of population.

In post-1947 West Bengal, output, land productivity and acreage of both food and non-food crops grew at higher rates. But, generally speaking, *per capita* food output was still declining. During this period, *per capita* acreage declined significantly indicating a near-exhaustion of the land frontier. Growth of land productivity was too slow to compensate for this. Growth of non-foodgrains output was again significant, but total agricultural output nevertheless failed to keep pace with population during a greater part of the period.

A persistent feature throughout the entire period (1891-1977) was a faster rate of growth of non-foodgrains output as compared to that of foodgrains output. This reflected a growing commercialisation of agricultural production. It is arguable that but for this, food output might have kept pace with population. But this should not detract attention from the fact that a stagnation or slow growth in land productivity was a major cause of the decline in *per capita* food output. It is also clear that in the seventies, when certain regions in India were experiencing the much trumpeted 'green revolution' characterised by a significant rise in yield of food crops, growth of yields in West Bengal showed little tendency to accelerate. Indeed, over the period 1970/1-1976/7, yield rates of food crops actually declined.[17] This fact clearly is disturbing, particularly if one recognises that growth of land productivity must be the major source of growth of food output in future.

The reason we focus on trends in food production is that *per capita* food output, and in Indian conditions local *per capita* food output, is a

Table 3.1: Growth of Output, Land Productivity, Acreage and Population in West Bengal

| | Average Annual Percentage Rates of Change in | | | |
	Output	Land Productivity	Acreage	Population
Bengal Presidency 1891-1947				0.65[a]
Foodgrains	−0.73[a]	−0.55[a]	0.00[a]	
Non-foodgrains	0.23[a]	0.59[a]	−0.41[a]	
All crops	−0.45[a]	−0.34[a]	−0.06[a]	
Bengal Presidency 1920-46				−
Foodgrains	−0.6[b]	−0.7[b]	0.3[b]	
Non-foodgrains	0.6[b]	0.8[b]	0.0[b]	
All crops	−0.2[b]	−0.5[b]	0.3[b]	
Bengal 1920-46				0.80[b]
Foodgrains	0.7[b]	−0.2[b]	1.0[b]	
Non-foodgrains	1.5[b]	1.2[b]	0.2[b]	
All crops	0.9[b]	0.1[b]	0.9[b]	
West Bengal 1952/3-1964/5				2.89[g]
Foodgrains	1.14[c]	0.88[c]	0.26[c]	
Non-foodgrains	3.77[c]	0.15[c]	3.61[c]	
All crops	1.94[c]	1.34[c]	0.59[c]	
West Bengal 1960/1-1964/5				−
Foodgrains	2.03[d]	1.55[d]	0.49[d]	
Non-foodgrains	4.91[d]	0.21[d]	4.69[d]	
All crops	2.87[d]	2.05[d]	0.83[d]	
West Bengal 1960/1-1970/1				2.42[h]
Foodgrains	2.36[e]	1.32[e]	1.02[e]	
Non-foodgrains	3.15[e]	1.32[e]	1.82[e]	
All crops	2.51[e]	1.37[e]	1.12[e]	
West Bengal 1971/2-1976/7				2.07[i]
Foodgrains	1.38[f]	−0.42[f]	1.80[f]	
Non-foodgrains	4.05[f]	4.42[f]	−0.37[f]	
All crops	2.01[f]	0.47[f]	1.54[f]	

Note: 1977/8 appears to have been an unusually good year, and consequently the growth rates for the period 1971/2-1977/8 turn out to be very high. They would, however, give a false picture in our opinion. The relevant data are presented in Appendix Table 8, so as to enable the reader to form his/her own judgement.
Bengal Presidency included present-day West Bengal, Bangladesh, Bihar and Orissa. Bengal included present-day West Bengal and Bangladesh.

major determinant of rural poverty.[18] Declining *per capita* food output, therefore, cannot but generate poverty (though it would be wrong to suppose that increasing food output *per capita* would necessarily solve the problem of poverty). It also means, in the final analysis, declining marketed surplus of foodgrains which acts as a brake on the process of industrialisation.

II.2 Technology

The trends in land productivity observed earlier already suggest that changes in production technology in West Bengal's agriculture could not have been significant. Constancy or decline of land productivity over the period 1891-1947 and a slow growth since 1947 leaves little room for doubt. Nevertheless, it would be useful to go into some detail, not so much for demonstrating the stagnation in technology as for isolating the underlying cause of stagnation.

First, let us consider the last half century of British colonial rule. During this period, very little effort was made to introduce improved seed varieties or to promote the use of chemical fertilisers.[19] Techno-logical change, such as there was, reflected itself mainly in the changes in incidence of irrigation and in intensity of cropping. As the figures in Table 3.2 show, incidence of irrigation increased appreciably over the period 1906-20 and remained constant afterwards. Somewhat surpris-ingly, the intensity of cropping does not show any corresponding rise over the same period.[20] The only plausible interpretation is that irriga-tion was extended to areas where double cropping was practised even in the absence of irrigation, perhaps because they were well served by rain-fall in normal years. It is likely, therefore, that the increase in the inci-dence of irrigation helped to reduce the variability of yields rather than to raise them. The evidence on the time trends of yields, examined earlier, is consistent with this hypothesis.

Coming to the post-1947 period, we find that there was some increase in the incidence of irrigation between 1948/9 and 1962/3, but

Sources: a. Blyn (1966). b. Islam (1978). c. *Indian Agriculture in Brief*, 8th edn (1967), p. 168. d. Estimated by fitting a semi-logarithmic equation to the data provided in *Indian Agriculture in Brief*, 7th edn (1965), pp. 144-6, and 8th edn (1967), pp. 162-4. e. Estimated on the basis of initial and terminal values pro-vided in the Government of West Bengal (1979b), pp. 43-6. f. Estimated by fitting a semi-logarithmic equation to the data presented in Appendix Table 8. g. Population Census (1961). h. Population Census (1971). i. Office of the Registrar General, 'Provisional Population Totals, Census of India, 1981' (New Delhi, Mar. 1981), quoted in Tim Dyson, 'Preliminary Demography of 1981 Census', *Economic and Political Weekly*, vol. XVI, no. 33 (15 Aug. 1981).

Table 3.2: Indices of Technological Change in West Bengal

Year	Irrigation[a]	Double Cropping[b]	Fertiliser[c]	HYV[d]
1891/2−1895/6	−	17.86		
1901/2−1905/6	1.7	21.64		
1916/17−1920/1	14.96	21.16		
1931/2−1935/6	14.32	20.90		
1941/2−1945/6	16.54	24.28		
1948/9	16.42	11.62		
1954/5	21.45	15.90		
1962/3	26.27	17.42		
1966/7	−	−		0.6
1967/8	26.54	19.46		
1968/9	−	−	8.9	
1970/1	−	−	−	16.0
1973/4	24.07	20.65	−	
1974/5	−	−	17.5	16.1 (22.0)
1975/6	−	−	17.3	19.4 (26.9)
1976/7	−	−	20.5	22.9 (29.8)
1977/8	−	−	21.8	25.5 (31.7)
1978/9	−	−	34.8	33.2 (39.0)

Notes: a. Net irrigated area as a percentage of net sown area.
    b. Net double cropped area as a percentage of net sown area.
    c. Kg. per hectare of gross area under crops (excluding tea).
    d. Gross area under HYV rice as a percentage of gross area under rice.
Figures in parentheses represent percentages of gross area under rice and wheat under HYVs.
Sources: Blyn (1966); Central Statistical Office, *Statistical Abstract*, various issues; Government of West Bengal (1981a), Statistical Appendix.

no increase thereafter. Net area under irrigation, however, expanded all through, albeit at a slow pace, since the net sown area was increasing. Till the early sixties, there was also a significant rising trend in the intensity of cropping which seems to have tapered off since then. Since the late sixties, however, certain important developments have been taking place, though again at a slow pace. These relate to the introduction of chemical fertilisers, and high-yielding varieties of rice and wheat. As can be seen from Table 3.2, the level of use of chemical fertilisers and that of adoption of high-yielding varieties have had a slow rate of growth and remain quite low.

Precisely in the seventies, however, the rate of growth of output and yield rate of foodgrains slowed down significantly (see Table 3.1 and Appendix Table 8). This seems somewhat curious at first sight, but a closer scrutiny suggests plausible explanations. Nearly the entire area under wheat had been under HYVs by 1976/7 and the yield rate of wheat had indeed risen significantly. But only about 8 per cent of the gross area under foodgrains is devoted to wheat in West Bengal. Rice accounts for about 80 per cent of the gross area under foodgrains, and even in 1978/9 only about 33 per cent of rice area was under HYVs. Moreover, it appears that there is a high degree of regional concentration in the adoption of HYVs; this is suggested by the fact that the yield rate of rice rose significantly in only 5 of the 15 districts[21] (it remained constant or declined in the remaining areas). Given the fact that these five districts also account for about 44 per cent of the wheat area of West Bengal, it seems probable that they also account for the bulk of the fertilisers and irrigation facilities. Under these conditions, progressive adoption of the new technology and deceleration in growth rate in the context of West Bengal as a whole are quite consistent.

A further development that took place during the seventies also deserves attention. This relates to private investment in irrigation. While such investments were negligible until the late sixties, they appear to have been encouraged by the introduction of chemical fertilisers and high-yielding varieties, since an assured and timely water supply is an essential precondition for the success of these innovations. The tendency is revealed by the figures in Table 3.3. The feature that stands

Table 3.3: Pumps and Tubewells in West Bengal Agriculture

			Tubewells	
			Private	Public
1961	3 637	256	neg.	neg.
1969	–	–	12 000	1 308
1974	–	–	60 500	1 940
1977	–	–	85 000	2 300
1978	–	–	95 000	2 400

Source: Dhawan (1979).

out is the remarkable growth in privately owned tubewells in recent years. The fact is significant, for it points to the success of the new technology in encouraging some private investment in agriculture, a precondition for capitalistic growth. The achievement, however, falls far short of the potential. In 1978, there still were about 13 tubewells for 1,000 hectares of gross cropped area (excluding the area under tea); the number of installed tubewells constituted only about 20 per cent of the total number that could be installed, given the ground water reserve; and the overall performance of private investment compared very unfavourably with that in the Punjab, Haryana and Uttar Pradesh.[22] Unfortunately, for lack of data, we cannot establish the extent to which growth in private tubewells may have led to an expansion of net irrigated area. Given the evidence on yields, the possibility that tube-well irrigation may have expanded in areas where other forms of irriga-tion already existed certainly cannot be ruled out.

On the whole, therefore, while the developments, in terms of tech-nological changes, during the post-1947 period, particularly in the seventies, did constitute a break with the past, they were clearly not significant enough to prevent a decline in *per capita* food output.

II.3 Agrarian Structure

British colonial intervention set in motion profound changes in the social relations of production in Bengal's agriculture. While it is impos-sible, and perhaps unnecessary for our purpose, to go into a detailed analysis of these changes here, a brief outline would place the current problems in perspective. The central question to which we shall address ourselves here concerns the direction of change in the organisational basis of production, since this is dialectically related to the technological and distributional changes within the agrarian sector. Accordingly, we shall confine attention to three basic variables: relative importance of hired labour, incidence of tenancy and the structure of landholding.

The widely held view[23] that in pre-British India agricultural produc-tion was carried out by small peasant cultivators existing within isolated, self-sufficient village communities has recently been seriously chal-lenged. It has been argued that various forms of agrestic servitude and even slavery existed in Mughal India.[24] However, it seems clear that landownership in the modern sense of the term did not exist; what existed was a hierarchy of obligations and rights concerning the use of land for production purposes and the distribution of output.[25] Popula-tion growth was very small and land was, generally speaking, available in plenty. The system could thus reproduce itself without much diffi-culty and economic differentiation, although considerable, was static in character.

This equilibrium was disturbed by two important changes intro-duced by the British. On the one hand, a system of legal landownership was introduced, and on the other hand successful efforts were made to promote a commercialisation of agricultural production mainly through the introduction of cash crops such as jute. As a consequence, cate-gories such as agricultural labourers, tenants and landlords made their appearance, and a dynamic process of stratification of the peasantry, with control of land as a central variable, was set in motion. By the turn of the century, the process was quite advanced in Bengal, and was proceeding apace thenceforth.

The estimates, from census data, of the proportion of agricultural workers surviving as sellers of labour-power reflect this. As can be seen from Table 3.4, by 1931 the proportion of agricultural labourers among agricultural workers had risen to 27 per cent, which was comparable to that in many European countries around the same period.[26] Moreover, the upward trend was consistent, though uneven, throughout the period 1901-41. There can be little doubt, therefore, that hired-labour-based production was displacing peasant production.

Table 3.4: Agricultural Labourers as Percentage of Agricultural Workers in Bengal

Year	Estimated Percentage
1901	13.63
1911	14.65
1921	15.93
1931	27.22
1941	26.38
1951	25.91
1961	28.24
1971	44.28

Note: In 1901, Bengal comprised present-day Bangladesh, West Bengal, Bihar and Orissa. During 1911-41, it comprised present-day Bangladesh and West Bengal and thereafter only West Bengal.
Source: Ghose (1979).

For lack of data, it is not possible to determine the nature of changes in tenancy relations and landholding structure that accompanied the growth of hired labour during this period. But there are reasons to believe, as will be clear from subsequent observations, that it was not accompanied either by a decline in the incidence of tenancy or by an increase in the relative importance of large farms.

As is evident from Table 3.4, growth in hired labour continued and even accelerated in post-1947 West Bengal. For this period, we can directly observe the nature of changes in tenancy relations and landholding structure. Table 3.5 contains estimates of percentage area under tenancy in West Bengal. It is clear that the incidence of tenancy has not declined significantly over quite a long period. The estimates for relatively recent periods may in fact be underestimates, since the existence of a threat of tenancy reforms by the government may have led to

74328

Table 3.5: Percentage of Operated Area under Tenancy in West Bengal

Year	Percentage
1938	23.45
1953/4	25.43
1960/1	17.65
1970/1	18.73

Source: Ghose (1979).

under-reporting.[27] But even if we accept the estimates at their face value, the decline in the recent period cannot be said to be sharp. Moreover, the decline in the incidence of tenancy occurred between 1953/4 and 1960/1, while the growth of hired labour was large between 1961 and 1971. Thus the growth of hired-labour-based production does not seem to have occurred at the expense of tenant-cultivation.

The structure of landholding shows an indisputable trend towards progressive disaggregation. As revealed by the data in Table 3.6 (see also Appendix Table 9), small farms have not only been growing in number but have also been increasing their share of the operated area.

It is known that no significant land redistribution programme was implemented by the government between 1953/4 and 1970/1. To the extent that a threat of land reform may have induced large landowners to report falsely parts of their family holdings as being independently operated by different members of their families, there may have been a bias in the data.[28] It is doubtful, however, if this would fully explain the observed tendency towards disaggregation. The effect of increasing subdivision of landholding in accordance with the laws of inheritance in response to growing pressure of population may conceivably have proved stronger than that of any force for concentration and consolidation that might have been operative.[29] Incidentally, the tendency is apparent not only in the case of operational holdings but in the case of ownership holdings as well (see Appendix Table 1).

It should perhaps be noted that one effect of this progressive disaggregation of landholding structure, to the extent that it occurred, was almost certainly an increase in fragmentation. This is suggested by the fact that smaller farms tend to have a greater relative degree of fragmentation (see Appendix Table 2). We shall return later to the significance of this fact.

The overall pattern of evolution of West Bengal's agrarian economy thus presents a somewhat curious picture. The growth of hired labour, instead of being accompanied by a growth of large farms, was actually

Table 3.6: Operational Distribution of Land (percentages), West Bengal

Size-class of Household Operational Holding (acres)	1938 House-holds	1953/4 House-holds	Area	1959/60 House-holds	Area	1960/1 House-holds	Area	1970/1 House-holds	Area
0.00	)		–	30.44	–	33.88	–	30.94	–
	) 48.5								
0.01–1.00	)		3.31	13.72	2.00	12.74	2.30	19.80	4.34
1.00–2.50		16.16	10.50	17.52	11.40	16.95	11.58	22.42	20.45
2.50–5.00		17.52	22.56	20.55	28.15	19.62	27.98	15.77	28.94
Below 5.00	74.60	82.23	36.37	82.23	41.55	83.19	41.86	88.93	53.73
5.00–7.50		8.03	18.61	9.14	21.12	9.12	21.54	6.48	20.03
7.50–10.00		4.12	13.20	3.82	12.29	3.19	10.78	2.47	11.02
5.00–10.00	17.00	12.15	31.81	12.96	33.41	12.31	32.32	8.95	31.05
Above 10.00	8.40	5.17	31.82	4.81	25.04	4.50	25.82	2.12	15.22
All	100.00	100.00	100.00	100.00	100.00	100.00	100.00	100.00	100.00

Sources: Government of Bengal, *Report of the Land Revenue Commission*, vol. II (Calcutta, 1940); NSS, 8th round, Report no. 66, p. 57; NSS, 16th round, Report no. 159, p. 79; NSS, 17th round, Report no. 144, p. 160; NSS, 26th round, Report no. 215 (West Bengal), vol. I, p. 70.

accompanied by a growth of small farms. There thus appears to have been a progressive substitution of hired labour for family labour on increasingly smaller farms. Nor did the growth of small farms discourage tenancy to any significant extent. The scenario becomes intelligible only if we suppose that two distinct tendencies operated simulta-neously. First, there was a growing pauperisation of a section of the peasantry. This seems plausible in view of the fact that neither acre-age nor productivity kept pace with population, reflecting both a fail-ure of the agrarian sector to grow and a failure of the non-agricultural (industrial) sector to effect a net withdrawal of labour from agriculture. Second, there probably was a significant transfer of land from the culti-vating to non-cultivating classes. Definitive evidence in support of this hypothesis is difficult to provide. But recent evidence shows that many small farms which could easily be operated with family labour are actually operated with wage labour.[30] Some illuminating examples from colonial periods are also available. It has been widely noted that rural moneylenders, taking advantage of peasants' indebtedness, often gained ownership of their land.[31] There is also clear evidence to show that in the course of the Bengal Famine of 1943, a large amount of land passed into the hands of non-cultivating classes.[32] There may also have been a tendency among solvent agriculturists to cease active participation in agricultural activities and to engage in moneylending, trading, etc.

Thus the basic cause of changes in agrarian structure was not growth but rather the lack of it. Growth of hired labour reflected acquisition of land by non-agriculturists and not by dynamic nascent rural capitalists, it reflected pauperisation of the peasantry, and not proletarianisation.[33]

III   Agrarian Structure, Agricultural Growth and Rural Poverty: An Analytical Perspective

It should be clear from the foregoing discussions that the specific pattern of structural evolution that characterised West Bengal's agriculture was underlined by a lack of growth and technological change.[34] We shall examine presently how the extant structural features relate to this historical process. But one issue must be commented on first.

The question remains: why was there a lack of growth and technological change? The British colonial government of course made little effort to promote technological innovations. Even in the post-1947 period, government investment in irrigation was meagre, and it is not until the late sixties that other important technological innovations were attempted. But why did individual landholders not invest in land improvement? With regard to the colonial period, the generally accepted answer is in terms of tenurial obstacles. The British-instituted permanent settlement tenure system created an atmosphere inimical to investment in agricultural production. A small class of landowner-cum-revenue collectors, often owning (theoretically) vast areas they were scarcely familiar with, found it convenient to live sumptuously on a portion of the rent extracted from the actual cultivators. The actual cultivators, on their part, could only view themselves as mere tenants, and in any case were left with little investible surplus. This mutually reinforcing 'disincentive structure' was further accentuated by the existence of a whole hierarchy of *rentier*-intermediaries between the landowners and the cultivators.[35] After 1947, the system of permanent settlement was abolished. But by then, one suspects, the long-term stagnation had given rise to other structural features which began to act as constraints on growth and technological change.

It is indeed this latter proposition that is of principal interest to us here, particularly since, as we noted earlier, the pace of growth and technological change continues to be sluggish. Let us take a look at some of the principal features of agrarian structure in post-1947 West Bengal. These may be summarised as follows. First, a large proportion of the cultivated area is operated in very small farms. As the data pertaining to 1970/1 (in Table 3.6) show, farms with less than 5 acres operate more than 50 per cent of the area. Second, farms, large and small alike, tend to have a high degree of fragmentation (see Appendix Table 2). Third, at least 18 per cent of the cultivated area is operated by sharecroppers.[36] It should also be noted that all landlords are not large landowners; nor are all sharecroppers small operators, as is often supposed.[37] Fourth, a significant proportion of the area is operated with hired labour.[38] But not all farms that are operated with hired labour are large. In fact, there seems to be little correspondence between the size of farm and the nature of labour process.[39] Fifth, a very large proportion of the cultivators do not have adequate access to the organised credit market and consequently fall victim to exploitation by private moneylenders.[40] A large part of cultivators' surplus is thus siphoned off by a group of people who are not directly involved in agricultural production.

The manner in which these features relate to the historical process of structural evolution, examined in the last section, is quite apparent. The process of progressive disaggregation of landholding structure has led to a predominance of small farms and a high degree of fragmentation. The process of growth of hired labour, through the operation of forces of pauperisation and land transfers from cultivating to non-cultivating classes, has led to a situation where farm size is not a reliable index for differentiating between various types of cultivators. Molecular landlords co-exist with medium or large sharecroppers; small hired-labour-based farms co-exist with medium or large peasant farms.

Although the possibility that structural factors may act as constraints on growth and technological change has often been recognised in development literature,[41] it has rarely been taken seriously in practice. The opposed view that growth can be promoted within the existing structure in almost all cases if only governments would make a conscious effort through providing modern inputs, extension services, etc.,[42] is so well entrenched that the sole purpose of agrarian reform is frequently thought to be a redistribution of income and wealth. The fallacy implicit in this view is perhaps best illustrated by the failure of the 'green revolution' to occur in many areas. West Bengal happens to be one such area.

There are indeed good reasons to believe that several features of agrarian structure in West Bengal pose serious obstacles to private investment, and hence to the adoption of a modern technology. We shall devote some space to an analysis of these.

The most obvious features of this nature are smallness and fragmentation of holdings. To the extent that elements of a modern technology may involve indivisibilities, their adoption becomes extremely difficult on small farms which are moreover scattered in several smaller plots. For instance, installation of a tubewell, to be economically viable, requires a minimum command area. Even such innovations as a bamboo tubewell would not prove economic on a farm which consists of widely scattered minuscule plots. When one recalls that more than 50 per cent of the cultivated area is operated in farms of less than 5 acres, and that a 5-acre farm may be scattered in 8 or 9 plots, it becomes difficult to imagine an appropriate technology other than the traditional one of simply carrying water manually from a nearby river or a tank or a government canal. There is no evidence either to suggest that cultivators respond to such problems of indivisibilities by resorting to joint cultivation.[43] In so far as an assured and timely water supply is a necessary precondition for the success of such innovations as use of chemical fertilisers and introduction of high-yielding varieties, smallness and fragmentation of holdings constitute serious obstacles to an adoption of modern technology.

But features such as smallness and fragmentation of holdings themselves need to be explained. There are obvious ways (joint cultivation, renting, etc.) in which land consolidation could take place in response to opportunities provided by new technology. If this is not happening, then one must suppose that the existing social relations of production militate against the adoption of the new technology and hence against land consolidation.

The institution of sharecropping has often been singled out as a major obstacle to the adoption of modern technology.[44] This view shifts attention from the traditional preoccupation concerning questions of efficiency of resource allocation under sharecropping,[45] and explicitly recognises that investment decisions are generally made by landlords and not by sharecroppers. It characterises sharecropping as a semi-feudal[46] institution, observing that landlords often have sources of income other than rent, the most important being usury. Sharecroppers are exploited through both rent and usury. Landlords, striving to maximise total income from rent and usury, may find adoption of modern technology unattractive. For increased income from rent may be cancelled out by a decline in income from usury, since technological development may raise sharecroppers' income and thus free them of perpetual debt bondage. In addition, landlords stand to lose personal power that they currently wield over their sharecroppers because of the latter group's indebtedness.

This view has been widely criticised on both theoretical and factual grounds. We do not wish to go into the details of these criticisms here, but shall mention them briefly. First, a generalised characterisation of cropsharing tenancy as semi-feudal is open to question unless hired-labour-based production, as it exists in West Bengal, is also characterised as feudal. For landlords also engage in cultivation with hired labour, and sharecroppers also hire out labour.[47] Moreover, landlords are not necessarily larger landowners than their sharecroppers.[48] Nor are they, generally speaking, professional moneylenders. They do, of course, provide loans to their sharecroppers; but essentially such loans are provided in order to keep the labour process going. And such loans are often made by employers to their labourers.[49] Second, even if the premiss of semi-feudalism is accepted, the argument that it impedes technological change still remains unconvincing.[50] In order to be valid, it requires very restrictive assumptions concerning the specification of debt situation and the determination of rental share. Third, it needs to be explained why landlords should prefer tenant-cultivation with primitive technology to hired-labour-based cultivation with modern technology.

In a situation where most landlords are also cultivators employing hired labour, and where most sharecroppers also work as hired labourers, attempts to analyse sharecropping in isolation from the hired-labour-based production are difficult to justify; they need to be analysed together. Sharecropping implies little or no involvement in production on the part of landowners, while hired-labour-based production requires landowners to assume managerial and supervisory roles. It may be supposed, therefore, that hired-labour-based production is worth while from the point of view of the landowners, only if it generates a profit for landowners over and above what they could get as rent had they opted for sharecropping. In other words, one expects to find either a superior technology in use in the case of hired-labour-based production or a higher rate of return to labour in sharecropping. Neither of these expectations, however, is borne out by facts. The available evidence seems to suggest that there are no significant differences between hired-labour-based production and sharecropping in terms of land productivity and returns to landowners and labour (see Appendix Table 10). From the point of view of the landowners, therefore, hired-labour-based production is somewhat inferior to sharecropping; the farmer obliges them to assume managerial and supervisory roles but yields what is in effect a ground rent. The implication is that hired-labour-based production exists because cultivation of the available area under a sharecropping system is not feasible, probably because a section of the peasantry has been dispossessed not only of land but also of such assets as draught animals, ploughs, etc.

Observed features of the market in hired labour[51] point in the same direction. The bulk of the hired labour is employed in the form of casual labour. Casual wage rate, however, is not determined within a demand-supply framework except during some short and busy periods. The minimum casual wage rate is determined on the basis of a norm which is derived from the annual wage paid to an attached labourer. This latter in turn is fixed on the basis of a notion of subsistence requirements. Thus it is that during most parts of a year, it is the level of employment of the casual labourers, and not the wage rate, which is determined by the forces of demand and supply. Furthermore, relationships between employers and hired labourers tend to be personalised, a principal mediating factor being indebtedness of the labourers.[52]

What all this suggests is not that sharecropping is a dominant feature of the agrarian economy, nor that sharecropping alone is semi-feudal, but that the agrarian economy as a whole operates within a framework defined by feudalistic notions and relations. The primary form of surplus is ground rent rather than profit, value of labour is determined on the basis of a notion of subsistence rather than by the forces of demand and supply, and economic relations are personalised rather than mediated by the 'invisible land' of the market. The difference between sharecropping and hired-labour-based production is one of form and not of essence.

In such a context, usury capital performs two specific functions: it acts as an instrument of personalisation of economic relationships and as an instrument of surplus appropriation. Most small (and many medium) cultivators, sharecroppers and owners alike, need consumption loans in order to sustain themselves from one production period to another. Such loans are provided by rich farmers, professional moneylenders, traders, etc., who charge an extraordinarily high rate of interest. As a consequence, many indebted cultivators are forced to hand over their entire surplus (above basic subsistence requirements) to the moneylenders in payment of interest.

Data presented in Table 3.7 clearly show the importance of moneylenders' capital in West Bengal's agrarian economy. Although credit from institutional sources (government, co-operatives, etc.) has grown in importance over the past years, private moneylenders remain a major

*Table 3.7: Percentage Distribution of Cash Credit Received by
Cultivators by Sources in West Bengal*

Source	1951	1961	1971
Government	1.8	25.0	16.3
Co-operatives	1.3	4.7	15.1
Commercial banks	0.0	0.0	1.3
Other institutions	0.0	0.0	0.3
Landlords	3.2	0.7	3.7
Agriculturist moneylenders	2.1	33.8	12.5
Professional moneylenders	58.9	6.8	15.6
Traders	0.5	10.8	9.4
Friends and relatives	32.1	9.4	24.1
Others	0.1	8.8	1.7

Sources: Reserve Bank of India (1969), p. 103; Reserve Bank of India (1977),
pp. 33-4.

source of credit in agriculture. Private moneylenders, it should be noted,
include landlords, agriculturist moneylenders, professional moneylenders, traders, and friends and relatives. The inclusion of friends and
relatives may seem somewhat improper in the present context. But in
fact the terms and conditions on which they provide credit are often
similar to those of the professional moneylenders.[53]

Furthermore, the importance of moneylenders' capital is actually
much greater than is suggested by its share in total credit. This is so
because institutional sources provide credit primarily for production
purposes, while moneylenders provide credit primarily for consumption
purposes.[54] This is why institutional credit tends to be distributed in
the same proportion as landholding among agricultural households,[55]
but credit from moneylenders tends to be distributed in inverse proportion as landholding. Thus, on the whole, total credit tends to be distributed in inverse proportion as landholding, as can be seen from Table 3.8.

The implication of this is that smaller landholders pay an extraordinarily high rate of interest (sometimes as high as 100 per cent) on a
large part of the credit received by them,[56] and often are left with bare
subsistence (and no investible surplus) or less as a consequence. Why
and how moneylenders are able to extract such high rates of interest
from the cultivators is a subject of controversy. Until recently, the
prevailing view was that private moneylenders in rural areas faced a high
risk of default in repayment by borrowers and hence were obliged to
charge high rates of interest.[57] Quite apart from the question of its

Table 3.8: Percentage Distribution of Land and Credit Received by the Cultivators by Size-class of Owned Area, and by Size-class of Operated Area in West Bengal, 1971

Size-class of Area Owned (acres)	Percentage of Area Owned	Percentage of Credit Received	Size-class of Area Operated (acres)	Percentage of Area Operated	Percentage of Credit Received
0.00	0.00	0.01	0.00	0.00	13.97
0.01–1.00	6.83	23.83	0.01–1.00	4.34	13.11
1.00–2.50	20.46	24.00	1.00–2.50	20.45	22.91
2.50–5.00	25.69	22.43	2.50–5.00	28.94	21.98
5.00–7.50	18.38	12.81	5.00–7.50	20.03	12.16
7.50–10.00	9.34	5.38	7.50–10.00	11.02	5.39
Above 10.00	19.30	11.54	Above 10.00	15.22	10.48

Sources: Data on land distribution are taken from Table 3.6 and Appendix. Table 1. The distribution of credit is based on data provided in Reserve Bank of India (1976).

factual basis, this view suffers from a theoretical weakness. A high rate of interest increases the risk of default so that we need to impose quite restrictive conditions on the functional relationship between the risk of default and the rate of interest in order to arrive at a determinate solution.[58]

In a recent contribution, Amit Bhaduri has rejected this view and has provided an alternative theory of the formation of usurious interest rates.[59] This analysis rests on three key propositions. First, small cultivators need consumption loans in order to survive from harvest to harvest. Second, small cultivators lack access to the organised credit market because the collateral they can offer is unacceptable to the organised credit institutions. Third, this in turn gives a certain monopoly power to rural moneylenders.[60] Bhaduri builds a model so as to demonstrate how these factors may combine to generate a situation where the rate of interest can be exceedingly high.

This formulation has its weaknesses, particularly in that it takes no account of the fact that continued survival of the poor peasantry is necessary for the survival of usury as a mode of exploitation,[61] but it certainly brings into focus some very relevant factors determining the rate of interest on private credit. It also focuses attention on the fact that a significant part of the potential investible surplus is extracted by private moneylenders in the form of interest.

The conclusions that emerge from foregoing discussions may be summarised as follows. There are three basic types of constraints on growth and technological change in West Bengal's agriculture: (a) a combination of extreme smallness and fragmentation of holdings, (b) the dominance of ursury capital, and (c) the existence of a group of parasitical landowners superimposed on the peasantry. *Rentiers* and usurers can hardly be expected to invest in productive capacity. In any case, any investor would need to generate a rate of profit comparable to the usurious rate of interest over and above the high rate of ground rent — a difficult task even with the new technology. Past history gives us no reason to believe that these constraints will be overcome by forces internal to the agrarian sector. If anything, they are likely to become stronger, given the fact that we cannot reasonably expect the industrial sector to be so buoyant in the foreseeable future as to effect a large-scale withdrawal of labour from agriculture and hence the pressure of population on land is likely to go on increasing.

Agrarian reforms in a broad sense, therefore, constitute necessary preconditions for agricultural growth in West Bengal. The basic set of problems that such reforms must tackle are also clear. But before we can visualise concrete programmes, we must look into the broad correlates, other than agricultural stagnation itself, of rural poverty and unemployment since reforms attempting to overcome the constraints through a process that either leaves unchanged or accentuates rural poverty and unemployment must be considered unacceptable.

Poverty of rural families corresponds roughly to their landholding status. This fact by now is well established,[62] and we need not labour the point here. Families who rely primarily on employment as hired labourers in agriculture for their living are the poorest. These include families of landless labourers, very small owner cultivators and share-croppers. These families are poor both because wage rates are low and because level of unemployment (or rather underemployment) is high, but the former is a more important reason. There are two facts which suggest this. First, an average casual agricultural labourer finds work for more than 200 days in a year.[63] Second, among the agricultural workers (including landowning peasants) level of poverty varies inversely, up to a point, with level of underemployment.[64] This is partly because those who work on their own land appropriate rent as well as reward for labour, and partly because the reward for a unit of labour itself varies inversely with the size of landholding.[65] Thus a family with land often has lower levels of both poverty and employment than a family which is landless. Similarly, up to a certain level, a family with more land has lower levels of both poverty and employment than a family with less land.

The problems of poverty and unemployment are therefore two separate problems, and it may not always be possible to tackle them simultaneously. If a choice has to be made, it seems reasonable to assign a
higher priority to the problem of poverty. The solution to this problem
would evidently involve a combination of the following policies: (a) providing land, to the extent possible, to those families who have little or
no land, and (b) raising wage rates for agricultural labour. This is not to
say that creation of extra employment opportunities is not important.
But unemployment is a privilege enjoyed by the relatively better off
and reduction of aggregate unemployment need not necessarily lead to a
decline in poverty. Attempts to reduce poverty, on the other hand, can be
so designed as to increase overall employment in agriculture. And they
may be desirable even if they do not increase aggregate employment.

Let us examine briefly the implications of the policies suggested
above. A redistribution of land from the relatively large owners to the
households owning little or no land would, in addition to reducing the
level of poverty of these households, increase employment in agriculture though it would probably reduce wage employment. This is
suggested by the fact that there is an inverse relationship between farm
size and the intensity of labour use.[66] But what is decisive in reducing
poverty is not the increase in employment (given the rather low marginal productivity of labour) but the fact that newly acquired landownership would enable the labour households to appropriate the rent
element in output. However, it must be recognised that the scope for
land redistribution under the present circumstances in West Bengal is
rather small. In 1970/1, the average cultivated area available per rural
household was estimated to be only 1.9 acres.[67] Given the fact that
inequality in land distribution cannot be completely eliminated in the
near future, the limitations of land redistribution as a method of
eradicating poverty are apparent. This does not mean of course that
this method should not be used, but it does mean that this method
cannot provide a comprehensive solution.

Formulating policies for raising agricultural wages requires a theory
of wage formation in agriculture. The foregoing discussion on this issue
suggests that there is little scope for increasing minimum casual wage
rates by creating new demand for labour (i.e. by creating fresh employment opportunities) unless conditions of full employment can be
created. What is required is a direct intervention designed to change the
norm for fixing wage rates. Such an intervention might be more effective if it is coupled with measures to increase employment, but measures to increase employment by themselves are unlikely to raise wage

rates substantially. Here it must be noted that abolition of landlordism and usury constitutes a prior condition for successful intervention in raising wages. For one thing, wage relations tend to be personalised through the medium of usury, as we have noted. For another, rent and usurious interest, being residual categories, can expand only at the cost of wages; hence a system where they form the major categories of surplus has a built-in tendency to depress the value of labour.

Let us pull the arguments together before concluding this section. Promotion of growth and technological change requires the creation of larger consolidated production units than are there at present. This involves negating the existing conditions of extreme smallness and a high degree of fragmentation. Two further prior requirements are: (a) the abolition of landlordism (involving, presently, both sharecropping and the employment of wage labour), and (b) the abolition of usury capital. Reduction or eradication of rural poverty, on the other hand, requires (a) reduction or abolition of landlessness in so far as this is possible, and (b) rise in agricultural wages.

To what extent are these objectives mutually consistent? There is one obvious contradiction between the requirements of growth and those of poverty eradication. In a land-scarce situation, the need to reduce landlessness contradicts the need to create reasonably large production units. Given private property rights in land, it is also difficult to visualise a process of consolidation of holdings which can proceed without generating further landlessness.

This contradiction, however, appears less serious if we view its resolution in terms of processes rather than of events. The following scenario is certainly conceivable. Suppose, in the first instance, there is a redistribution of land in favour of the landless and the very small cultivators. Suppose further that ownership rights in sharecropped land are transferred to the sharecroppers. The immediate impact of these measures will be one of reducing poverty and of weakening the hold of landlordism and usury. Production in the short run is unlikely to suffer and may even increase, given the evidence on the inverse relationship between farm size and land productivity, and on the production effect of sharecropping.[68] At the next stage, groups of small cultivators could be induced to pull their resources together and organise production on a co-operative basis. This, if achieved, would largely solve the problems of extreme smallness and fragmentation of landholdings, and would thus facilitate technological change. Once the need for technological change and the imperatives of modern technology are apparent to the cultivators, such a change might be feasible. In theory, therefore, it is possible to think of policies that can resolve the contradiction. Whether or not the policies of the present government in West Bengal are on these lines is a matter to which we now turn.

IV The Current Programme of Agrarian Reform in West Bengal: Objectives, Achievements and Limitations

The agrarian reform programme currently being implemented[69] by the government of West Bengal has the following basic components:

(i)   acquisition of estates and implementation of a ceiling on land-holding and distribution of land thus acquired to the landless and small owners;

(ii) abolition of absentee landlordism through the acquisition of all lands owned by persons or individuals who live away from the areas in which these lands are located and who do not bear direct responsibilities for their cultivation;

(iii) recording of sharecroppers ('bargadars') with a view to preventing their unlawful eviction and securing their rights concerning cultivation of the land and distribution of the produce;

(iv) provision of special credit facilities for the poorer section of the cultivators; and

(v)   implementation of a minimum wage rate for casual agricultural labour and of programmes designed to boost employment.

Except for the last two items, the present government is implementing a programme which was in fact formulated long ago in the form of the West Bengal Estates Acquisition Act, 1953, and the West Bengal Land Reforms Act, 1955. Some amendments have, however, been made so as to close certain loopholes which could be taken advantage of by the landholders for frustrating the basic objectives of the programme.

The innovative aspect of the process of implementation is the involvement of the Gram Panchayats, the elected bodies at village level.[70] The members of these bodies, being residents of the village(s) they represent, have in general a better grasp of existing realities of village life than government officials. This can of course be both a help and a hindrance to the implementation of substantive reforms, depending on the extent to which these bodies are truly representative, in structure and composition, of the village population. Table 3.9 gives some indication of their representative character.

Table 3.9: Percentage Distribution of Members of Gram Panchayats by Occupation

Occupation	Percentage
Owner cultivators: All	50.7
Owning below 2 acres	21.7
Owning between 2 and 5 acres	14.3
Owning more than 5 acres	14.7
Sharecroppers	1.8
Landless labourers	4.8
Others (non-agriculturists)	42.7

Source: Government of West Bengal (1981a), pp. 20-1.

The table is based on a survey of 100 sample Gram Panchayats conducted in early 1979 by the Development and Planning Department of the government of West Bengal. Comparing this to the pattern of landownership (Appendix Table 1), it can be said that while the owner cultivators have a fair representation, sharecroppers and landless labourers are seriously under-represented. Power of the large landowners, however, is obviously limited.

Let us examine the progress made in implementing the programme in some detail. The ceilings on landholding[71] are as follows:

(1) For a single adult person, 2.5 hectares (= 6.2 acres) of irrigated or 3.5 hectares (= 8.6 acres) of unirrigated agricultural land;
(2) for a family of up to 5 members, 5 hectares (= 12.4 acres) of irrigated or 7 hectares (= 17.3 acres) of unirrigated agricultural land; and
(3) for a family of more than 5 members, an additional 0.5 hectares (= 1.2 acres) of irrigated or 0.7 hectares (= 1.7 acres) of unirrigated land for each additional member subject to a maximum of 7 hectares (= 17.3 acres) of irrigated or 9.8 hectares (= 24.3 acres) of unirrigated agricultural land.

Total amount of agricultural land taken into possession by the government up to December 1980 through the application of the Estates Acquisition Act and the ceiling laws amounted to 1,211,617 acres. Of this, however, 179,207 acres were hit by court injunctions and about 1,032,410 acres, which constituted approximately 6.8 per cent of net sown area, were free for distribution. Up to December 1980 about 673,453 acres, constituting approximately 4.8 per cent of net sown

area, were actually distributed to 1,194,176 beneficiaries.[72] The beneficiaries were, as stipulated by the Act, residents of the locality where
the land was situated who owned either no land or less than 1 hectare
(= 2.47 acres).[73] The size of the parcel given to a single beneficiary was
not fixed, but did not generally exceed 1 hectare. On average, each
beneficiary received about 0.56 of an acre.

With regard to sharecroppers, laws have long existed forbidding arbitrary eviction as well as stipulating the manner in which the produce of
sharecropped land is to be shared by sharecroppers and landlords. These
laws, however, have not been seriously applied until recently. For
example, a sharecropper and his landlord are, by law, supposed to share
the produce in the proportion of 50:50 in a case where plough, cattle,
seeds and manure necessary for cultivation are supplied by the landlord,
and in the proportion of 75:25 in all other cases.[74] But in reality, the
widespread practice is to share the produce in the proportion of 50:50
where a sharecropper provides plough, cattle and a certain proportion
of seeds and manure. In cases where plough, cattle, seeds and manure
are provided by the landlord, he takes 75 per cent of the produce.[75]
Similarly, arbitrary eviction of sharecroppers has continued unchecked.
A major reason for this state of affairs is that in most cases formal
sharecropping contracts simply did not exist, so that no law could be
made applicable.

With a view to creating conditions for an enforcement of the existing
laws, the government has launched a campaign (called 'Operation
Barga') for recording sharecroppers and for regularising sharecropping
contracts. This campaign has so far met with considerable success. Up
to December 1980, 1,001,986 sharecroppers had been officially
recorded.[76]

As regards the provision of special credit facilities for poor cultivators, the government succeeded in convincing some nationalised
commercial banks to advance credit exclusively to poor cultivators and
sharecroppers. The government has also undertaken to shoulder the
entire burden of interest payment, provided that the loans are repaid
within the specified time period. This is designed both to encourage the
poor cultivators to use the loan productively and to lighten their
burden of repayment. Progress made so far under this programme, however, is unsatisfactory, although it is still of great significance. According to available estimates, 59,114 sharecroppers and assignees of redistributed land had received loans under the scheme in 1979; the corresponding figure for 1980 was 71,054.[77]

The attempt to raise agricultural wages has involved two types of activities. First, serious efforts are being made to ensure a minimum wage rate for agricultural work. The minimum wage rate for an adult worker has been fixed at Rs 8.10. In the case of workers employed on a day-to-day basis, the rates of cash wages can be reduced by Rs 1.25 for each principal meal (lunch or dinner) provided. In the case of workers employed on a monthly basis, two principal meals a day and accommodation will have to be provided in addition to cash wages.[78] The administrative machinery being relied upon for enforcing these rates consists of labour commissioners at district and subdivisional levels, and labour inspectors at block level. It is admitted, however, that administration of minimum wages is very difficult in practice. This is the basic reason for undertaking the second type of activity, which involves encouraging organised movements of agricultural labourers for better wages. Such movements were virtually non-existent in West Bengal before.

Alongside these attempts to raise agricultural wages, the government has undertaken to implement certain schemes to develop rural infrastructure which have helped to generate additional employment. Three types of programmes are being implemented: rural works programme, food for work programme, and composite rural restoration programme. Those employed under these schemes receive payments partly in cash and partly in foodgrains (wheat). Table 3.10 presents data relating to resources that were made available for these programmes. It is estimated that about 21.84 million man-days of employment were created through these schemes in 1977/8.[79] The corresponding figures for 1978/9 and 1979/80 were 53.34 and 54.05 million man-days respectively.[80]

Table 3.10: Availability of Resources for Rural Employment Schemes
(cash in million rupees, wheat in tonnes)

Programme	Resource	Amount		
		1977/8	1978/9	1979/80
Rural works	Cash	87.72	102.12	98.40
programme	Wheat	36 226	55 928	95 712
Food for work	Cash	10.20	37.04	39.49
programme	Wheat	25 721	49 072	30 388
Composite rural	Cash	–	285.15	25.01
restoration				
programme	Wheat	–	43.997	18 566

Sources: Government of West Bengal (1979a), p. 29; Government of West Bengal (1980), p. 32; Government of West Bengal (1981b), pp. 28-9.

The involvement of Gram Panchayats in the process of implementation of these programmes has been significant and has proved to be a distinct advantage. They have helped in identifying surplus land, in mobilising sharecroppers for recording, in choosing the recipients of loans from commercial banks, and in organising movements of agricultural labourers for higher wages. But perhaps the most sigificant aspect of participation of these elected village councils is their involvement in selecting and administering projects implemented under the rural works programme, the food for work programme and the composite rural restoration programme. In most cases, the government merely provided the resources, and left the responsibility for selecting and implementing projects with the Gram Panchayats. This is unprecedented and the government clearly took a risk, but it seems to have paid off. From all accounts, imaginative suggestions reflecting the felt needs of rural population are forthcoming. There are also indications that benefits of employment have gone primarily to those who need them most (landless labourers, marginal and small cultivators, etc.).[81]

As is evident from the above description, agrarian reform is still an on-going process in West Bengal. This makes an evaluation of it somewhat difficult. It is too soon to determine factually the long-term impact of various programmes on rural poverty, unemployment and growth. The success of much of what has been done is contingent on what remains to be done. We cannot, therefore, attempt a definitive assessment at this point. Some observations on the short-term consequences may, however, be made. First, no serious disruptions in production have occurred. Second, a significant redistribution of income in favour of the poor has obviously taken place. This outcome is expected in view of such measures as land redistribution, special credit facilities for the poorer peasants and special employment schemes for the landless. It can also be presumed, on this basis, that the level of rural poverty has been reduced. Third, it is difficult to make any judgement concerning changes in employment. Although the level of employment of rural labourers may have increased because of the special employment schemes, changes in employment in the sphere of agricultural production remain unknown.

It is probably more useful to attempt to evaluate the general direction of change in the agrarian structure since this is what will determine the long-term impact of the reforms.[82] One observation can be made straight away. The programmes being implemented do have a coherent structure, and are based on a correct appraisal of the problems of agriculture in West Bengal. Land redistribution in favour of the landless and small cultivators, tenancy reform, provision of easy credit, raising agricultural wages, boosting employment — all these are necessary first steps towards evolving an agrarian economy which is not only capable of growth but can also ensure a basic minimum level of living for all members of the rural community.

Reservations begin to crop up when one looks into the details. First, the scale of land redistribution appears to fall short of the requirement. As was stated earlier, it has so far involved only 4.8 per cent of net sown area benefiting about 1,194,176 households or about 27.8 per cent of the estimated 4,297,500 rural households who are either landless or own less than one hectare.[83] This does not appear very surprising when one considers that according to the 26th round of the National Sample Survey only 11.15 per cent of the land area was owned by households owning more than 12.5 acres. But it raises two questions. First, is the ceiling on landholding perhaps too high? The question arises because given the enormity of the problem, marginal changes are unlikely to solve it and may indeed be quickly undone. Inequality, beyond a certain level, generates further inequality. Second, do ceiling laws provide an adequate basis for land redistribution programmes in West Bengal? It has been observed that size of landholding is a poor indicator of the real position of a landholder within the production system; many small landholders are actually landlords.[84] In this context, an objective such as 'land to the tiller' would have been more desirable.

The issue can be viewed from another perspective. Given the twin facts that the average land area available per rural household amounts to only 1.9 acres and that prospects of withdrawal of a significant proportion of rural population in the near future must be considered dim, it is difficult to visualise any effective alternative to a co-operative organisation of production in agriculture if the problems of growth, poverty and unemployment are to be simultaneously resolved. Co-operation does not flourish in an unequal environment. And the best way of promoting co-operation may be to make a majority of cultivators individually non-viable to begin with.

Coming to the question of tenancy reform, it cannot be doubted that the objectives of Operation Barga are commendable in so far as they aim to ensure a fair deal for sharecroppers, a large majority of whom are poor. In terms of landholding structure, however, it merely institutionalises the existing operational distribution of land by removing the possibility of large-scale eviction of sharecroppers. This does help the poorer cultivators in many cases by effectively transferring the possession of land to them from larger landowners. But, more importantly, it transfers land from non-cultivating landlords to peasants, thus promoting a sort of *repeasantisation* of the economy.[85] This must be considered desirable; it reverses an unfortunate historical trend and increases both the incentive and capacity of sharecroppers to invest in production.[86] By the same token, however, it risks encouraging a new type of differentiation – that within the group of cultivating peasants.[87] For while Operation Barga certainly benefits landless and near-landless peasants, it benefits the medium-sized peasants even more. According to National Sample Survey data (see Appendix Table 5), about 36 per cent of the sharecropper households, each owning between 1.0 and 5.0 acres of land, account for about 58 per cent of the total sharecropped area. Provision of security of tenure and a strict enforcement of the laws on rent would draw them close to the peasant households owning between 2.5 and 7.5 acres. Operation Barga thus serves to strengthen the economic position of the group of households operating between 2.5 and 7.5 acres, constituting approximately 22 per cent of all rural households and accounting for approximately 49 per cent of the total operated area (see Table 3.6). These households would obviously be better able to take advantage of special credit facilities than the landless or near-landless households. Consequently, a new process of differentiation may be set in motion whereby this group (medium-sized peasants) may continue gaining economic benefits at the expense of both large landowners, and landless and near-landless households. To be sure, this is healthier than the peasant-landlord type of differentiation in so far as it is predicated on productive investment, but it may nevertheless defeat the objective of eradicating rural poverty.

With regard to the government's attempt to provide easy credit to poor cultivators, the only comment that can be made is that its achievement so far has been far too inadequate. It is perhaps worth recalling that the major cause of dependence of poor cultivators on usurious moneylenders is the former group's need for consumption loans (and not only production loans) for which, moreover, they can offer as collateral only such things as personal security, commitments for labour, etc. Until and unless such facilities can be developed as would meet this type of need of the poor cultivators, no amount of law-making is likely to remove the moneylenders from the rural scene. This is also a problem which is hard to tackle through organised movements of the poor.

The government's efforts to raise agricultural wage rates appear to have largely failed (see Appendix Table 9). In the first place, the minimum wage rate fixed by the government was not much higher than the ruling market rate in 1976/7. Second, in 1979/80, the market wage rate was only 83 per cent of the minimum wage rate. Third, the real wage rate showed no significant rise between 1976/7 and 1979/80. All this is perhaps not very surprising in view of the earlier observation that the dominance of landlordism and usury over the agrarian economy, though considerably weakened, has by no means been destroyed by the reforms so far implemented. So long as this dominance remains, attempts to raise wage rates cannot effect more than marginal changes.

With regard to the overall programme, one palpable gap is the absence of any programme for consolidation of holdings. The West Bengal Land Reform Act of 1955 in fact included a programme for consolidation of holdings through a process of redistribution, but the present government has made no attempt as yet to implement this. Fragmentation, as we noted earlier, constitutes a fundamental constraint on growth and technological change in West Bengal agriculture. Indeed, it has already proceeded so far that one cannot help being sceptical about the feasibility of tackling it through a process of redistribution. This would be a massive task and would involve enormous administrative problems. Furthermore, this method cannot remove basic forces generating fragmentation, namely, increasing pressure of population on land and the laws of inheritance. Progress in consolidation will always be liable to be cancelled out by the progress of fragmentation. The only effective solution to the problem lies in the development of joint cultivation by groups of families. This, unfortunately, is not part of the tradition in West Bengal and has to be promoted consciously.

Apart from facilitating investment and technological change, the incidental advantages of joint cultivation are very considerable. It can effectively check the progress of differentiation (i.e. growth of landlessness on the one hand and expansion of a rich peasant stratum on the other); it can facilitate the development of an appropriate institutional credit system; and, above all, it can enhance the spirit of co-operation among cultivators and thereby further weaken the stranglehold of landlordism and usury.

It is our view, therefore, that while the present efforts should certainly be continued as far as possible, additional efforts should be made in future to promote joint cultivation, especially among poorer cultivators. An analysis of the specific problems that might arise in the course of such efforts falls outside the scope of this chapter. These may be considerable, but would probably be less than those of any conceivable alternative, and the benefits would be incomparably larger. We cannot fail to recognise that unless mechanisms are developed to

counter the forces, operating within the agrarian economy, that obstruct growth and generate inequality, the fruits of all other efforts will
remain perishable.

V Concluding Observations

The perspective we have developed for judging agrarian reforms in West
Bengal focuses on the imperatives of growth, technological change, and
reduction in poverty and unemployment within the agrarian sector.
This perspective is derived from our analysis of the specific situation in
West Bengal, and no generality is claimed for it. Given the perspective,
our broad conclusions may be stated as follows. The agrarian reform
programme, currently being implemented by the government of West
Bengal, does address itself to the fundamental problems facing the
agrarian economy. It aims to curb landlordism, and thereby aims to
promote a repeasantisation of the economy. It aims to improve the lot
of the rural poor through such measures as a reduction in landlessness,
an expansion of employment opportunities, and an increase in rural wage
rates. It aims to free the small cultivators from the clutches of usurious moneylenders through the creation of special credit facilities. In
the short run, these measures are both necessary and desirable. However, the concrete achievements so far, though considerable, have been
inadequate. To say this is not to underrate the achievements, for problems accumulated over decades cannot be solved overnight, but rather
to underline the need for continued efforts. Furthermore, if the benefits of the current measures are to be consolidated and built upon in the
longer run, conscious efforts must be made to develop co-operative or
joint farming at the next stage. This is necessary not only for pre-
empting a renewed process of polarisation of income and wealth, but
also for facilitating the adoption of modern technology which is essential for growth.

Before concluding this chapter, we should comment on a gap, albeit
deliberate, in our analysis. It will be noted that throughout the chapter,
we have not once addressed ourselves to the important question concerning the manner in which agrarian reforms might impinge on the
growth of the non-agricultural sectors, the urban-industrial sector in
particular. The crucial linkage between the two sectors is provided by
'marketed surplus' (primarily of foodgrains, but also of industrial raw
materials). In much of the writings on agrarian reform, marketed surplus occupies the centre of the stage, as it were. The issue first came to

the fore in the course of the Soviet 'industrialisation debate',[88] and remains controversial to this day. Some economists take the view that the only valid criterion for judging agrarian reforms is provided by the imperatives of industrialisation, i.e. the possible movements of the marketed surplus as a consequence of agrarian reforms.[89] At the other end of the spectrum of opinions, there are those who insist that the welfare of agricultural producers should receive priority over the imperatives of rapid industrialisation.[90]

Whatever views one holds, it is difficult to be indifferent to the problem in a discussion of agrarian reform. Our indifference, therefore, warrants an explanation. A part of our justification lies in the fact that West Bengal is not a nation-state. Industrialisation in India is influenced by the marketed surplus available in India as a whole. Since marketed surplus from West Bengal constitutes only a small part of the marketed surplus of India as a whole, the latter is unlikely to be highly sensitive to changes in the former.

Our second reason needs elaboration. Agrarian reforms can hardly be guided by the principle of maximising marketed surplus (except in the long run) in a situation where a part of the marketed surplus exists precisely because a large section of the rural population starves. It has to be recognised that in the short run, it may not be possible to eliminate rural poverty without reducing the marketed surplus. A choice, therefore, may have to be made. If elimination of rural poverty is considered desirable, then short-run difficulties for the urban-industrial sector should be expected. There is nothing to be gained from a preservation of the imbalances inherited from the past. For buoyant growth, growth of marketed surplus, and hence the growth of the industrial sector, must depend on the growth of agricultural output and not on the growth of deprivation. By the same reasoning, agricultural growth must be one of the basic objectives of any agrarian reform programme, for rapid industrialisation is a necessary condition for development in the long run. To view agrarian reform as primarily a distributive mechanism is to ignore the lessons of history.

Appendix Tables

Appendix Table 1: Ownership Distribution of Land (percentages),
West Bengal

Size-class of Household Ownership Holding (acres)	1953/4		1959/60		1960/1		1970/1	
	House-holds	Area	House-holds	Area	House-holds	Area	House-holds	Area
0.00	20.54	–	13.92	–	12.56	–	9.78	–
0.01–1.00	36.29	4.29	40.04	4.74	38.58	4.11	46.74	6.83
1.00–2.50	16.64	11.61	18.14	13.14	18.11	13.43	21.10	20.46
2.50–5.00	12.61	18.59	14.33	22.82	16.81	25.97	12.64	25.69
5.00–7.50	5.43	14.11	6.22	16.70	7.04	18.39	5.38	18.38
7.50–10.00	3.14	11.40	2.66	10.10	2.77	10.42	1.92	9.34
Above 10.00	5.35	40.00	4.69	32.50	4.13	27.68	2.44	19.30
All	100.00	100.00	100.00	100.00	100.00	100.00	100.00	100.00

Sources: National Sample Survey (NSS), 8th round, Report no. 66, p. 54; NSS, 16th round, Report no. 159, p. 38; NSS, 17th round, Report no. 144, p. 125; NSS, 26th round, Report no. 215 (West Bengal), vol. I, p. 66.

Appendix Table 2: Estimated Number of Parcels per Operational
Holding and Average Area per Parcel: West Bengal

Size-class of Operational Holding (acres)	No. of Parcels per Holding	Average Area per Parcel (acres)
0.01–0.50	1.64	0.14
0.50–1.00	2.90	0.26
1.00–2.50	5.26	0.33
2.50–5.00	7.48	0.48
5.00–7.50	10.02	0.60
7.50–10.00	13.38	0.65
10.00–12.50	12.68	0.80
12.50–15.00	16.36	0.83
15.00–20.00	16.12	1.08
20.00–25.00	22.29	1.01
25.00–30.00	35.75	1.00
Above 30.00	26.17	1.71

Source: NSS, 17th round, Report no. 144, Table 10.15, p. 177.

Appendix Table 3: Distribution of Households Leasing Out and Area Leased Out (percentages), West Bengal

Size-class of Household Ownership Holding (acres)	1953/4		1959/60		1960/1		1970/1	
	House-holds	Area	House-holds	Area	House-holds	Area	House-holds	Area
0.01–1.00	19.78	2.08	26.98	5.25	36.97	4.11	27.75	4.90
1.00–2.50	22.25	8.13	26.16	12.50	20.04	12.32	30.45	19.19
2.50–5.00	17.75	12.10	16.35	10.92	23.83	31.96	21.19	27.49
5.00–7.50	10.56	9.70	13.62	16.81	5.12	8.66	6.44	14.86
7.50–10.00	8.09	10.14	2.45	1.26	3.56	6.44	4.56	9.49
Above 10.00	21.57	57.85	14.44	53.26	10.48	36.51	9.61	24.07
All	100.00	100.00	100.00	100.00	100.00	100.00	100.00	100.00

Sources: NSS, 8th round, Report no. 66, p. 55; NSS, 16th round, Report no. 159, p. 60; NSS, 17th round, Report no. 144, p. 142; NSS, 26th round, Report no. 215 (West Bengal), vol. I, p. 67.

Appendix Table 4: Distribution of Household Leasing In and Area Leased In (percentages), West Bengal, 1970/1

Size-class of Household Ownership Holding (acres)	Distribution of	
	Households	Area
0.00	27.62	7.82
0.01–1.00	38.97	29.56
1.00–2.50	20.32	47.90
2.50–5.00	9.39	9.29
5.00–7.50	2.57	3.01
7.50–10.00	0.51	1.17
Above 10.00	0.62	1.25
All	100.00	100.00

Source: NSS, 26th round, Report no. 215 (West Bengal), vol. I, p. 68.

Appendix Table 5: Distribution of Households Leasing Out and Leasing In on a Cropsharing Basis and of Sharecropped Land (percentages), West Bengal, 1970/1

Size-class of Household Ownership Holding (acres)	Distribution of			
	Households Leasing Out	Area Leased Out	Households Leasing In	Area Leased In
0.00	–	–	12.63	7.07
0.01–1.00	20.72	4.00	46.21	29.14
1.00–2.50	32.98	20.29	25.33	49.85
2.50–5.00	22.76	29.11	11.14	8.33
5.00–7.50	7.68	16.10	3.46	3.16
7.50–10.00	5.28	10.06	0.60	1.14
Above 10.00	10.58	20.44	0.63	1.31
All	100.00	100.00	100.00	100.00

Source: NSS, 26th round, Report no. 215 (West Bengal), vol. I, pp. 67-8.

Appendix Table 6: Hooghly, 1972/3, Employment in 8-hour Days

Size-class of Household Operation Holding (acres)	Peasant Owners		Total	Peasant Tenants		Total
	Own Farm	Hired Out		Own Farm	Hired Out	
Below 1.0	46.9	101.5	148.4	53.2	170.5	223.7
1.0–2.5	91.3	49.5	140.8	155.2	114.1	269.3
2.5–5.0	157.0	0.0	157.0	224.0	47.1	271.1
5.0–7.5	380.0	0.0	380.0	398.2	43.9	442.1
7.5–10.0	783.6	0.0	783.6	–	–	–
Above 10.0	–	–	–	568.9	180.9	749.8

Source: Author's estimates on the basis of disaggregated farm level data. It can be observed that for the households operating between 0.01 and 5.0 acres, total employment (on-farm and off-farm) remains roughly constant. Number of adult workers per household, however, increases with the size of the farm operated by a household. Thus, the amount of employment per adult worker actually declines as farm size increases (over the relevant range).

Appendix Table 7: Yield (kg.) per Hectare of Rice and Wheat, 1976

Country/Region	Rice	Wheat
West Bengal	1 265	2 100
India	1 262	1 410
Japan	5 503	–
Egypt	5 227	3 342
USA	5 244	2 039
USSR	4 008	1 630
Canada	–	2 112
France	–	3 760
United Kingdom	–	3 877

Sources: Government of West Bengal (1979b), p. 56; *Indian Agriculture in Brief*, 17th edn (1978), pp. 244-5.

Appendix Table 9: Daily Wage Rates of Casual Agricultural Labourers in West Bengal

Year	Money Wage Rates	Consumer Index Numbers for Agricultural Labourers	Wage Rates in 1976/7 Prices
1976-7	5.65	100	5.65
1979-80	6.75	115	5.87

Sources: Money wage rates are from Government of West Bengal (1981a), p. 41. Consumer index numbers are from *Agricultural Situation in India* (a monthly journal published by the Government of India), various issues.

Appendix Table 8: Indices of Area, Production and Productivity in West Bengal, 1971/2-1977/8
Base: Triennium ending crop year 1971/2 = 100

Year	Foodgrains			Non-foodgrains			All crops		
	Area	Production	Productivity	Area	Production	Productivity	Area	Production	Productivity
1971/2	100.60	105.05	104.42	105.13	108.98	103.66	101.17	105.87	104.65
1972/3	99.86	91.10	91.23	89.36	98.37	110.08	98.53	92.62	94.00
1973/4	103.62	92.49	89.26	103.31	111.34	107.77	103.58	96.42	93.09
1974/5	108.77	105.28	96.79	95.41	111.85	117.23	107.08	106.65	99.60
1975/6	112.00	113.72	101.54	93.45	118.20	126.48	109.66	114.65	104.55
1976/7	105.52	98.69	93.53	101.32	129.48	127.79	104.99	105.11	100.11
1977/8	107.33	120.53	112.30	113.35	142.95	126.11	108.09	125.21	115.84
1978/9	97.34	107.73	110.67	127.42	164.50	129.10	101.14	119.57	118.22
1979/80	99.33	94.89	95.53	118.68	144.83	122.03	101.77	105.31	103.48
Average annual growth rate, 1971/2-1979/80	−0.15	0.93	1.09	3.04	5.70	2.59	0.28	2.11	1.83

Sources: Government of West Bengal (1979b), pp. 33-41; Government of West Bengal (1981b), pp. 39-47.

Appendix Table 10: Hired-labour-based Cultivation and Sharecropping: A Comparison of Some Production Characteristics (Hooghly, West Bengal)

	Hired-labour-based Cultivation				Sharecropping[a]			
	1955/6	1956/7	1971/2	1972/3	1955/6	1956/7	1971/2	1972/3
1. Value of output per acre (Rs)	198	390	1,780	1,656	273	331	1,503	1,986
2. Labour days per acre	71	75	123	98	71	72	90	106
3. Net income per acre (Rs)	50	209	936	985	–	–	–	–
4. Rent paid per acre (Rs)[b]	–	–	–	–	107	155	528	843
5. Income per family labour day (Rs)	–	–	–	–	1.71	1.31	5.49	7.07
6. Market daily wage rate (Rs)	–	–	–	–	1.60	1.70	3.50	3.50

Notes: a. All the sharecroppers were observed to rely wholly or primarily on family labour. Net income per acre = gross output per acre – wages of hired labour per acre – costs of material inputs per acre – imputed (at the market wage rate) value of family labour per acre. Sharecroppers' income per family labour day = (gross output per acre – wages of hired labour per acre – costs of material inputs per acre – rent paid per acre) ÷ family labour days employed per acre.

b. It may be noted that in the two later years, *rent* falls short of *net income* and, correspondingly, income per family labour day is higher than the market wage rate. This is because purchased inputs (fertilisers and HYVs) have been introduced, and the costs are usually shared between landlords and sharecroppers in equal proportion. In effect, therefore, landlords and sharecroppers are sharing the returns on investments in purchased inputs. From the landlords' point of view, the equivalence between hired-labour-based cultivation and share-cropping is maintained if they can mop up the sharecroppers' excess income through the medium of usury. This is a very likely situation.

Source: Author's estimates on the basis of disaggregated data from the Farm Management Surveys.

Notes

1. Government of West Bengal (1981a), Statistical Appendix, p. 2.

2. Data from the 17th round (1961/2) of the National Sample Survey shows the average levels of consumption expenditure per person per month to be Rs 38.42 and Rs 20.83 in urban and rural areas respectively. See *National Sample Survey*, Report no. 184. Data from the 19th round (1964/5) show these levels to be Rs 41.13 and Rs 23.18 respectively. See *National Sample Survey*, Report no. 208.

3. Ahluwalia (1978).

4. Ibid.

5. See Government of West Bengal (1979a), p. 4, Table 1.5.

6. See Appendix Table 7.

7. See Table 3.2.

8. These figures have been worked out by the author from the information provided in *National Sample Survey*, Report no. 215.

9. Structural parameters of the agrarian economy, for the purpose of this chapter, include indicators of patterns of landholding, tenancy relations and labour process.

10. Cf. Sanyal (1977b).

11. *National Sample Survey*, Report no. 215. This figure, however, is generally believed to be an underestimate.

12. See Ghose (1979).

13. Cf. Ahluwalia (1978). Ahluwalia's estimates show an unambiguous rise in the incidence of rural poverty and a virtual constancy in the inequality of distribution of consumption expenditure. The latter estimates are, however, based on data relating to expenditure in money terms. Given the fact that over the same period food prices increased faster than other prices, a constancy in the inequality of distribution in money expenditure actually implies a deterioration in the distribution of real consumption, since the proportion of expenditure on food declines as the level of total expenditure rises. It is also worth noting that in general the inequality in the distribution of income is higher than that in the distribution of consumption expenditure.

14. Stagnation or slow growth, it is true, is one of the causes but by no means the only cause of increasing poverty and inequality, and in this sense causes of stagnation are also among the causes of increasing poverty and inequality.

15. Famine Commission (1880).

16. Cf. Blyn (1966), pp. 172-7.

17. For the purposes of Table 3.1, we have taken the period up to 1976/7 because the present government assumed power in 1977. If one takes the period 1970/1-1978/9, yield rate of foodgrains does show a positive growth rate (1.09 per cent per annum; see Appendix Table 8). However, it remains lower than that during the preceding period (1960/1-1970/1).

18. Cf. United Nations (1975), Ch. 1.

19. Cf. Blyn (1966), Ch. VIII.

20. Since the increase in net sown area was small, the increase in net double cropped area must have been small too, the ratio of the latter to the former remaining roughly constant.

21. These five districts are Burdwan, Birbhum, Hooghly, 24-Parganas and Murshidabad.

22. Cf. Dhawan (1979).

23. Cf. Maine (1871), Dutt (1940), Patel (1952) and Marx (1964a, 1964b).

24. Cf. Moreland (1962, 1968), Habib (1963), Kumar (1965) and Saradamoni (1973).

25. Cf. Moreland (1962, 1968), Habib (1963), Grover (1963) and Moore (1966).

26. See Howard (1935), pp. 45-8.

27. In all probability, the decline between 1953/4 and 1960/1 was real. Cf. Joshi and Narain (1969) and Sanyal (1972).

28. Cf. Ghosh (1976) and Sanyal (1977a, 1977b).

29. Such tendencies were observed in other countries too. See, for example, Geertz (1963) and Radwan (1977).

30. See section III.

31. See, for example, *Report of the Bengal Provincial Banking Enquiry Committee, 1929-30* (Calcutta, 1930) and Chaudhuri (1969).

32. See Ghose (1979a).

33. The significance of the growth of wage labour in Indian agriculture was a subject of serious controversy (perhaps not surprisingly) in the Indian debate on the question of characterising production relations in Indian agriculture. See, in particular, Banaji (1975), Byres (1972), Chattopadhyay (1972a, 1972b), Frank (1973), Rudra (1970, 1971), Patnaik (1971a, 1971b), and Alavi (1975).

34. A contrasting example is provided by Punjab where growth and technological change ensured structural stability over a long period. See Ghose (1979).

35. We cannot go into the details of these arguments here. They have been discussed by several authors. See, for example, Moore (1966), Frykenberg (1979) and Islam (1978). For very interesting discussions of the arguments and motivations that led to the introduction of permanent settlement, see Stokes (1959) and Guha (1963).

36. Area under sharecropping constituted 79.26 per cent of the total area under tenancy in 1953/4 (see *National Sample Survey*, 8th round, Report no. 59, p. 71) and 92.19 per cent in 1970/1 (see *National Sample Survey*, 26th round, Report no. 215, West Bengal, vol. 1, p. 69).

37. We shall return to this later.

38. No estimates for the State are available. My own estimate on the basis of farm management survey data from Hooghly District shows about 40 per cent of the total cultivated area being operated with wage labour. This estimate is of course subject to sampling error, and Hooghly, in view of its proximity to Calcutta, is not perhaps a typical district of West Bengal.

39. See Ghose (1979b) for evidence and discussion.

40. To this issue we shall return later.

41. See, for example, UN (1951), Jacoby (1953) and Myrdal (1968).

42. The major source of inspiration for this view is Schultz's work. See Schultz (1964). For critiques of Schultz's basic premisses, see Sen (1967) and Lipton (1968). For modified statements of the Schultzian position, see Hayami and Ruttan (1971) and Mellor (1976).

43. In 1953/4, 3.36 per cent of the holdings and 3.78 per cent of the area were operated jointly (see NSS, 8th round, Report no. 66, p. 57). In 1970/1, the corresponding figures were 0.11 per cent and 0.07 per cent respectively (see NSS, 26th round, Report no. 215, West Bengal, vol. I, p. 89).

44. For a forceful exposition of this view, see Bhaduri (1973).

45. The preoccupation dates back to Adam Smith. Johnson (1950) and Cheung (1969) provide excellent summaries of the views of classical economists. They also provide rigorous formulations of the problem. The issue has been hotly debated in recent times. See Bardhan and Srinivasan (1971), Newbery (1974), Stiglitz (1974) and Bell and Zusman (1976).

46. The main features of semi-feudalism are: (i) sharecropping, (ii) perpetual indebtedness of the tenant to his landlord, and (iii) lack of accessibility of the tenant to commodity or capital markets. For a lucid account, see Bhaduri (1973).

47. See Ghose (1980c).

48. See Appendix Table 5. Households owning between 1.00 and 5.00 acres predominate among both landlords and sharecroppers.
49. See Bardhan and Rudra (1978).
50. For detailed discussions, see Griffin (1974), Newbery (1975), Ghose and Saith (1976), Banaji (1977) and Rudra (1978b, 1978c).
51. Cf. Ghose (1980a).
52. See Labour Bureau (1978), Tables 3.1(a) and 6.1(a).1.
53. See Kurup (1976).
54. Incidentally, household expenditure absorbed 61.1 per cent and 44.9 per cent of total cash credit received by cultivators in West Bengal in 1961 and 1971 respectively. See Reserve Bank of India (1977), p. 85.
55. We do not have any direct evidence on this. But Prof. K. N. Raj provided some relevant evidence in his Radhakrishnan Memorial Lecture delivered at the University of Oxford in 1976.
56. See Bardhan and Rudra (1978) for some evidence.
57. See Bottomley (1963a, 1963b, 1964a, 1964b) and Long (1968a, 1968b).
58. See Bhaduri (977).
59. Ibid.
60. It is of interest to note that 67.0 per cent and 27.1 per cent of total cash credit received by cultivators was on the basis of personal security in 1961 and 1971 respectively. See Reserve Bank of India (1977), p. 121.
61. See Ghose (1980b).
62. Cf. Bardhan (1970, 1973), Minhas (1970), Dandekar and Rath (1971a, 1971b), ILO (1976) and Ghose (1980a).
63. Labour Bureau (1973, 1978).
64. Some relevant estimates pertaining to Hooghly District of West Bengal are presented in Appendix Table 6.
65. This is the obverse of the celebrated inverse relationship between farm size and land productivity. See Ghose (1979b).
66. See Ghose (1979b).
67. *National Sample Survey*, 26th round, Report no. 215, West Bengal, vol. I, p. 70.
68. See Chakravarty and Rudra (1973), Dwivedi and Rudra (1973), Ghose (1979b) and Appendix Table 10.
69. The programme was launched in stages: land redistribution and rural works programme in 1977, Operation Barga in 1978, and credit programme in 1979.
70. There are 3,242 such bodies in West Bengal.
71. Source: Legislative Department, Government of West Bengal.
72. For details, see Government of West Bengal (1981c).
73. In the case of a sharecropper, only 50 per cent of the area operated by him was taken into account for this purpose.
74. Source: Legislative Department, Government of West Bengal.
75. See Rudra (1975).
76. See Government of West Bengal (1981c). The actual total number of sharecroppers in West Bengal is a matter for speculation. The prevailing view is that there are around 2 million of them.
77. See Government of West Bengal (1981c).
78. Information received through personal communication with the Department of Labour, Government of West Bengal.
79. See Singh (1979).
80. See Government of West Bengal (1981a), p. 40.
81. These features are indicated by the results of the aforementioned sample survey of Gram Panchayats conducted by the Development and Planning Department, Government of West Bengal.

82. For some partial assessments, see Bandyopadhyay (1979), Bandyopadhyaya (1981), Dutt (1981), Ghosh (1981), Khasmabis (1981), Rudra (1981) and Sengupta (1981).

83. Estimated number of rural households owning either no land or less than 1 hectare is taken from NSS, 26th round, Report no. 215 (West Bengal), vol. I, p. 66.

84. See Ghose (1979b).

85. This follows from our earlier observation that land is generally leased out by non-cultivating owners and is leased in by family farmers.

86. Recall, in this context, our earlier observations on the character of the non-cultivating landowners and on the production effects of sharecropping. Note further that security of tenure is needed before a sharecropper can venture to invest in production.

87. This is the classical type, analyses of which are available in Marx (1967) and Lenin (1960).

88. For comprehensive discussions, see Lewin (1968), Carr and Davies (1974) and Mitra (1977).

89. See, for example, Byres (1974).

90. See Warriner (1969) and Lipton (1974). Lipton's more recent work (1977), however, emphasises the need for solidarity among rural classes and hence the need for avoiding measures that might create tension among them. Since agrarian reforms would necessarily create such tensions, he appears to be arguing against agrarian reforms.

References

Ahluwalia, M. S. (1978) 'Rural Poverty and Agricultural Performance in India', *Journal of Development Studies, 14* (3)

Alavi, H. (1975) 'India and the Colonial Mode of Production', *Economic and Political Weekly, 10*, Special Number

Banaji, J. (1975) 'For a Theory of the Colonial Mode of Production', *Economic and Political Weekly, 10* (49)

—— (1977) 'Capitalist Domination and the Small Peasantry: Deccan Districts in the Late Nineteenth Century', *Economic and Political Weekly, 12*, Special Number

Bandyopadhyay, D. (1979) 'West Bengal Experience in Land Reform', *Mainstream*, Annual Number

Bandyopadhyaya, N. (1981) 'Operation Barga and Land Reforms Perspective in West Bengal: A Discursive Review', *Economic and Political Weekly, XVI* (25-6), Review of Agriculture

Bardhan, P. K. (1970) 'On the Minimum Level of Living and the Rural Poor', *Indian Economic Review, 5* (1)

—— (1973) 'On the Incidence of Poverty in Rural India', *Economic and Political Weekly, 8*, Annual Number

—— and Srinivasan, T. N. (1971) 'Cropsharing Tenancy in Agriculture: A Theoretical and Empirical Analysis', *American Economic Review, 61* (1)

Bardhan, P. and Rudra, A. (1978) 'Interlinkage of Land, Labour and Credit Relations: An Analysis of Village Survey Data in East India', *Economic and Political Weekly, 13*, Annual Number

Bell, C. and Zusman, P. (1976) 'A Bargaining Theoretic Approach to Cropsharing Contracts', *American Economic Review, 66* (4)

134   Agrarian Reform in West Bengal

haduri, A. (1973) 'A Study of Agricultural Backwardness under Semi-Feudalism',
Economic Journal, 83 (1)
—— (1977) 'On the Formation of Usurious Interest Rates in Backward Agriculture', *Cambridge Journal of Economics, 1* (4)
Blyn, G. (1966) *Agricultural Trends in India, 1891-1947*, University of Penn-
sylvania Press, Philadelphia
Bottomley, A. (1963a) 'The Cost of Administering Private Loans in Underdeveloped Rural Areas', *Oxford Economic Papers, 15* (2)
—— (1963b) 'The Premium for Risk as a Determinant of Interest Rates in Underdeveloped Rural Areas', *Quarterly Journal of Economics, 77* (4)
—— (1964a) 'The Structure of Interest Rates in Underdeveloped Rural Areas',
Journal of Farm Economics, 46
—— (1964b) 'The Determination of Pure Rates of Interest in Underdeveloped
Rural Areas', *Review of Economics and Statistics, 46* (3)
Byres, T. J. (1972) 'The Dialectic of India's Green Revolution', *South Asian
Review, 5* (2)
—— (1974) 'Land Reform, Industrialisation and the Marketed Surplus in India:
An Essay on the Power of Rural Bias' in D. Lehmann (ed.), *Agrarian Reform
and Agrarian Reformism*, Faber and Faber, London
Carr, E. H. and Davies, R. W. (1974) *Foundations of a Planned Economy – 1*,
Penguin, Middlesex
Chakravarty, A. and Rudra, A. (1973) 'Economic Effects of Tenancy: Some
Negative Results', *Economic and Political Weekly, 8* (28)
Chattopadhyay, P. (1972a) 'On the Question of the Mode of Production in Indian
Agriculture: A Preliminary Note', *Economic and Political Weekly, 7* (13)
—— (1972b) 'Mode of Production in Indian Agriculture: An Anti-Kritik',
Economic and Political Weekly, 7 (53)
Chaudhuri, B. B. (1969) 'Rural Credit Relations in Bengal 1859-1880', *Indian
Economic and Social History Review*, vol. 6, no. 3
Cheung, S. N. S. (1969) *The Theory of Share Tenancy*, University of Chicago
Press, Chicago
Dandekar, V. M. and Rath, N. (1971a) 'Poverty in India', *Economic and Political
Weekly, 6* (1)
—— (1971b) 'Poverty in India', *Economic and Political Weekly, 6* (2)
Dhawan, B. D. (1979) 'Trends in Tubewell Irrigation, 1951-78', *Economic and
Political Weekly, 14* (51 and 52)
Dutt, K. (1981) 'Operation Barga: Gains and Constraints', *Economic and Political
Weekly, XVI* (25-6), Review of Agriculture
Dutt, R. P. (1940) *India Today*, Victor Gollancz, London
Dwivedi, H. and Rudra, A. (1973) 'Economic Effects of Tenancy: Some Further
Negative Results', *Economic and Political Weekly, 8* (29)
Famine Commission (1880) *Report of the Indian Famine Commission*, Calcutta,
Part I
Frank, A. G. (1973) 'On Feudal Modes, Models and Methods of Escaping Reality',
Economic and Political Weekly, 8 (1)
Frykenberg, R. E. (ed.) (1979) *Land Control and Social Structure in Indian
History*, enlarged India edn, Manohar Publications, New Delhi
Geertz, C. (1963) *Agricultural Involution: The Process of Ecological Change*
University of California Press, Berkeley
Ghose, A. K. (1979) 'Institutional Structure, Technological Change and Growth
in Poor Agrarian Economies: An Analysis with Reference to Bengal and
Punjab', *World Development, 7*
—— (1979a) 'Short-term Changes in Income Distribution in Poor Agrarian
Economies: A Study of Famines with Reference to Indian Subcontinent',
Working Paper, ILO, Geneva

—— (1979b) 'Farm Size and Land Productivity in Indian Agriculture: A Reappraisal', *Journal of Development Studies, 16* (1)
—— (1980a) 'Wages and Employment in Indian Agriculture', *World Development, 8,* 413-28
—— (1980b) 'The Formation of Usurious Interest Rates', *Cambridge Journal of Economics, 43* 169-72
—— (1980c) 'Production Organisation, Markets and Resource Use in Indian Agriculture', Unpublished PhD dissertation, University of Cambridge
—— and Saith, A. (1976) 'Indebtedness, Tenancy and the Adoption of New Technology in Semi-feudal Agriculture', *World Development, 4* (4)
Ghosh, Ratan (1976) 'Effect of Agricultural Legislations on Land Distribution in West Bengal', *Indian Journal of Agricultural Economics, 31* (3)
—— (1981) 'Agrarian Programme of Left Front Government', *Economic and Political Weekly, XVI* (25-6), Review of Agriculture
Government of West Bengal (1979a) *Economic Review, 1978-79,* Calcutta
—— (1979b) *Economic Review, 1978-79,* Statistical Appendix, Calcutta
—— (1980) *Economic Review, 1979-80,* Calcutta
—— (1981a) *Economic Review, 1980-81,* Calcutta
—— (1981b) *Economic Review, 1980-81,* Statistical Appendix, Calcutta
—— (1981c) *Land Reforms in West Bengal,* Statistical Report V, Calcutta
Griffin, K. B. (1974) *The Political Economy of Agrarian Change,* Macmillan, London
Grover, B. R. (1963) 'Nature of Land Rights in Mughal India', *Indian Economic and Social History Review, 1* (1)
Guha, R. (1963) *A Rule of Property for Bengal,* La Haye, Paris
Habib, I. (1963) *The Agrarian System of Mughal India,* Asia, London
Hayami, Y. and Ruttan, V. W. (1971) *Agricultural Development: An International Perspective,* Johns Hopkins Press, Baltimore
Howard, L. E. (1935) *Labour in Agriculture: An International Survey,* Oxford University Press, London
International Labour Office (1976) *Poverty and Landlessness in Rural Asia,* Geneva
Islam, M. M. (1978) *Bengal Agriculture, 1920-1946,* Cambridge University Press, Cambridge
Jacoby, E. H. (1953) *Interrelationship between Agrarian Reform and Agricultural Development,* FAO, Rome
Johnson, D. G. (1950) 'Resource Allocation under Share Contracts', *Journal of Political Economy, 58* (2)
Joshi, P. C. and Narain, D. (1969) 'Magnitude of Agricultural Tenancy', *Economic and Political Weekly, 4* (39)
Khasmabis, R. (1981) 'Operation Barga: Limits to Social Democratic Reformism', *Economic and Political Weekly, XVI* (25-6), Review of Agriculture
Kumar, D. (1965) *Land and Caste in South India,* Cambridge University Press, London
Kurup, T. V. N. (1976) 'Price of Rural Credit: An Empirical Analysis of Kerala', *Economic and Political Weekly, 11* (27)
Labour Bureau (1973) *Rural Labour Enquiry, 1963-65,* Final Report, Sinla
—— (1978) *Rural Labour Enquiry, 1974-75,* Final Report on Wages and Earnings of Rural Labour Households, Chandigash
Lenin, V. I. (1960) 'Development of Capitalism in Russia', *Collected Works,* vol. 3, Progress, Moscow
Lewin, M. (1968) *Russian Peasants and Soviet Power: A Study of Collectivization,* Allen and Unwin, London
Lipton, M. (1968) 'The Theory of the Optimizing Peasant', *Journal of Development Studies, 4* (3)
—— (1974) 'Towards a Theory of Land Reform' in D. Lehmann (ed.), *Agrarian*

Reform and Agrarian Reformism, Faber and Faber, London
—— (1977) *Why Poor People Stay Poor*, Temple Smith, London
Long, M. (1968a) 'Interest Rates and the Structure of Agricultural Credit
   Markets', *Oxford Economic Papers, 20* (2)
—— (1968b) 'Why Peasant Farmers Borrow', *American Journal of Agricultural
   Economics, 50*
Maine, H. S. (1871) *Village Communities of East and West*, John Murray, London
Marx, K. (1964a) *On Colonialism*, Progress, Moscow
—— (1964b) *Pre-capitalist Economic Formations*, ed. E. J. Hobsbawm, Lawrence
   and Wishart, London
—— (1967) *Capital*, Penguin, Middlesex, vol. 1
Mellor, J. W. (1976) *The New Economics of Growth*, Cornell University Press,
   Ithaca
Minhas, B. S. (1970) 'Rural Poverty, Land Redistribution and Development',
   Indian Economic Review, 4
Mitra, A. (1977) *Terms of Trade and Class Relations*, Frank Cass, London
Moore, B. Jr. (1966) *Social Origins of Dictatorship and Democracy*, Penguin,
   Middlesex
Moreland, W. H. (1962) *India at the Death of Akbar*, Atma Ram, Delhi
—— (1968) *The Agrarian System of Moslem India*, Atma Ram, Delhi
Myrdal, G. (1968) *Asian Drama*, Pantheon, New York, vol. 2
Newbery, D. M. G. (1974) 'Cropsharing Tenancy in Agriculture: A Comment',
   American Economic Review, 64 (6)
—— (1975) 'Tenurial Obstacles to Innovation', *Journal of Development Studies,
   11* (4)
Patel, S. J. (1952) *Agricultural Labourers in Modern India and Pakistan*, Current
   Book House, Bombay
Patnaik, U. (1971a) 'Capitalist Development in Agriculture – A Note', *Economic
   and Political Weekly, 6* (39)
—— (1971b) 'Capitalist Development in Agriculture – A Further Comment',
   Economic and Political Weekly, 7, Annual Number
Radwan, S. (1977) *Agrarian Reform and Rural Poverty: Egypt, 1952-1975*, ILO,
   Geneva
Reserve Bank of India (1969) *Report of the All India Rural Credit Review Com-
   mittee*, Bombay
—— (1975) *All India Rural Debt and Investment Survey, 1970-71*, Bombay, vol. I
—— (1977) *Indebtedness of Rural Households and Availability of Institutional
   Finance*, Bombay
Rudra, A. (1970) 'In Search of Capitalist Farmers', *Economic and Political
   Weekly, 5* (26)
—— (1971) 'Capitalist Development in Agriculture: A Reply', *Economic and
   Political Weekly, 6* (45)
—— (1975) 'Sharecropping Arrangements in West Bengal', *Economic and Political
   Weekly, 10* (13)
—— (1978a) 'Class Relations in Indian Agriculture – I', *Economic and Political
   Weekly, 13* (22)
—— (1978b) 'Class Relations in Indian Agriculture – II', *Economic and Political
   Weekly, 13* (23)
—— (1978c) 'Class Relations in Indian Agriculture – III', *Economic and Political
   Weekly, 13* (24)
—— (1981) 'One Step Forward, Two Steps Backward', *Economic and Political
   Weekly, XVI* (25-6), Review of Agriculture
—— and Mukhopadhyay, M. M. (1976) 'Hiring of Labour by Poor Peasants',
   Economic and Political Weekly, 8 (39)
Sanyal, S. K. (1972) 'Has There Been a Decline in Tenancy?' *Economic and*

Political Weekly, 7 (19)
—— (1977a) 'Trends in Some Characteristics of Landholdings – An Analysis for a Few States – I', *Sarvekshana*, *1* (1)
—— (1977b) 'Trends in Some Characteristics of Landholdings – An Analysis for a Few States – II', *Sarvekshana*, *1* (2)
Saradamoni, K. (1973) 'Agrestic Slavery in Kerala in the Nineteenth Century', *Indian Economic and Social History Review*, *10* (4)
Schultz, T. W. (1964) *Transforming Traditional Agriculture*, Yale University Press, New Haven
Sen, A. K. (1967) 'Surplus Labour in India: A Critique of Schultz's Statistical Test', *Economic Journal*, 77
Sengupta, S. (1981) 'West Bengal Land Reforms and the Agrarian Scene', *Economic and Political Weekly*, *XVI* (25-6), Review of Agriculture
Singh, P. N. (1979) 'Rural Unemployment and Food for Work Programme', *Khadi Gramodyog*, *26* (1)
Stiglitz, J. E. (1974) 'Incentives and Risk Sharing in Sharecropping', *Review of Economic Studies*, *41* (1)
Stokes, E. (1959) *The English Utilitarians and India*, Oxford University Press, Oxford
United Nations (1951) *Land Reform: Defects in Agrarian Structure as Obstacles to Economic Development*, New York
—— (1975) *Poverty, Unemployment and Development Policy: A Case Study of Selected Issues with Reference to Kerala*, New York
Warriner, D. (1969) *Land Reform in Principle and Practice*, Clarendon Press, Oxford

PART THREE

ON THE ROAD TO COLLECTIVISM

4 AGRARIAN REFORM, STRUCTURAL CHANGES AND RURAL DEVELOPMENT IN ETHIOPIA

Alula Abate and Fassil G. Kiros

Introduction

The present study is only a preliminary attempt to identify the different policies and modalities underlying the process of agrarian transition in Ethiopia since 1974, and to assess the extent to which they have initiated a process of transformation of the institutional and social bases of production in Ethiopian agriculture. To those who have been following events in Ethiopia over the last few years, the reasons for choosing 1974 as a starting point should be obvious. That year had a great significance in the history of the country, since it was then that the Ethiopian revolution overthrew the feudal order and ushered in a new era of social and political change. In the course of this study, we have tried to focus on problems that are being faced in the post-revolutionary period of transition to a collective agriculture. In this endeavour, if we have succeeded in posing the right kind of questions, then our efforts will have been justified.

The study begins by providing an outline of the socio-political conditions that prevailed in the pre-revolution period and then describes the manner in which a limited development that took place within that context led to certain basic contradictions, which in turn led to the 1974 revolution. The revolutionary response, particularly in the form of agrarian reform, is then assessed mainly in terms of the fundamental structural changes and their impact on the economic conditions of the rural population. This is followed by an analysis of the likely process, direction and pattern of future developments in rural Ethiopia.

Thus, in the first part of the chapter the land tenure arrangements that tended to stifle the economic development of Ethiopian agriculture before 1974 are reviewed. The features of the 1975 proclamation concerning land tenure and the basic changes in the agrarian structure which followed its implementation are then assessed. In order to provide a view of the dynamics of the process of agrarian transition, changes in the size distribution of landholding and in the organisation of production are analysed. Attempts are also made to indicate changes in employment, income distribution and consumption patterns. However, for many of these items, owing to a paucity of statistical data, only fragmentary and often indirect assessments have been possible. Finally, we have endeavoured to identify some of the main issues of long-term rural and national development and indicated the types of strategies that might be appropriate given the aims adopted, changes already initiated and the constraints that might conceivably be encountered in future.

Traditional Polity and Superimposed Modernisation: A Contradiction

I.1 A Preview

Land was the source of all wealth and political power in pre-revolution Ethiopia. The maintenance and enlargement of political power, therefore, depended on the perpetuation of the traditional landholding system. The regime reinstated in 1941, however, also saw the advantages of a policy of modernisation. Modernisation would increase the power of the ruling oligarchy over its subjects, would help to burnish its image abroad, and would allow access to the coveted styles of living in the Western world.

Traditionally, rent, tribute and tax sustained the political power structure. The policy of modernisation which required the superimposition of a new infrastructure and bureaucratic machinery could also only be served by the same system of exploitation. The class of traditional landlords was joined by the new elements of the bureaucracy and the military in plundering the peasantry. Behind the veil of modernisation, therefore, the traditional mechanisms of surplus extraction widened and deepened. The temptation of higher standards of living based on modern and imported luxuries, on the other hand, pulled the new establishment increasingly into the orbit of international capital, and towards commercial and capitalist modes of organisation of production and extraction of a surplus.

Gradually, the traditional power structure of the oligarchy was eroded by its own integument, this veil of modernisation. What was intended to increase its power and prestige served to alienate it from its power base. Contradictions began to appear. Peasants were evicted by landlords who had turned capitalist. Pastoralists were pushed away from their traditional grazing land by government turned entrepreneur. Finally, unprepared to adjust to the new economic conditions and unable to cope with the increasing political pressures, the old order shook at its foundation and collapsed.

I.2 The Traditional Mechanism of Exploitation and its Fortification

The reconstruction of the old system of oligarchic rule was achieved almost immediately after the return of Emperor Haile Selassie I from exile in 1941. The oligarchy consisted of the Emperor and his extended family relations, provincial governors, ministers, and the military and church aristocracy. Among them were divided vast territories of agricultural land by inheritance, imperial grant and outright confiscation from the weak peasantry and pastoralists. The members of the oligarchy, with the Emperor at the apex, were united by blood, marriage and tutelage.

Essentially, the members of the oligarchy were absentee landlords who extracted from one-quarter to two-thirds of the produce of a large segment of the peasantry, mainly in the form of rent, to provide for standards of living unequalled in the past. Taxes of various sorts levied on the peasantry helped to finance the symbols of modernisation. But this simultaneously required the introduction of a new administrative set-up as well as the adoption of the vogue of 5-year economic planning.

A new bridgehead of administrative structure was therefore constructed. Appointment to the new hierarchy was made on the basis of tutelage and loyalty to the power centre. If the administrative set-up served and protected the interests of the ruling class, it became at the same time by nature a new burden on the peasantry. The position of district governor (or other positions lower still) was prized not for the salary that it offered, which was quite low, but for the access to land which it afforded. As the new administrators increased their wealth by acquiring land or by a variety of corrupt methods, they helped fill the coffers of the government with tax money for financing the new programmes of controlled modernisation. Modernisation meant largely the gradual introduction of limited economic and social infrastructures which did little to raise the level of living of the rural population.

I.3 Exploitation in the Market-place

To these categories of absentee landlords was soon added an increasing army of merchants and speculators who profiteered at the expense of the peasantry. The merchants bought at prices which they dictated the small quantities of grain brought to the market-place by tenant and small farmers. Trade in food became more and more profitable as the urban population swelled because of rural-to-urban migrants who moved in search of employment in the nascent modern sector. From this new form of exploitation, no one except the landed rich could escape.

The commercial form of exploitation affected the peasantry as much when they sold as when they bought necessities. The local dealers of essential manufactures such as clothing, kerosene, cooking oil, etc., sold to the peasants at prices which included the high profits of the chain of middlemen as well as the relatively high costs of production in industry. This meant essentially that the peasantry subsidised the high levels of living of those involved in the production and distribution of manufactured articles. Many merchants of grain and manufacturers used their profits to acquire agricultural land and joined the landlord class, further increasing the burden of exploitation on the peasantry.

I.4 The Symbols of Modernisation

As urbanisation increased, the ruling oligarchy, the bureaucracy and the merchants saw a new opportunity for exploitation based on the control of urban land and housing. Already, the land in cities such as Addis Ababa, Asmara and Dire Dawa was controlled by the members of the oligarchy. Speculation became widespread as the rich began to invest in urban land. The rich also took advantage of the loans provided by a new mortgage bank to construct houses, apartments and office buildings for renting. Over the period of a generation, the government continued to encourage such symbols of modernisation.

The effect of this development was to increase the contrast between the urban centres and the rural periphery. Not that the government had a well-defined strategy of economic development. The example of economic planning is revealing. The adoption of planning was deemed necessary in order to attract international capital. The plan documents were prefaced with high-sounding statements, and explained priority areas of investment. The actual government measures introduced and projects undertaken, however, had little relation to what was promised in the plan documents; these only served the purpose of boosting the image of the government abroad and of attracting international finance which helped to initiate the development of capitalistic enterprises.

I.5 The Need for Increased Food Production: a New Challenge

Because agriculture was ignored, food production could not keep up with the rate of population growth (see section III.2 below) which rose from a low of roughly 1.5 per cent in the mid-1960s to about 2.5 per cent in the early 1970s. (This general rate of increase was, of course, exceeded by the rate of increase of the urban population, which had reached a rate of 6.6 per cent by 1973.[1]) As a consequence, the country's need for imports of cereals, particularly of wheat, continued to increase. Experiments with 'green revolution' technology were started with assistance from international agencies. Agricultural investment was moreover supported by a liberal policy towards the importation of farm machinery. Some landowners, aware of the benefits to themselves of the capitalistic organisation of production, began to replace men by machines.

The effects of the new technology of production, though restricted, generated new pressures in the rural economy. The new technology had come into direct conflict with the traditional landlord-peasant relations of production. As more tenant peasants were evicted by landlords, and as more pastoralists were pushed away from their traditional grazing land the condition of the rural population deteriorated. Open unemployment, already rapidly increasing in urban areas, now became a serious problem among the rural population.

I.6 The Ferment of Revolution

These developments became the seedbed for revolution. The undercurrents of revolutionary activity, originating in the university campuses, had become widespread in schools by the late 1960s. An educated elite which was to be a major symbol of modernisation became a source of dissent as the exploitation of the masses increased. Students and teachers openly and directly challenged the power base of the oligarchy by raising the slogan, 'Land to the Tiller!' The oligarchy altogether failed to respond to the demands for reform. This became clear in 1971 when the draft of a 'land-reform' proclamation which would only have enabled tenants to claim a legal right for their tenancy failed to pass through Parliament.

Despite this economic and political situation in Ethiopia, however, hardly anyone could predict, even in late 1973, that the end of oligarchic rule was approaching. The proximate causes of revolutionary change — economic crises — so conspired as to result in widespread and continued strikes in Addis Ababa and in other urban centres. One symptom of a regime in serious difficulty appeared when the entire cabinet of the government, overcome by the mounting economic and political pressures, resigned in February 1974. The months that followed saw the armed forces organise a common front and take the leadership in sustaining the revolutionary movement. The united military forces adopted the strategy of gradually and cautiously dismantling the old structure.

II The Pre-revolution Agrarian Structure

II.1 Land Tenure

Land tenure systems have a long and intricate history in Ethiopia, intimately linked to political power structures and social class lines. While an extensive discussion of this history is beyond the purview of the present study,[2] it is important to sketch briefly, as a backdrop, the most important features of Ethiopia's agrarian structure before the revolution in February 1974. The multifarious taxonomy — running into well over 100 terms — used to classify and explain Ethiopia's archaic land tenure system is an indication of the complex pattern of production relations characteristic of the then feudal agrarian structure of the country. The intricacy and complexity of nomenclature associated with tribute extraction is partly responsible for the difficulties encountered in understanding the system. However, many of the categorised types are only marginally different, for these names are merely a reference to forms of tribute that must be surrendered for the right to use land.

It has already become an established tradition among some Ethiopianists to classify tenures into the so-called 'communal' holdings of the north (comprising Begemeder, Tigrai, Eritrea and parts of Shewa, Wello and Gojam provinces), and the private ownership system of the south. There is obviously sufficient justification for treating the north and south separately. This should not, however, be understood to mean that there were two different types of 'feudalism', one manifesting itself in the north and the other in the south. The south presents only a variant of the feudal mode of production prevalent in the north.

The most common form of landholding in the north, often misleadingly termed communal, was known as *rist* land. It is a genealogically based system of land distribution, by which a person could claim land through both male and female ancestors, and, if married, could also gain access to the *rist* rights of his wife. *Rist* rights are essentially land-use rights which are in principle hereditary and inalienable. While ultimate ownership resided in an extremely remote and often legendary

ancestor, actual cultivation and distribution of production were rigo-
rously individualistic. What was inherited was not a particular plot of
land with a permanently fixed location, but rather the right to a share
of a larger tract of land held corporately by the descendants of the first
holder of the estate as *rist*. Once a claim was established, parcels of land
were allotted by the community elders who were sanctioned to adjudi-
cate such claims. The amount of land a would-be owner was likely to
end up with was a direct function of his acumen in litigation, political
power and social standing. Because claims lodged by descendants did
not refer to particular plots with fixed boundaries, *rist* rights encouraged
endemic conflict and interminable litigation among kinsmen and neigh-
bours, sometimes ending in bloody family feuds and vendettas that
could go on through several generations.

Normally, land held under the *rist* system was cultivated by the
peasant as he pleased, so long as he met his tax and other service obligations associated with it. The tenancy system did not flourish in areas of
the *rist* system (as in the south) primarily because land was not considered a commodity to be disposed of or exchanged at will. However,
due to the practice of claiming land through both parents, a person's
rist rights were not only very extensive but spatially scattered in several
other descent groups as well. As a result, limited tenancy arrangements
developed, though almost everyone cultivated some plots rent-free,
however small these plots might be. Only young men who were certain
of eventually obtaining land, and comparatively better-off farmers, with
sufficient oxen and labour, were likely to lease in land.

Particularly in Eritrea, a form of tenure known as *diesa* — perhaps
best translated as 'village' ownership — was to be found. Here, residence
in a village provided the basis for an individual's claim to use a portion
of the available land. The village land was classified in three fertility
grades, and every family was given a share from each category at the
time of redistribution, which would be at regular intervals of anything
ranging from a three-year to a twenty-year cycle.

Superimposed over these tenures, which formed the base of a large
hierarchical structure, was a class of claimants to the peasant's output
and services. The ruling classes, composed in the main of the aristocracy
and the higher echelon of the ecclesiastical hierarchy, extracted the
peasants' agricultural surplus, labour and loyalty through a system of
'fiefs' or 'service tenements', known as *gult*. *Gult* rights were granted to
members of the ruling elite — both lay and clergy — in lieu of salaries,
and to churches and monasteries as endowments. As supreme arbiter,

the monarch delegated a small portion of his powers to his warlords, who in turn passed on to their subordinates some of their delegated powers in a tiered system of sub-infeudation. The area held under a *gult* right varied in the case of an individual according to his rank in the feudal 'pecking order', and in the case of a church or a monastery, according to its fame.

Gult holders had the right to collect taxes and/or extract corvée labour from the *rist* holders who farmed the land; they also had judicial and administrative authority and thus a system of control of the peasantry. In the course of exercising these wide-ranging powers, the *gult* holders systematically drained the surplus of the producers through 'a plethora of fees, fines and taxes imposed on practically everything that lived and grew on the land, and on many forms of activity engaged in by the peasants'.[3] *Gult* as a medium of surplus appropriation enabled the traditional elite to collect a good part of the smallholder's crop. The collected tax and tribute were quickly dissipated by the elite in the financing of their sumptuous lifestyles, characterised by a minimum of investment and personal effort, and consisting of frequent feasts for their large retinue and other hangers-on. Wasteful as they were, these feasts and gestures were nevertheless essential to the maintenance of the political power and legitimacy of the elite in the eyes of the peasantry.

In the process of the formation and consolidation of the Ethiopian empire in the second half of the last century, the traditional mode of production of the north was imported into the south by Menelik's warlords, in a region ethnically, linguistically and culturally different from the north.[4] Immediately after the conquests, the warlords directly confiscated two-thirds of the land, in many cases in close collaboration with the indigenous ruling elite, and left the rest for the local population. As governors of the conquered territories, they were given a free hand to subdivide the land among their officers, soldiers, retainers and other settlers who followed in the wake of the armies, intending to escape land pressures in their home areas. The church as an institution, and individual clergymen, were also allotted land. Traditional chiefs, given the title of *balabat*, were incorporated in the feudal polity as the lowest-level functionaries by grants of *gult* and private tenure rights over the one-third share of the local population. To carry out the division, a very poorly standardised form of land measurement and classification based on fertility was introduced. In accordance with this measure, the land, together with the peasants, was divided on the basis of rank and service, the imperial family, the aristocracy and the influential warlords, who gained huge latifundia in the process.

A good part of the land was granted as a permanent and hereditary possession while on the rest the grantee enjoyed only temporary rights, which reverted to the state when the holder was removed from office or transferred to another region. Prior to the development of a salaried bureaucracy, the political-military administration relied for its sustenance on collection of rent, tribute and services. With the advent of a functionally differentiated bureaucracy under the direct control of the central government, however, this was rendered unnecessary. This in effect turned the majority of the indigenous population into tenants, and most of the grants into freehold. Thus emerged a class of landlords superimposed upon the direct cultivators of the soil.

II.2 The Pattern of Land Distribution in Pre-revolution Ethiopia

Although gross disparity in landownership was all too evident in pre-revolution Ethiopia, reliable statistics on landholding were not available. The reasons are not far to seek. Since the previous regime was basically controlled and run by major landowners or their allies, every obstacle was created to frustrate attempts to document information on land-ownership. The reliability of the surveys conducted by the Department of Land Tenure in the Ministry of Land Reform and Administration was compromised by the absence of standard measurements in many areas, an utterly disorderly system of registration and the equivocal nature of some forms of traditional land tenure. The sketchy data presented below are incomplete and should thus be taken only as rough indicators of prevailing conditions at the time.

The process of alienating the basic means of production from the peasantry resulted in an oppressively unequal distribution of land in southern Ethiopia, which is reflected in the extremely high rate of tenancy (Table 4.1). In northern Ethiopia, where the kinship form of tenure predominated, pure tenants were insignificant, amounting to only a little over 10 per cent of the population. Figures related to distribution of total cultivated land between owners and tenants corroborate the overall picture of a severely skewed distribution of land. The overwhelming majority of the cultivated land in the south was either rented, or partly owned and partly rented. Ownership of land was highly concentrated. For example, a survey conducted in the very fertile district of Harerghe administrative region revealed that 44 per cent of the landowners owned only 3.4 per cent of the total 'measured' land, while a tiny minority of 0.2 per cent owned 75 per cent; out of this vast area, 95.4 per cent was owned by two people only. Again, 0.01 per cent of

Table 4.1: Percentage of Tenants in Total Farm Population and Percentage of Total Area Cultivated by Tenants

Administrative Region	Wholly Rented		Partly Rented		Total	
	Tenancy	Area	Tenancy	Area	Tenancy	Area
Arssi	45	51	7	11	52	62
Begemeder	9	1	6	1	15	2
Gamo Gofa	43	46	4	6	47	52
Gojam	13	–	7	–	20	–
Harerghe	49	46	5	15	54	61
Illubabor	73	62	2	4	75	66
Keffa	59	67	3	4	62	71
Shewa	51	55	16	17	67	72
Sidamo	37	35	2	1	39	36
Tigrai	7	7	18	6	25	33
Wellega	54	49	5	5	59	54
Wello	16	14	16	25	32	39

Source: Ministry of Land Reform and Administration, *The Major Features of the Prevailing Land Tenure System in Ethiopia* (Addis Ababa 1971), pp. 35 and 39; *Report on Land Tenure Survey of Begemeder and Simien Province* (Jan. 1970), Tigrai Province (Jan. 1969) and Gojam Province (Jan. 1971).

the population held about 64.4 per cent of the total *kuter gebar* (unmeasured) land, of which 63.9 per cent was owned by one person. This same person also held 21.7 per cent of the measured land.[5] As the figures in Table 4.2 indicate, a substantial percentage of private land was owned by absentee landlords. In fact, at local levels, the share of absentee landlords may have been well above 50 per cent.

Besides this obviously lopsided distribution of land, the prevailing feudal production relations encumbered the tenants with a plethora of payments. The most extensively used tenancy arrangement was share-cropping where, as indicated, tenants were obliged to pay between one-quarter and three-quarters of their net harvest.[6] Except on rare occasions, tenancy agreements were oral, and it was not uncommon for tenants to be evicted at very short notice, with little or no justification. In addition, tenants were at times charged a premium to start their tenancy, and also had to pay fees and services to the landlord to renew tenancy.

The power of landlords lay in the manner in which payment of various taxes and tributes was made, that is, in kind. This system of payment opened the way for all kinds of extortion.

Table 4.2: Percentage Distribution of Absentee Landlords to Total Number of Landowners and Area of Land Owned

Administrative Region	Absentee Owners	Area Owned by Absentee Owners
Arssi	28	27
Gamo Gofa	10	42
Harerghe	23	48
Illubabor	42	42
Keffa	18	34
Shewa	35	45
Sidamo	25	42
Wellega	29	28
Wello	26	13
Bale	15	12

Source: Ministry of Land Reform and Administration, *The Major Features of the Prevailing Land Tenure System in Ethiopia* (Addis Ababa, 1971), p. 33.

In the country as a whole, 57.6 per cent of the holdings were less than 1 hectare in size and accounted for only 18.4 per cent of the cultivated area. At the other end of the spectrum, holdings of more than 5 hectares accounted for only 0.8 per cent of the total number, but covered 29.3 per cent of the area (see Table 4.3). The problem of small holdings was aggravated by the fragmentation of holdings into smaller parcels, scattered over a wide area. Rapid increase in rural population, combined with the prevailing inheritance and other land-right laws, required the breaking up of property upon the death of the owner/operator and its distribution among the heirs. Fragmentation occurred in all administrative regions but it posed the greatest problem in the northern *rist* areas.

Table 4.3: Percentage Distribution of Number and Area of Operational Holdings by Size

Holdings	Holdings by Size of Total Area (in hectares)						
	Under 0.10	0.11– 0.50	0.51– 1.00	1.01– 2.00	2.01– 5.00	5.01– 10.00	10.01 and above
	Cumulative percentage of total						
Number	3.5	31.1	57.6	81.0	96.2	99.0	100.00
Area	0.2	6.0	18.4	40.7	70.7	82.8	100.00

Source: PMAC Ministry of Agriculture, *Agricultural Sample Survey, 1974-75* (report) (Addis Ababa, July 1975), vol. 1, p. 54.

III The Picture of Uneven Development in Pre-revolution Ethiopia

The modernisation policy of the government since the early 1940s gradually led to the creation of a small modern enclave; but this did not form an integral component of the national economy which continued to be dominated by subsistence agriculture. The reason for this development should be obvious from earlier discussions. The uneven pattern of economic development that has resulted as a consequence will be discussed in the following pages.

Reliable statistical data on the Ethiopian economy are scarce. The first systematised statistical report was published by the Central Statistical Office in 1963, but no national accounts/statistics were included in it. National income data available since 1963 are of uncertain quality; among the main reasons for this is the difficulty of estimating production in the subsistence sector.

Some attempts have nevertheless been made to estimate the production in the monetary and non-monetary sectors of the economy. Based on one such set of estimates for the 1963-9 period, as well as on other figures (though strictly speaking not always comparable) made available by the Central Statistical Office up to 1973, highlights of the economic picture during the period of the last decade of rule by the old regime are presented here.

III.1 Gross Domestic Product (GDP) and Employment

Thirty years ago, the *per capita* income of Ethiopia was among the very lowest to be found anywhere. On the eve of the revolution of 1974, Ethiopia was again ranked among the least developed of the developing countries.[7] The GDP at constant factor cost of 1969 was estimated at US$1,156 million in 1963, growing to US$1,544 million by 1969. This implied an average annual rate of growth of about 5 per cent over the period 1963-9. Agriculture contributed 67 per cent to total GDP in 1963, declining to 57 per cent by 1969. The share of manufacturing in the GDP, including small-scale industry and handicrafts, grew from a low of 9 per cent in 1963 to 14 per cent in 1969. Of the total GDP of 1963, 52 per cent was non-monetary; this proportion declined to 44 per cent in 1969. This reflected the importance of the subsistence sector in agriculture, about which more will be said below.[8] GDP at current factor cost was estimated to have grown at an estimated average rate of around 4 per cent between 1969/70 and 1972/3.[9] Thus, over the period of the decade 1963-73, the average annual rate of growth of GDP was less than 5 per cent.

Roughly 90 per cent of the workforce was employed in agriculture around 1963, declining to about 85 per cent by the end of the decade. Around 80 per cent of the work-force in agriculture was estimated to be employed in the subsistence sector in the mid-1960s, showing the relatively small part played by commercial agriculture in providing employment.

The population of Ethiopia was estimated at about 22 million in 1963, and at about 27.1 million in 1973. The average annual rate of growth therefore was approximately 2.1 per cent. Rough estimates of GDP *per capita* yielded about US$64 for 1963 and about US$83 for 1973, implying an average annual rate of growth of about 2 per cent.

The foregoing presents the picture of a predominantly subsistence agricultural economy growing at a moderate pace in the decade 1963-73. It does not, however, show the underlying situation of stagnation of the most important sector, i.e. agriculture, and the poverty of the agrarian population.

III.2 Neglected Agriculture

In addition to the archaic and extremely unwieldy land tenure system, rural Ethiopia was also characterised by crude and inefficient methods of cultivation and a very limited infrastructure. Every conceivable epithet has been used to describe the backward nature of Ethiopian agriculture, demonstrated, among other things, by the rudimentary nature of the traditional farm implements. Except on the recently introduced commercial farms (starting from about 1965), farming on the small and extremely fragmented plots, particularly in the northern highlands, was carried out with the help of a wooden plough, fitted with a small iron tip, drawn by a pair of oxen. This plough, of ancient vintage, was light and did not turn the soil but merely broke it. Much time and effort was wasted by ploughing the same plot several times in order to prepare a seedbed adequate for local conditions. In areas where draught animals were not employed, either the hoe or another similar digging-stick was used to break the ground.

Whatever little grain or other crop the peasant was able to save from his subsistence production, he brought to the market — a task which often involved a day's journey. He was a victim of price fluctuations and was obliged to sell at harvest time, when prices were lowest. Low *per capita* productivity, income, consumption and savings, characterised the small-scale peasant agriculture in Ethiopia.

A substantial quantity of grain was lost due to inefficient harvesting and storage. It has been estimated that between 30 and 35 per cent of the grain could not usually be recovered from the packed-earth threshing floor, particularly because it was trampled on by the oxen which were used for threshing.[10] The loss through improper storage was no less substantial. Traditionally, produce which was not disposed of immediately was stored in mud-plastered wicker containers or in covered pits. A study carried out in the south-eastern part of the country came out with alarming figures: the rate of damage to crops from rodents, insects and mould could be 30 per cent or more, depending on how full the container was.[11] The fuller the container, the lower was the loss.

While total GDP grew by nearly 5 per cent between 1963 and 1969, agricultural GDP grew by about 2.2 per cent during the same period. A relatively high proportion of the agricultural growth was accounted for by growth of production of commercial crops, the vast non-commercial sector growing at the average rate of about 1.9 per cent per annum. The industrial GDP, on the other hand, grew at the much higher rate of 12.7 per cent per annum between 1963 and 1969.[12]

It might be assumed, superficially, that the higher rate of growth of industry and its increasing share in the GDP showed a normal pattern of progressive industrialisation. The reality in Ethiopia, however, was that the industrial sector was in its infancy even in 1973. The relatively higher rate of growth of the industrial sector largely reflected the narrow base from which it started.

The low rate of growth of agriculture generally, and of subsistence agriculture particularly, was the result of the relative neglect of that sector. As indicated above, labour productivity in agriculture remained extremely low. One estimate made for 1967, for instance, revealed that the output per worker in the large subsistence sector was as low as about 23 per cent of that in the monetary sector of agriculture.[13]

The malaise of agriculture is further underlined by the worsening situation of grain supply during the period under discussion. The figures of agricultural GDP cited above included, *inter alia*, raw materials supplied to industry, since Ethiopian industry was (and still is) predominantly of the processing type. It must also be recognised that nearly all of Ethiopia's foreign exchange earnings came from agriculture. Production in agriculture was in good part accounted for by export produce such as coffee, hides and skins, oil seeds, etc. This meant that the component of food production reflected in agricultural GDP was much lower than might at first appear. Assuming, generously, that the rate of growth of food production was slightly less than 2 per cent per year between 1963 and 1969, it may be concluded that food production was falling behind population during that period (given a population growth rate of over 2 per cent). Indeed, it would seem that food production in the subsistence sector was hardly sufficient to meet the needs of the growing population within that sector (assuming no change in improvements of *per capita* consumption), let alone to supply increasing quantities of food to the urban population.

As a matter of fact, it was since the late 1950s that Ethiopia had been importing wheat, the total quantity imported amounting to over 68,000 metric tons by 1972.[14] In 1972/3 and in a few subsequent years, of course, food imports increased further because of the famine conditions prevailing in a number of regions of the country.

The relative stagnation of agriculture was recognised by the government by the late 1960s, as evidenced by the identification of agriculture as a primary area of emphasis after 1968. However, the government's intention was not the creation of a long-term socio-economic basis for bringing about general and sustained increase in productivity in agriculture, but rather the rapid increase of food production to meet the deficit by promoting commercial agriculture. Though not necessarily implemented in a disciplined planning framework, this encouraged a limited growth of capitalistic methods of organisation of production. This development tended to aggravate the conditions of the peasantry for it led to the eviction of tenants by landlords who mechanised their farms and thereby initiated a process of proletarianisation of the peasantry.

III.3 The Rural-urban Contrast

The neglect of agriculture was not astonishing to those who recognised that the pattern of development in Ethiopia was biased against the rural sector. Urban centres grew at the expense of the rural sector. To be sure, urban poverty too had been deep and dehumanising in Ethiopia. Behind the skyscrapers and the luxury living along the main roads of the major cities were found the unsightly conditions of life of large segments of the urban population. We are only too conscious of this fact when we present the contrasting picture between rural and urban sectors that has developed over a period of a generation or more. Yet, in economic and social terms, the relative position of the urban population is found to be better than that of the rural population.

In economic terms, the rural population became poorer as the urban population (except for the poor) benefited from the government's drive towards modernisation. Whatever industry the country had was concentrated in the major cities. In 1969, income *per capita* of the urban population was estimated at about US$273;[15] the figure for the rural population was about US$44 during the same year. The wide gap in income levels was basically the result of the systematic extraction of 'surplus' from the peasantry to finance investment in the large urban centres.

The contrast becomes even more glaring when we consider the distribution of educational and health services between the rural and the urban sectors. At the beginning of the 1970s, 50 per cent of all educational benefits went to the 10 per cent of the children living in urban regions. The remaining 90 per cent living in rural regions obtained only 50 per cent of the benefits. The participation rate in government primary schools in 1971/2 (enrolment as a proportion of the population aged 7-12) was 70 per cent in urban areas and only 8 per cent in rural areas.[16] These figures suffice to indicate the serious imbalance that existed in the distribution of educational services. It must, however, also be appreciated that attendance at elementary school on the part of those rural children who could be spared from work on the land did not mean that they would eventually replace their fathers in agricultural occupations, nor that they would be otherwise employed in the rural areas. Education in Ethiopia was an alienating factor, since those capable of higher levels of education had to move to the main urban centres, and then rarely returned to their place of origin. Even those students who dropped out at the elementary level would generally turn their back on the rural communities, hoping for higher pay and less grinding work in the cities and small towns. Thus, the physical presence of educational establishments in rural regions and attendance by rural children, in the absence of a policy of general rural development, could hardly be said to be of much benefit to the country areas.

The need for health services is great in rural Ethiopia not merely because 90 per cent of the people live there, but also because disease is much more rampant in rural areas. The basic reason underlying the prevalence of communicable diseases is the general poverty of the population, resulting in poor nutrition and insanitary conditions of living. The pattern of distribution of health services in the years prior to the revolution was, however, heavily biased against the rural regions. Of the 14 administrative regions, three (Shewa, Eritrea and Harerghe, containing four of the largest cities of the country) received over 63 per cent of all health expenditures between 1970/1 and 1972/3. Of the total expenditures for Shewa administrative region during the same period, about 80 per cent went to Addis Ababa alone, a city containing less than 18 per cent of the population of the entire region. In the administrative region of Bale, only one doctor was stationed amid a population of over three-quarters of a million during the period under discussion, whereas in the relatively highly urbanised Shewa region, the ration was 1:16,000.[17]

While the rural Ethiopians received few educational and health services, they were nevertheless made to pay taxes levied specifically for these purposes. Much of the tax funds found their way to financing modernisation programmes in the already privileged urban centres. The urban population's access to relatively better housing, clean water and electricity further emphasised the contrast between rural and urban areas.

IV The Cornerstone of the Ethiopian Revolution

It should be obvious from the foregoing profile of pre-revolution Ethiopian society and economy that the country's future development depended in the first place on agrarian reform. Where a small number of big landlords played a dominant role in agriculture and intensified the exploitation of the peasantry, and where new forms of exploitation gained ground and deepened the peasantry's impoverishment, land reform came to be regarded more and more as an essential precondition for development. The overall performance of the economy was inadequate to ameliorate the social and political problems. The steady increase in population, in the face of the slow growth of output and the inequitable distribution of income and wealth generally, aggravated the country's problems. The regime, however, stubbornly resisted change even as the contradictions sharpened and popular pressure increased. It was left for the revolution that erupted in 1974 to undo the archaic and complex system of landholding, and the political, economic and social structure that it supported. The Proclamation to Provide for the Public Ownership of Rural Land was issued in March 1975, one year after the beginning of open and widespread popular challenge to the old regime, and six months after the removal of the Emperor from the throne.

IV.1 The Key Elements of the Proclamation

The key provisions of the Proclamation will be briefly outlined below so as to identify the ways in which the government sought to respond to the social, political and economic problems of Ethiopia. A fuller appreciation of the objectives and character of agrarian reform can be gained in the context of the specific socio-economic philosophy adopted, and this latter aspect will be discussed in the next section.

The Proclamation banned private ownership of rural land and declared that land would be distributed to the tiller. The area of land to be allotted to any single family was at no time to exceed 10 hectares, no part of which could be sold, leased or mortgaged. The relationship between landowners and tenants was abolished and the latter were freed from payment not only of rent but also of any debt or other obligation to the former. A tenant was furthermore given the right to retain agricultural implements and a pair of oxen, for which the landowner was to be compensated, provided however that the latter was not thereby left without such implements and oxen for cultivating the land to which he/she was entitled.

Specific additional provisions were included in the Proclamation regarding 'communal and nomadic lands'. Peasants in *rist* and *deisa* areas were provided 'possessory rights' over the land they cultivated, and no new claim of land right could be entered in these areas as per tradition. Nomadic people were also given 'possessory right' over their customary grazing land.

The provisions of the Proclamation were to be implemented by peasant associations to be organised with a minimum command area of 800 hectares. These associations were authorised to distribute land to former tenants and landowners, to evicted tenants, to persons with no other means of livelihood in the area, to migrants, to pensioned persons and to organisations requiring land 'for their upkeep'.

The Proclamation thus sought to eliminate the most fundamental problem of Ethiopian society once and for all, by means of reforms leading to egalitarian distribution. Land was not only to be distributed to those who actually tilled it, but the concentration of land in a few hands was also effectively barred: a ceiling was established on landholding, and the enlargement of holdings through purchase or lease was prohibited. Not only was tenancy abolished, but so also was the use of hired labour by individuals (except in special circumstances stipulated by the law), thus firmly closing the door for any possibility of exploitation of the many by the few. These measures could also be expected to stimulate agricultural production and growth in so far as they removed some of the fundamental institutional constraints. The Proclamation thus not only aimed at reducing the inequality in land distribution and destroying the power base of the ruling classes, but also at replacing the former system by radically different social relationships of production.

The formulation and articulation of the socialist organisation of production were to come only later, in 1979, with the issue of the

Directives on Agriculture Producers' Co-operatives. Those who appreciated the depth of the problems associated with land in Ethiopian society may have doubted the realisability of this objective; but it came as no surprise to them that a 'half-way' measure of agrarian reform had been rejected.

IV.2 The Ideological Underpinning of the Revolutionary Change

As implied in the preamble to the Proclamation, and as articulated in the 1974 Declaration of the Provisional Military Government of Ethiopia[18] and in the Programme of the National Democratic Revolution of 1976,[19] the goal in Ethiopia is to build *HibretteSebawinet* − a socialist society. Rural transformation in Ethiopia was therefore expected to occur within the institutional framework of a socialist organisation of production. The 1974 Declaration interpreted the motto of *Ethiopia Tikdem* ('Ethiopia First'), adopted in the early days of the revolution, in terms of *HibretteSebawinet*. The Programme of the National Democratic Revolution, which provided broad guidelines for development in the various sectors, constituted a strategy for transition to socialism, during which phase the material and institutional bases of socialism were to be developed. In the period of transition, therefore, individual production in agriculture and continued participation of local private capital in defined areas of industry and trade were not precluded. The longer-term aim clearly was the building of a socialist society in which the state and co-operative sectors would dominate all economic activity.

The background of the ideological development, its unveiling in the post-1974 period, and the events that influenced it may be of great interest to students of political change. For our purposes, it suffices to indicate the major reform measures taken by the government. Thus, in July 1975, just four months following the Proclamation on rural lands, another Proclamation was issued to nationalise urban land and extra houses:[20] all urban land became the collective property of the Ethiopian people, and all extra houses (except single family dwellings and places of business) were made public property. The Proclamation also prohibited any rental of residential or other buildings on the part of individuals, again reflecting the basic aim of eliminating exploitative relationships and promoting an egalitarian society. Urban dwellers' associations (*Kebelles*) were organised in all towns and cities, *inter alia*, to administer and rent small houses and buildings; the large villas and other buildings were to be managed and rented by a central government authority.

The nationalisation of the major industries as well as of banks and insurance companies was also among the significant measures taken by the Provisional Military Administrative Council (PMAC). In addition, the government took control of a substantial part of local and international trade.

Guided by the ideology of socialism, these measures aimed to respond to the central socio-economic problems of Ethiopia. Thus, the quasi-monopoly of control of urban land and housing by the traditional ruling class and the *nouveau riche* was dissolved. Urban land, which had been practically inaccessible to the great majority of town dwellers was now obtainable by any citizen desiring, perhaps with a building loan, to construct a dwelling, whereas, in the pre-reform period, close to 70 per cent of housing in, for example, Addis Ababa, was rented.[21] The amassing of wealth through the construction and renting of villas and high-rise apartment and office buildings was no longer permissible.

As pointed out above, the modern sector of the Ethiopian economy had increasingly become a new source of economic inequality. This sector, though small, was already relatively strong and without appropriate measures it may have perpetuated some of the exploitative relations in the rural sector. Agrarian reform alone would not have been sufficient for the realisation of the aim of balanced rural and national development.

IV.3 Land Reform Implementation: The Testing Ground

The implementation of land reform was inseparably associated with the policy of organising peasant associations in rural Ethiopia and these associations were also expected to play a large role in the initial stage of rural transformation. The Proclamation on rural lands charged the peasant associations with a number of other tasks in addition to the distribution of land, including, *inter alia*, the administration and conservation of public property, the establishment of judicial tribunals within their jurisdiction, the creation of marketing and credit cooperatives, the construction of schools and clinics in co-operation with the government, and the undertaking of villagisation programmes.[22] The powers and duties of peasant associations were further detailed in a second Proclamation[23] issued in December 1975 to reinforce the first. Some of the major aims were as follows:

(1) to enable peasants to secure and safeguard their political, economic and social rights;
(2) to enable the peasantry to administer itself;

(3) to enable the peasantry to participate in the struggle against feudalism and imperialism by building its consciousness in line with *HibretteSebawinet*;

(4) to establish co-operative societies, women's associations, peasant defence squads and any other associations that may be necessary for the fulfilment of its goals and aims;

(5) to enable the peasantry to work collectively and to speed up social development by improving the quality of the instruments of production and the level of production.

The newly established peasant associations would evidently have been helpless in fulfilling their expected functions if it had not been for the active role played initially by participants of the Development Through Co-operation Campaign (the 'Zemecha') and by the government cadres. Indeed, the speedy creation of the peasant associations in many parts of the country was itself made possible mainly by the campaign participants. It was reported that during the campaign period (1974/5-1975/6) the participants organised over 20,000 peasant associations, with a total membership of some 5 million members,[24] close to 90 per cent of what was reported in early February 1980 (Table 4.4).

The most immediate task to be undertaken by the peasant associations in 1975 was the implementation of the land reform law. It was the campaign participants who in many areas played the role of vanguard in this great undertaking. In the rural areas in which they were stationed, the participants (and particularly the students) took it upon themselves to carry out political agitation and to wage the struggle against former landowners and their supporters. Many campaign participants also pushed towards higher levels of institutional development and in some regions began to promote 'communal' cultivation of land. In several *weredas* (administrative districts), for example Ada, Lume, Shashemene and Dodota, peasant association members were stirred into cultivating the major part of the land 'communally' with only small plots (*yeshet*) left for individual production. It was impressive to observe peasant association members (often exceeding 100 men) ploughing the land *en masse* during the period just before planting in 1975.

It appears that many of the campaigners were primarily motivated by the objective of promoting a socialist organisation of production. Others saw 'communal' cultivation as the only practical measure at the time, since it would have been extremely difficult to divide up the land among the peasants before the critical ploughing and planting season

Table 4.4: Number of Peasant Associations and their Membership[a]

Administrative Region	Number of Peasant Associations[b]	Number of Members[c]
Illubabor	980	198 546
Eritrea	281	60 750
Keffa	1 645	535 474
Wello	1 703	1 022 482
Wellega	2 210	449 792
Gamo Gofa	777	211 976
Gondar	1 137	427 209
Gojam	1 920	456 242
Arssi	1 119	290 000
Harerghe	1 608	420 957
Sidamo	1 482	601 043
Shewa	6 044	1 703 662
Bale	696	170 000
Tigrai	1 064	499 076
Total	22 666	7 047 209

Notes: a. Table excludes over 200 settlements.
b. Number of peasant associations has been reduced due to consolidations.
c. The number of members appeared to be overestimated.
Source: *KeYekatit Iske Yekatit*, Report of the Ministry of Agriculture, 1980 (1972 EC).

immediately following the agrarian reform Proclamation. As it was, it took the commitment and youthful energy of the campaigners to accomplish the task of delimiting the areas of 800 hectares within which peasant associations were to be organised. Rough estimates, heavily relying on traditional measures, had to be applied for this purpose. In those areas where 'communal' cultivation was not introduced, those peasants were allowed to continue private cultivation of their earlier holdings, provided that the sizes of such holdings, in the estimates of the campaigners, did not exceed the upper limit of 10 hectares stipulated by the law.

For subsequent discussions of institutional development in rural Ethiopia, it would be of particular interest to take a closer look at what took place in areas where communal activities were pushed forward. As the harvest season approached, campaigners in areas such as Ada and Lume became busy working out formulae to be applied in the distribution of produce. The key problem was how to divide the produce among the *domegnas* ('those with the hoe') and those who had pooled oxen for ploughing the land. Should the *domegnas* and those with oxen share the produce equally? There were campaigners who felt that any arrangement to compensate animal power contributed would tend to perpetuate inequality within the peasant community. This became a matter of a good deal of debate among the campaigners, and was naturally a serious point of contention between the *domegnas* and those with oxen, the latter often arguing that they not only contributed animals but also did most of the work. Compensation for oxen hired, however, had been even in former times a well-known factor among the peasantry. Thus, the procedure often applied by the peasant associations was to settle liabilities of seeds borrowed and to set aside a reserve of seeds for the following planting season, then to compensate for animal power by such amounts as decided by the general assembly, and finally to share the rest of the produce equally among the members.

The question may be asked as to what lessons were learned from the 1975 measures taken in such areas as Ada, Lume and others. The most common difficulties faced in Ada and Lume had to do with the fact that there was then no fully developed socialist organisation of production, no administrative set-up to plan the work that had to be done, to monitor the contribution of each peasant towards the common endeavour, or to devise a satisfactory procedure for the distribution of the produce. There was no procedure to guarantee that each peasant would be compensated according to the socialist distributive principle: from each according to his ability, to each according to his work.

Anyone reading the June 1979 Directives on Agricultural Producers' Co-operatives, which will be discussed below, will readily appreciate that the initial moves towards mass 'communal' cultivation were premature. The new Directives emphasise the Leninist principle of voluntariness, and go to great lengths to elaborate the stages of development of producers' co-operatives and the conditions that must be fulfilled in each stage.[25] It will come as no surprise to learn that all peasant association land in the *weredas* referred to above, except for much smaller plots left for 'communal' cultivation, was in 1976 divided up among the peasants.

The students indeed often exceeeded their mandate and assumed the role of final arbiter with regard to land distribution, a power which belonged to the peasants themselves. Where the students prematurely pushed communal cultivation of land, it was without explicit authority to do so. Given the low level of political consciousness of the peasants, and their individualistic tendencies, this move undoubtedly led to suspicion about what the future might hold for them. It would be fair to recognise, however, that these early heroic attempts were not without advantage. They played a role in awakening the peasantry and in challenging the tradition-bound mode of petty production. Limited 'communal' cultivation of land has gained acceptance in many parts of rural Ethiopia since 1975.

V Institutional Developments in Rural Ethiopia

V.1 The Peasant Associations as Transitional Institutions

The process of organising peasant associations (and for that matter the implementation of land reform) was not carried out evenly throughout the country, because of differences in the traditional landholding systems, the revolutionary struggles, and the war situation that prevailed in many parts of the country. On the whole, progress in the establishment of peasant associations was relatively rapid in those areas of southern and south-western Ethiopia where tenancy and absentee landlordism had dominated. In the traditional *rist* areas of northern Ethiopia, progress was slow, and in the nomadic regions of the country even slower.

The agrarian structure in the north placed singular constraints on the implementation of the policy laid down by the government — a policy which had met with considerable success in the south. It was easier to isolate the landed elite in the south for many of them were northerners with large estates whereas, in the north, class differentiation was less pronounced, and at the same time the landlords were able to resist far more effectively what they considered as encroachment by the government on their political 'clientele'.[26]

As of early 1980, there were nearly 23,000 peasant associations reported as established throughout the country, including an estimated total membership of over 7 million household heads (Table 4.4). Progress in the implementation of the diverse roles of the peasant associations, however, could hardly be expected to be as rapid as their establishment. It should be recognised that the peasant associations came into being as a result of external impetus; they were mass organisations formed very rapidly while the revolution was yet in a very early stage. They had to face many problems resulting from the very low level of administrative capacity which proved inadequate even for managing day-to-day affairs, let alone for realising the range of economic, social and political objectives with which they were charged. As mass organisations, they stand apart from the governmental administrative apparatus, and they have yet to be integrated into the national planning and policy-making system.

The important point to bear in mind is that despite the understandable problems and difficulties they have encountered, the peasant associations have gradually acquired firm control of the land and have greatly contributed to the consolidation of the gains of the revolution in rural Ethiopia. Moreover, the peasant associations are not politically regarded as permanent institutions; even though the All-Ethiopia Peasant Association has now come into being, the institution of the peasant association itself is, as we shall see below, a transitional development in the process of the long-term institutional transformation envisaged in Ethiopia.

V.2 The Service Co-operatives

An important institutional development in rural Ethiopia, allied with the development of peasant associations, has been the rapid expansion of service co-operatives. The establishment of service co-operatives was provided for in the Proclamation to Provide for the Organisation and Consolidation of Peasant Associations, cited above. That Proclamation stipulated that service co-operatives be formed, each with not less than three and not more than ten peasant associations. The objectives and duties of the service co-operatives included the following:

(1) to procure crop expansion services;
(2) to market the produce of members at fair prices;
(3) to give loans at fair interest rates;
(4) to give storage and savings services;
(5) to supply consumer goods to the members according to their needs;
(6) to give education in socialist philosophy and co-operative work in order to enhance the political consciousness of the peasantry;
(7) to supply improved agricultural implements and provide tractor services;
(8) to collect contributions;
(9) to give flour-mill services;
(10) to organise craftsmen in order to promote cottage industry; and
(11) to provide political education with a view to establishing agricultural producers' co-operative societies.[27]

It was estimated that nearly 3,000 service co-operatives representing over 15,000 peasant associations, were formed as of early 1980. Nearly 90 per cent of the members of peasant associations benefit from the services of the co-operatives, whose estimated total capital of about US$15 million was contributed largely by the peasantry (Table 4.5).

Table 4.5: Service Co-operatives: Statistical Report

Administrative Region	Number of Service Co-operatives	Number of Members' Peasant Associations	Peasant Associations: Members Served	Capital (Birr)[a]
Gamo Gofa	85	385	216 493	430 187
Sidamo	156	702	444 346	728 656
Harerghe	231	1 121	959 848	3 043 990
Bale	15	420	86 735	797 715
Shewa	988	4 813	1 182 864	11 307 975
Illubabor	118	608	98 943	722 159
Keffa	94	481	205 740	852 976
Gojam	460	2 319	318 938	2 575 274
Wello	218	694	379 577	1 549 181
Gondar	206	1 807	124 132	928 016
Tigrai	129	204	95 663	914 980
Wellega	85	433	122 960	537 688
Arssi	141	1 078	15 000	5 225 572
Total	2 996	15 065	4 368 239	29 614 369

Note: a. US$1 = 2.07 Birr.
Source: *KeYekatit Iske Yekatit.*

Most of the service co-operatives have given priority to the establishment of co-operative shops, and have not yet been in a position to fulfil the other economic functions adequately. The shops are established to make available to the peasantry essentials such as cooking oil, sugar, salt, matches and to some extent clothing, footwear and agricultural implements. The shops have served to provide the peasantry with a protective shield against the exploitative market relations involving the hierarchy of wholesalers and retailers of industrially produced consumption items.

As might be expected, the service co-operatives have also had problems of institutional development to face. Their activities require planning and financial organisation for which they are not yet adequately equipped. Despite their present shortcomings, the fact that they have been established throughout rural Ethiopia, and the fact too

that they are managed by the peasantry, constitute significant developments. And, as we shall see below, the service co-operatives, like the
peasant associations, are to form an important but transient step
towards the future institutional development envisaged by the government.

V.3 The Producers' Co-operatives

The establishment of producers' co-operatives was envisaged by the
1975 Peasant Associations Consolidation Proclamation. The organisation and stage-by-stage development of such institutions were fully
elaborated in the 1979 Directives on Agricultural Producers' Co-
operatives referred to above.[28]

According to the Directives the initial step is the establishment of
the type of co-operative referred to as *malba* (elementary producers'
co-operative). The basic requirement for the establishment of the *malba*
is the transfer of private holdings of land to communal holding, leaving
plots of up to one-fifth of a hectare for individual cultivation. Draught
animals and implements are to remain private property in this stage, for
the use of which the co-operative would pay rent to the owners.

The stage of *welba* (advanced producers' co-operative) is reached
when all land becomes communal holding and when all draught animals
and implements are transferred to the co-operative. Land for individual
cultivation of up to one-tenth of a hectare would, however, be permitted. Compensation is to be paid by the co-operative for all draught
animals and implements that become communal property. All members
of a peasant association, or a minimum of 30 of them, may initiate a
malba or they may directly establish a *welba*.

The service co-operatives discussed above are to become 'associations
of producers' co-operatives' (instead of 'associations of peasant associations'), when most of the peasantry belonging to the member peasant
associations become members of producers' co-operatives. The functions of the new associations are to be similar to those of the institutions that they succeed. They also become the basis for yet another
level of institutional development. Thus, the place that service co-
operatives are expected to occupy in the process of rural institutional
change is unique in Ethiopia.[29]

At a later stage of development, several *welbas* already associated
through the Association of Producers' Co-operatives, may become
united to form a new body named *weland*, with an average landholding
of 4,000 hectares and a membership of 2,500 peasants. Each *welba*, by
this time including all peasant association members, is to be designated

as a *habre* or brigade, united under a *weland*. The latter stage maks a high level of institutional and technological development. Thus, the service co-operatives of today are to be gradually transformed into a kind of 'commune' (*weland*), and the present-day peasant associations are to be converted into their brigades (*habres*).

A total of 89 producers' co-operatives are today in various stages of development in rural Ethiopia (Table 4.6). It must be noted that many

Table 4.6: Agricultural Producers' Co-operatives Established (as of early 1980)

Administrative Region	Number of Co-operatives	Number of Members
Gamo Gofa	1	616
Sidamo	5	1 005
Harerghe	4	256
Bale	1	801
Shewa	23	2 583
Keffa	2	139
Gojam	19	862
Wello	9	2 608
Gondar	6	570
Tigrai	12	787
Wellega	7	1 091
Total	89	11 318

Source: *KeYekatit Iske Yekatit*.

of the producers' co-operatives were already in existence prior to the issue of the Directives; these are now in a process of reorganisation in accordance with the new provisions. Furthermore, not all of them are peesant producers' co-operatives; some are co-operatives of former agricultural workers, or of the urban unemployed or other interested groups.

The Directives provide built-in mechanisms for encouraging the development of producers' co-operatives. For example, when a producers' co-operative comes into being within a peasant association area, the key leadership posts of the association are to be held by persons elected from among the members of the co-operative. It is also stated in the Directives that land, draught animals and implements contributed by members of the co-operatives shall not be returned when and if for some reason a member leaves the co-operative. It is, however, left to the discretion of the members of a co-operative to compensate only for implements contributed. Furthermore, service co-operatives are required to provide loans amounting to 25 per cent of their surplus to producers' co-operatives established in their area and to give priority to meeting the needs of the co-operatives for agricultural inputs such as fertiliser. It can be inferred from these provisions that a relatively rapid development of producers' co-operatives is intended.

It is, however, realised that the process of institutional development is likely to require time and may encounter difficulties, as the experience of other socialist countries attests. Furthermore, it seems to be appreciated by the government that, to be successful, institutional development cannot be considered in isolation, but must be accompanied by economic and technological changes in order to overcome the multiple restraints that impede primitive agricultural production. The capacity to plan, finance and execute viable rural and national development in Ethiopia has also yet to be developed.

VI The Effects of Agrarian Reform in Ethiopia:
A Preliminary Assessment

VI.1 *The Basic Yardstick of Achievement*

A distinction is often made between land reform and agrarian reform, the latter being taken to mean not merely changes in tenure and terms of landholding but also, more broadly, the introduction of new institutional arrangements and economic and technological conditions leading to increases of production and income, and generally to rural and national development. Such a distinction between land reform and agrarian reform may, however, be spurious, since it is hardly conceivable that land reform is undertaken in developing countries without regard to the effects of such a reform on agricultural production and on economic development generally. Since it is also true that governments of developing countries which implement land reform are to one degree or another motivated by the aim of raising the level of living of the rural populations, it may not be justifiable to ascribe too restrictive a definition to land reform as opposed to agrarian reform.

From our point of view, a more fundamental distinction that must be made is between socialist agrarian reform and capitalist agrarian reform. The former aims to promote co-operative or collective organisation of production, while the latter seeks readjustments that merely modify the prevailing production relations based on private property in land.

The Ethiopian reform, as has already been shown, is a socialist agrarian reform. While it is evident that the 1975 Proclamation regarded development as the long-term challenge, its most fundamental and immediate aim was to 'alter .. the agrarian relations so that the Ethiopian peasant masses ... may be liberated from age-old feudal oppression, injustice, poverty'. The Proclamation further stated that 'the development of the future can be assured, not by permitting the exploitation of the many by the few ... but only by instituting basic change in agrarian relations which would lay the basis upon which, through work by co-operation, the development of one becomes the development of all'.[30] The realisation of the aim 'land to the tiller', as a necessary step towards a socialisation of production, has therefore been pre-eminent in Ethiopia. Institutional changes are considered to be fundamental not only for the realisation of this aim but also for boldly challenging the impediments to development inherited from the past. In this context, it is the revolution's achievement, in terms of its success in changing the feudal relations of production and in terms of the progress made in consolidating the gains of its reform, that should supply the basic yardstick for evaluating past efforts. The achievements in this regard have been remarkable, especially since they have been attained in the face of economic and social disruption caused by the continuing wars.

As we already indicated, the implementation of the reform pro- gramme has not been uniform throughout the country, nor has it been adequately documented. These factors explain the gaps that exist in our understanding and evaluation of the full impact of the reform, parti- cularly on rural development. An attempt will be made, however, in the following paragraphs to indicate some of the effects of the agrarian reform, based on field data on a number of *weredas* and other informa- tion available from official sources, and the inferences about the future that can reasonably be drawn.

VI.2 *The Effect of the 1975 Agrarian Reform on Land Distribution*

The fact that landownership in Ethiopia can no longer be concentrated has not brought about a substantial change in the size of average 'land- holdings'. However, in addition to eliminating ownership and control of large areas of agricultural land by a handful of landlords, the reform has initiated a trend towards narrowing the disparities in landholdings among the peasantry.[31]

Individualistic tendencies among the peasantry are strongly dis- couraged, although at present individual peasant cultivation continues to dominate in agriculture. Since co-operative development is likely to be gradual, the present pattern of land distribution is of interest. As indicated above, the Proclamation set the ceiling of 10 hectares for an individual or family holding. This upper limit has not, however, been reached, at least not in those areas surveyed so far. Data available for *weredas* mentioned earlier indicate that average landholding is significantly below the limit specified by the law (Table 4.7). In the four *weredas* surveyed, the highest average for any peasant association was found to be 4.6 hectares, and the lowest 1.7 hectares per holding, both in Dodota *Wereda*. There is no doubt that average holdings vary between different *weredas*, between peasant associations within a *wereda*, and between individual holdings within a given peasant association.

It is pertinent to ask why there are such variations in the land distribution pattern today, and indeed why average holdings have not increased significantly when compared with those of the pre-reform period in the areas of the survey (Table 4.2). Let us consider the second question first. It will be recalled that peasant associations were formed within areas of 800 hectares each, so that in general the peasantry who cultivated land within these areas prior to the reform essentially redistributed that amount of land among themselves. Where uncultivated land belonging to former rich landowners was brought under the control of peasant associations the total land size within a given peasant association was still kept within the legal limit. Furthermore, the landless had to be allotted land, and the very small holdings had to be raised in size. Therefore, there was little likelihood for the average holding within a peasant association area to differ significantly from that of the pre-reform period. It must, however, be recognised that there are areas of land in certain regions potentially available for cultivation; if these were distributed among the peasantry, they could significantly increase the average size of holdings. This would necessitate the planning and implementation of resettlement programmes and the movement of a large segment of the peasantry from the regions of high population density to regions of low population concentration. Resettlement of population in large numbers can be considered only as a long-term proposition, in view of the economic and social problems that it is likely to entail.

In answer to the question about variation in land distribution pattern, the variation of holdings within a given *wereda* and within a given peasant association area in a *wereda* have resulted from a number of factors. First, the area of 800 hectares within which a peasant

Table 4.7: Average Size of Landholding in Four Weredas (in hectares)

Administrative Region	Peasant Association	Pre-reform	Post-reform
A. Ada (Shewa)[a]	Aderie	2.1	2.0
	Algay	6.7	2.4
Aderie	Choba	3.4	NA
	Kerfie	3.0	2.4
	Keta-Weregenue	3.1	2.7
	Tulu-Dimtu	2.2	2.3
	Turura-Chello	3.3	2.1
	Yerer-Buti	2.1	2.6
B. Welmera (Shewa)[b]	Dewana Tebel	3.0	2.9
	Dilu Tulu Hora	2.2	3.0
	Fiota Mintilie	2.3	2.3
	Hide Idil Ber	2.0	2.0
	Ilu Berga	3.1	2.6
	Kawo	2.8	2.5
	Liben	2.5	2.4
	Minjaro	2.1	3.3
	Sokoru Golelie	2.0	2.7
	Wajitu Fiche	2.5	2.3
C. Dangla (Gojam)[c]	Semalta Gabriel	4.0	3.6
	Abela Lideta	1.8	1.8
	Kuancha Michael	4.1	3.9
	Kebesa Mariam	4.3	3.5
	Gumdere Abo	2.5	1.9
	Sahra Kidanemihret	3.6	2.6
	Sengure Ghiorgis	2.4	2.2
	Shangana Giorgis	2.1	2.1
	Gayta Sellassie	3.2	2.3
	Demsa Mariam	4.7	4.2
	Zebura Abo	3.5	3.2
D. Dodota (Arssi)[d]	Kobbo Memelle	2.9	3.3
	Berkuale Degaga	2.4	1.7
	Bueyenu Kiltu	4.3	4.1
	Bekka	3.7	2.6
	Dodota Alem	3.3	3.3
	Genjo Kuye	3.6	3.9
	Bestella Bedasso	3.7	3.6
	Messo Shokoksa	3.6	3.5
	Meleboma	5.3	4.6
	Koremeni	4.7	3.2
	Huluko	3.4	3.4

Sources: a. Fassil G. Kiros, Survey Data on Ada Wereda (in process of analysis) (IDR, 1979). b. Fassil G. Kiros and Asmerom Kidane, *Socio-Economic Baseline Survey of Welmera Wereda* (Ethiopian Science and Technology Commission, United Nations University, International Development Research Center, and IDR, Addis Ababa, 1979). c. Tesfaye Teklu, *Socio-Economic Conditions in Dangla*, Research Report no. 23 (Institute of Development Research, Addis Ababa University, 1977). d. Tesfaye Teklu, *Socio-Economic Conditions in Dodota*, Research Report no. 27 (Institute of Development Research, Addis Ababa University, 1977).

association was established was not, to the best of the authors' knowledge, actually measured. The Proclamation itself stated that peasant associations should be established using traditional settlement boundaries such as *chika*, *debr* and *got* as the bases. As a result, the actual sizes of peasant association areas have varied. And within a given *wereda*, population densities have also varied between the various peasant association areas (Table 4.8).

Thus, even if we assume that areas held by individual peasant associations scatter closely around the 800-hectare limit, average holdings would vary because of the differences in population densities. It may also be noted in this connection that variations are bound to exist in the levels of fertility of land held by the various peasant associations and even within a given peasant association area. In the case of many peasant associations, land is in fact distributed in such a way that each household obtains its share of land, taking full account of different levels of fertility. Although motivated by considerations of equity, this method of land distribution has again resulted in the fragmentation of holdings in a number of areas surveyed.[32] The ultimate solution to this type of problem is the development of producers' co-operatives which is necessarily a slow process.

Inequality in landholding within a particular peasant association may also occur because land is allotted in accordance with the policy of differentiating between the land needs of single or married persons and those with children. Lastly, it must be pointed out that even if the size of a peasant holding may now be about the same as in the pre-reform period, the majority of the peasantry still benefit because of the fact that the entire produce of the land accrues to them and because of other factors which will shortly be discussed.

Table 4.8: Size of Membership of Eight Peasant Associations in Ada Wereda

Peasant Association	Number of Members
Aderie	217
Algay	257
Choba	154
Kerfie	192
Keta-Weregenue	168
Tulu-Dimtu	202
Turura-Chello	266
Yerer-Buti	184

Source: As in Table 4.7, Source a., and based on interviews with peasant association leaders.
Note: Number sometimes includes a few pensioners supported by a peasant association.

VI.3 Effect of Agrarian Reform on Production and Employment

It was thought by many people that the immediate effect of the reform would be a chaotic situation in rural Ethiopia, resulting in severely depressed production and shortages of food supplies. Such fears were partly based on the reasonable assumption that a degree of disruption would be inevitable when revolutionary changes in production relations, such as have occurred in Ethiopia, took place.

It nevertheless turned out that the fears were exaggerated. A report of the Ministry of Agriculture indicates that production had not fallen as much as was feared (Table 4.9). There is no denying that there were shortages (usually temporary) of food items in the urban areas, but this may in good part be explained by the disruption in the marketing and distribution system, increase in rural consumption and by exogenous factors, in particular the wars in the Ogaden and in Eritrea, and the recurrent drought in several parts of the country.

In addition, the incomes policy of the government has contributed to an increased effective demand for food in urban areas. The establishment of a minimum wage and salary increments for low-income groups in industry and the tertiary sectors brought about effective increments in the demand for food. In a situation of rising inflation, together with fixed prices and food shortages, the evolving black market (although as yet minor) tends to confuse the data and its interpretation for the design of a coherent price policy. When the official price for grain fixed by the government is normally half of what it could fetch on the black market, it is not surprising to encounter dissatisfaction among farmers who are not able to sell at the higher price. Obviously, an inflationary situation and shortages of basic commodities constitute sources of worry for any government, because they tend to erode the political

Table 4.9: Estimates of Production of Major Crops for the Whole of Ethiopia, 1974/5-1978/9

Year	Estimates (in thousand quintals)[a]
1974-5	43 572.9
1975-6	52 900.5
1976-7	50 669.9
1977-8	45 866.1
1978-9	46 303.7

Note: a. 1 quintal = 100 kg.
Source: *Area, Production and Yield of Major Crops for the Whole Country and by Region, 1974-5 - 1978-9*, (Ministry of Agriculture, Provisional Military Government of Socialist Ethiopia, Addis Ababa, July 1979), p. 38.

support gained through the initiation of a structural transformation. Stimulating industrial and agricultural production is vital for the maintenance of this support.

Indeed, the implementation of agrarian reform, particularly in the south, and the readjustment to the new economic and political order were remarkably rapid. More important, it is the long-term impact of agrarian reform rather than its immediate effects on production that should be the primary basis for assessment.

Production increases have also resulted from the state farm sector, in large part because of the new lands that were brought under cultivation (Table 4.10). A good deal of emphasis has been placed on the expansion of this sector, which initially consisted of the large-scale commercial farms that were nationalised. Although land brought under cultivation by state farms increased steadily, rationalisation of production in this sector is still required. Without including depreciation costs for machinery and other fixed assets, the estimated loss between 1975/6 and 1978/9 is quite considerable, as indicated in Table 4.10. The important point to note, however, is that the vast expansion of the areas cultivated was made possible because of the agrarian reform. The major constraints on the development of state farms are inadequate managerial capacity and a shortage of capital required to bring the new lands under cultivation.

Table 4.10: Land Under Cultivation, Employment, Total Revenue and Expenditure, and Estimates of Major Crops in State Farms, 1975/6 – 1978/9

	A	B		C		D
Year	Land under Cultivation (in ha.)	No. of Workers Employed (in 000) Perm-anent	Temp-orary	Expenditure[a]	Revenue[a]	Various Crops[b] (in 000 quintals)[c]
1975/6	57 738	8 485	57 709	58 129 634	62 112 760	986 016
1976/7	53 433	18 482	55 934	84 994 144	84 666 348	1 044 168
1977/8	62 524	21 362	64 584	104 978 664	94 497 467	1 371 441
1978/9	82 634	24 039	76 868	158 258 799	116 947 386	2 028 588

Notes: a. US$1 = Ethiopian Birr 2.07.
Sources: A, B and D, 1975/6-1977/8, *State Farms Development Authority*, no. 1, 25 Yekatit 1972 (Eth. Cal.). A, B and D, 1977/9-1978/9 and C information is obtained from the Ministry of State Farms.

b. These farms are engaged in the production of cotton, maize, coffee, teff, sorghum, wheat, pulses and different fruit and vegetables.

c. 1 quintal = 100 kg.

The effect of the agrarian reform on employment is a subject of great interest. Unfortunately, data are hard to obtain in this area at the present time. There seems to be little doubt, however, that the effect of the implementation of the reform has been to increase employment opportunities in agriculture. In the past, rural and urban migration was in good part attributed to landlessness, eviction of tenants, and to the exploitative relations of production in agriculture. Such factors no longer exist in rural Ethiopia. It is the responsibility of every peasant association to allot land to everyone willing and able to work it. Young men and women who have come of age in the years since 1975 and who would have out-migrated in large numbers, were most probably absorbed in the agricultural sector. Many peasant associations redistribute the available land periodically, precisely to accommodate the needs of their new members. There are, of course, problems associated with this procedure. Aside from the time and the logistics required, the frequent redistribution of land may also result in a disincentive to invest on the part of an individual cultivator, since he/she does not expect to continue to hold a given plot of land.

The potential for wider employment in agriculture is also being realised to some extent by the expansion of the state farms and by resettlement programmes. The planned level of employment in the state farm sector in 1979/80 was estimated to be 30,500 permanent and 180,000 seasonal workers.[33] Although further increase of employment can be expected in this sector, this will very much depend on the type of production technology adopted. Resettlement programmes of the Relief and Rehabilitation Commission and the Settlement Authority (agencies now amalgamated) have provided productive employment to a great many people of different origins, including drought victims, the urban unemployed, etc. The number of families resettled by the Settlement Authority had exceeded 20,000 by 1977, and that of the Relief and Rehabilitation Commission was probably no less than this number up to about the same period, and this was expected to increase significantly in the subsequent years.[34]

It should not be inferred from the foregoing that there is at present 'full employment' in agriculture and that there is no longer rural to urban migration. The full utilisation of available manpower in agriculture depends, *inter alia*, on the technical conditions of production. Besides seasonal underemployment and unemployment resulting from socio-cultural factors, land redistribution based on family size irrespective of the actual work-force in the household is contributing to this problem. Rural to urban migration is inevitable in the process of development and to the extent that it now exists it must be explained by factors other than those mentioned above.

VI.4 The Effect of the Agrarian Reform on the Level of Living of the Peasantry

The concept of the level of living reflects the economic, social and political conditions in the life of a society. The liberation of the mass of the peasantry from age-old social degradation and political oppression has directly resulted from the 1975 Proclamation. There is hardly any need to dwell upon this matter, except to note that the peasantry can in the future be expected progressively to realise its own development potential, and to participate more fully in national and state affairs.

At the level of material well-being of the peasantry, the contributions of the land reforms have undoubtedly been significant, although data needed for a detailed assessment are lacking. The egalitarian land distribution, the ban on the employment of hired labour on private and communal farms (where the latter existed) operate to restrict differences in income among housholds. However, because of resource endowments, types of production and population density, differences in income between peasant associations and regions continue to exist.

As has been stated above, the produce of the peasantry now accrues to the peasantry. So far, the level of taxation of the peasantry has been very low. In the years following the Proclamation, a peasant family was required to pay, on an average, a tax of only about US$3.50 (approximately 3-4 per cent of average annual income) per annum. (However, the level of taxation can be expected to change when institutional conditions permit the implementation of a tax law already promulgated, based on the assessment of the income of individual farm families.[35])

The changing terms of trade between the agricultural and industrial sectors have also benefited the peasantry since 1975. It is recognised that movements in the terms of trade between food and agricultural raw materials on the one hand and industrially produced items on the other, had in the past been against the former. A trend in the opposite direction has been apparent in Ethiopia in recent years (Table 4.11). Prices of major staples such as teff rose significantly following the agrarian reform, although some of the price increases may be attributed to a fall in production. Generally, however, prices had remained on the higher levels even after production had recovered. Although there may be several other factors contributing to this situation, part of the explanation may have to do with the fact that the peasantry has now

Table 4.11: Retail Price Index for Addis Ababa (excluding rent) –
1963=100

Major Groups	1974	1975	1976	1977	1978
General index	159.7	170.1	218.7	255.1	291.6
Food	167.6	175.1	248.4	290.1	339.5
Household items	157.0	159.8	205.5	261.0	280.4
Clothing	175.2	190.6	205.9	223.7	245.2
Medical care	156.8	179.6	194.9	209.3	239.3

Source: National Bank of Ethiopia, *Annual Report 1978*, p. 14.

gained the upper hand in influencing the level of prices for food items.
The result of this has been to increase the money income of the
peasants, and therefore their capacity to acquire necessities such as
clothing and other industrially produced items whose prices have
increased at a relatively slower pace because of effective control of
prices of the nationalised industrial establishments. However, although
a central distribution centre has been created, and rural service co-
operatives have been established, the marketing of industrially produced goods in the rural sector continues to be difficult.

All in all, it can be judged without prejudice that the agrarian reform
has indeed greatly benefited the peasantry; and a survey made in Ada
Wereda shows that the peasantry there concur. Because of the difficulty
of obtaining reliable figures on production and income, the peasants
were asked general questions that might provide indications of the
changes in the level of living. Out of a random sample of 295 peasants
from eight peasant associations interviewed, a vast majority indicated
that they were materially better off than before (Table 4.12).

Table 4.12: Indications of Economic Change since the Land Reform –
Interview Responses of Ada Peasants (total no. of respondents: 295)

Responses	Number	Respondents Percentage of Total Respondents
(1) Consuming more of own produce	276	93.6
(2) Buying more manufactured goods since land reform	201	68.1
(3) Saving more grain now	204	69.2
(4) Selling more grain now	200	67.8
(5) Selling less grain now	33	11.2
(6) Producing less grain now	12	4.1

Source: Fassil G. Kiros, study undertaken in Ada for the purpose of this chapter
(IDR, 1979).

It must, however, be said that a systematic assessment of the impact of agrarian reform in Ethiopia has yet to be undertaken; and again, it is not only the immediate effects but also the long-term impact on national and rural development that will prove its success.

VII Issues of National and Rural Development in Transition: Concluding Observations

Ethiopia has now turned its attention to development. In the past few years, efforts have been focused on major campaigns of reconstruction to raise production to full capacity, and on institutional development in agriculture, industry and social sectors. Its current development policy is based on the Programme of National Democratic Revolution, which basically aims to 'lay the foundations for the transition to socialism'. What, then, are the main issues and how might development strategy in the transition be influenced by them? We consider only three broad questions, namely, the question of institutional change in agriculture, the challenge to overcome the diverse economic, social and technological conditions that impede development in the agricultural sectors, and the basic problem of increasing the rate of accumulation.

The collectivisation of agriculture is generally the second step of socialist agrarian reform.[36] As was explained above, Ethiopia has already adopted a long-term policy of developing co-operatives, regarded as pivotal under the socio-economic philosophy adopted not only for rural development but also for overall national development. There is an apparent major dilemma which needs to be resolved. Left to their own devices, the peasants, in the early post-distribution situation, are particularly vulnerable to alienation from socialist aims, being unable effectively to mobilise themselves without outside support, or often, without outside leadership. The problem is how to control and direct from above effectively, without stifling mass initiatives and self-reliance in local matters from below. There is a need to stimulate a change from within peasant mentality and attitudes in order to bring about democratic participation at every level.

Petty subsistence production continues to dominate in Ethiopian agriculture, and is likely to impede the immediate government efforts to promote technological change and to provide essential services to the peasant sector. The marketing of agricultural produce is also likely to be handicapped if deliveries are to be based on whatever small surplus the individual peasant household brings to the market. Petty production needs to be replaced by relatively large-scale production so as to realise the full potential of modern technology; subsistence production has to give way to specialisation and production for the market. This is expected to result from the development of producers' co-operatives, but this can only be gradual. Government efforts in the transition will be faced with economic and social problems of development inherited from the pre-revolution period, but the problems can be ameliorated if the drive towards collectivisation does not at the same time tend to diminish the important economic role that has been played by the peasant associations and their service co-operatives, until such time as producers' co-operatives become widespread in agriculture. Forcing the speed of collectivisation may endanger forever the prospects of a transition to socialism.

Initially, peasant farming should not be viewed as an insurmountable stumbling block to be avoided at all costs in the period of transition, so long as relevant measures are taken against the forces of class different-iation which are likely to create renewed inequalities among the peasantry. The experience of other socialist countries has taught two important lessons: that collective agriculture by and in itself is not a sufficient defining characteristic of socialism, and that the 'individualist' stage can realistically be skipped in favour of collectivisation only in the case of a relatively advanced capitalistic form of agriculture.[37] Attention to the strengthening of peasant associations and service co-operatives is therefore desirable in the transition in order to increase agricultural production and to promote self-sustained rural development.

The development of state farms can, of course, play an important role in increasing production. However, expansion of this sector is possible largely by bringing new lands under cultivation and it demands high levels of organisational and technical capacity. Mainly because of the pressure to increase production and the inability to enforce work discipline, state farms have so far been running at a substantial loss in Ethiopia. The investment requirements of state farms are always likely to be heavy, constraining development especially if mechanisation is emphasised, with an increased demand for scarce foreign exchange resources.

Agricultural development strategy needs to be recast in Ethiopia. In the past, emphasis was placed on increasing the production of main staple grains, generally in selected regions, by the use of inputs such as fertiliser, improved seed varieties, insecticides, etc. The features that characterise Ethiopian agriculture are the diverse agro-ecological conditions and production patterns, heavily influenced by local socio-economic traditions. These diverse conditions of production are little understood. Long-term development strategy calls for the institution of a national system of research, experimentation and dissemination of technologies appropriate for the various regions of the country, necessitating an elaborate organisational system and a much enlarged technical capacity. It may become necessary in present circumstances to emphasise maximisation of the marketable surplus by overcoming already identified problems associated with production, storage and the marketing of the produce of the peasant sector. It is well known, for example, that in many parts of Ethiopia a significant proportion of potential produce is lost because of untimely planting, inefficient weeding, and other conditions already indicated. With the deployment of low-level technical personnel, including trained peasant production cadres (as has already been modestly begun), significant returns could be obtained. This may appear to be a rather obvious necessity; yet it is felt that aspiration towards much higher levels of technological development (which are not attainable in the immediate future) can sometimes obscure and obstruct simple measures which may contribute importantly in increasing yield, productivity and marketable surplus.

Although the peasantry would have to bear an increasing share of the cost of development, initial government investment is required; indeed, a modest programme is at present under way to make a net transfer to agriculture through public-sector subsidies (for credit, etc.). Since 1974 an increase of about four times the pre-revolution fiscal budget for agriculture has been attained so as to bring about an intersectoral redistribution in its favour. A fundamental issue from the point of view of national development as a whole has to do with the magnitude of surplus that can be generated. The country aims at raising the level of living of the masses as rapidly as possible; at the same time the need for immediate sacrifice is evident in order to increase investment not only in agriculture but also in industry and in the tertiary sectors. If it is expected that much of the surplus will come from the agricultural sector, as indeed it must, how can this be achieved without at the same time sacrificing to some degree the goal of improving the level of living of the mass of the peasantry? It must also be noted that neither institutional conditions of production nor marketing mechanisms have yet developed sufficiently to facilitate the siphoning off of whatever potential surplus may exist in the agricultural sector.

In the transition, therefore, the desirable option might be one that favours a relatively self-reliant approach to development in the rural area, based on the promotion of economic and social development projects by the peasantry and for the peasantry. With this approach, the peasantry might be able to bear, without excessive hardship or resentment, an increased share of rural and agricultural development costs, to an extent that would permit investment of a substantial proportion of available national resources in other sectors of the economy. The development strategy would moreover be in consonance with the aim of spreading the development base, avoiding uneven development in the rural and urban sectors. With the increase in overall capacity of the peasant sector to generate surplus development and the mechanisms for its appropriation, investment in the non-agricultural sectors can be increased gradually.

This approach may be taken by some as an argument for the well-known 'agriculture first' strategy of development; others may argue that it implies a gradual process of change. Yet industrial development can be fostered in rural areas and based on local resources and initiatives much more constructively than is apparently appreciated. The approach outlined above is based on the assumptions that Ethiopia desires both immediate increases in the level of living of the masses and a relatively rapid overall rate of development of the national economy, and that such a combination is feasible. Its relevance depends, *ipso facto*, on the validity of these assumptions.

Notes

1. R. Pankhurst, 'Nineteenth, Early Twentieth Century Population Guesses', *Ethiopia Observer 5*, no. 2 (1961); also *Urbanization in Ethiopia*, Bulletin no. 9 (Central Statistical Office, Addis Ababa, 1972).

2. There are a number of sound studies of the agrarian question and land tenure systems of Ethiopia. See in particular Allan Hoben, *Land Tenure among the Amhara of Ethiopia: The Dynamics of Cognitive Descent* (University of Chicago Press, Chicago, 1973); John Markakis, *Ethiopia: Anatomy of a Traditional Polity* (Clarendon Press, Oxford, 1974); Addis Hiwet, 'From Autocracy to Revolution', *Review of African Political Economy* (London), Occasional Paper no. 1 (1975); Michael Stahl, *Ethiopia: Political Contributions in Agricultural Development* (Raben and Sjojrent, Stockholm, 1974): John M. Cohen and Dov Weintraub, *Land and Peasant in Imperial Ethiopia* (Van Gorcum, Assen, Netherlands, 1975); Patrick Gilkes, *The Dying Lion: Feudalism and Modernization in Ethiopia* (Julian Friedman, London, 1975); S. F. Nadel, 'Land Tenure on the Eritrean Plateau', *Africa*, vol. XVI (1946); Mahtama Sellassie W/Meskel, 'Land Tenure and Taxation from Ancient to Modern Times', *Ethiopia Observer 4*, no. 5 (1964); John M. Cohen, 'Peasants and Feudalism in Africa: The Case of Ethiopia', *Canadian Journal of African Studies*, vol. VIII, no. 1 (1974).

3. J. Markakis and Nega Ayele, *Class and Revolution in Ethiopia* (Spokesman Press, Nottingham, 1977).

4. For an interesting discussion of class formation in the southern regions of Ethiopia, see Markakis, *Ethiopia*, pp. 104-40; Hiwet, 'From Autocracy to Revolution', pp. 30-8.

5. Seleshi Wolde Tsadik, *Land Ownership in Hararghe Province* (Dire Dawa, HSIU, Ethiopia, June 1966), p. 21. This situation was quite widespread in almost all the southern provinces. Data collected for four *weredas* in Jimma *awraja* show that before the revolution 0.25 per cent of the total population owned about 18 per cent of the privately held land (Mesfin Kinfu, in *Proceedings of the Social Science Seminar* (IDR, 1974), p. 159).

6. Net harvest is what remains after the seed for the next harvest is set aside, and after tithes and sometimes taxes are paid on behalf of the landowner. An interesting insight is provided by Amit Bhaduri, who argues that the practice of deducting the required seed for the next harvest before the sharing of the produce implies that the tenant is in effect lending the landowner a part of the working capital, free of interest. See Amit Bhaduri, 'Agricultural Backwardness under Semi-feudalism', *Economic Journal*, vol. 83, no. 1 (1973), pp. 120-4.

7. Branislav, Gossovic, UNCTAD, *Conflict and Compromise* (A. W. Sijthoff, Leiden, n.d.) pp. 163-4, 282-3.

8. Fassil G. Kiros *et al*, 'The Urban Bias of Ethiopia's Development' in Fassil G. Kiros (ed.), *Introduction to Rural Development* (Institute of Development Research, Addis Ababa University, 1977), p. 41.

9. Planning Commission Office, Ethiopia, *Report on Current Economic Situation in Ethiopia* (Addis Ababa, 1976), Appendix I.1.

10. G. Gill, 'Improving Traditional Ethiopian Farming Methods: Misconceptions, Bottlenecks and Blind Alleys', *Rural Africana*, no. 28 (1975), p. 109.

11. The concentration of oxygen in a grain pit varies according to the level of the grain: the higher the concentration of oxygen, the greater the infestation of grain by insects. Wubshet Teferra, *Land Use Study of Ten Farmers at Gende Maya* (Dept. of Geography, HSIU, Addis Ababa, 1972).

12. Kiros *et al.*, *Introduction to Rural Development*, p. 41, and 'Ethiopia Economic Survey', *Africa Development* (Nov. 1971).

13. Kiros *et al.*, *Introduction to Rural Development*, p. 43.

14. FAO, *Food Security Mission to Ethiopia: Report* (FAO, Rome, 1974), p. 22.

15. Kiros *et al.*, *Introduction to Rural Development*, p. 42.

16. A. H. ter Weele, *Distribution of Education in Ethiopia* (Ministry of Education and Fine Arts, Addis Ababa, 1974).

17. Kiros *et al.*, *Introduction to Rural Development*, p. 46; Ministry of Health, *Health Services in Socialist Ethiopia* (Addis Ababa, 1978), pp. 18-19.

18. Provisional Military Government of Socialist Ethiopia, *Declaration of the Provisional Military Government of Ethiopia* (Addis Ababa, 20 Dec. 1974).

19. *Idem*, *Programme of the National Democratic Revolution of Ethiopia* (1976).

20. *Idem*, *Government Ownership of Urban Lands and Extra Houses*, Proclamation no. 47 (July 1975).

21. Markakis and Ayele, *Class and Revolution in Ethiopia*, p. 141.

22. Provisional Military Government of Socialist Ethiopia, *Public Ownership of Rural Lands Proclamation*, Proclamation no. 31 (1975).

23. *Idem*, *Peasant Association Organization and Consolidation Proclamation*, Proclamation no. 71 (1975).

24. *Idem*, *Development through Co-operation Campaign Report, 1967-68 E.C.*, vol. I (1976), p. 70.

25. *Idem*, *Directives on Agricultural Producers' Co-operatives* (June 1979).

26. The process of land distribution and the different sets of problems encountered in the south (Shashemene *Wereda*) and in the north (Dangella

Wereda) are described at length by Alula Abate and Tesfaye Teklu, 'Land Reform and Peasant Associations in Ethiopia: A Case Study of Two Widely Differing Regions' (ILO, Geneva, 1979; mimeographed World Employment Programme research working paper; restricted).

27. Provisional Military Government of Socialist Ethiopia, *Peasant Association Organization and Consolidation Proclamation*.

28. *Idem*, *Directives on Agricultural Producers' Co-operatives*.

29. For a distinction of the roles of service co-operatives, see Md. Anisur Rahman, 'Transition to Collective Agriculture and Peasant Participation – North Viet Nam, Tanzania and Ethiopia' (ILO, Geneva, 1980; mimeographed World Employment Programme research working paper; restricted).

30. Provisional Military Government of Socialist Ethiopia, *Public Ownership of Rural Lands Proclamation*.

31. There now appears to be a positive correlation between family size and landholdings. See Tesfaye Teklu, *Socio-economic Conditions in Shashemene* (RR No. 26, IDR, Addis Ababa University, 1979); *Idem*, *Socio-economic Conditions in Dangella* (RR no. 28, IDR, Addis Ababa University, 1979).

32. Fassil G. Kiros and Asmerom Kidane, *Socio-economic Survey of Welmera (II)* (Ethiopian Science and Technology Commission, UN University, International Development Research Center, and IDR, 1979).

33. Paper prepared for a seminar on agricultural mechanisation, Ministry of State Farm Development, Feb. 1980, Table 2.

34. Ministry of Agriculture and Settlement, *Settlement Authority at a Glance* (UNDP/FAO Project, Aug. 1978), Annex II. Data in population settled by the Relief and Rehabilitation Commission are hard to obtain, part of the reason being the disruption of settlements as a result of the war in some parts of the country.

35. *Negarit Gazeta*, Proclamation no. 77 of 1976, lists the tax rate to be paid for the different agricultural income brackets. It may be noted here that a segment of the peasantry obtains income from other sources, in addition to food production; an example is the supply of firewood to urban communities, the price of which has soared in recent years.

36. Some policy options open to the government are discussed by John M. Cohen, Arthur A. Goldsmith and John W. Mellor, 'Rural Development Issues Following Ethiopian Land Reform', *Africa Today 23*, no. 2 (Apr.-June 1976), pp. 13 and 15.

37. David Lehmann (ed.), *Agrarian Reform and Agrarian Reformism: Studies of Peru, Chile, China and India* (Faber and Faber, London, 1974), p. 22.

5 THE AGRARIAN REFORM IN PERU: AN ASSESSMENT

Cristóbal Kay

I. The Geographical Setting

Peru has an extremely diversified topography; its land altitude ranges from sea level to over 6,000 metres, its landscape from desert to tropical rainforest. As a consequence, agriculture is very varied making any overall generalisation difficult. For this reason, three distinct geographical regions need to be distinguished: the Coast, the Highlands and the Jungle. It is only the first two that are relevant for a study of recent agrarian reforms.

The Coast (Costa) is a narrow desert strip bordering the Pacific Ocean on one side and the foothills of the Andes on the other. It makes up about 144,000 square kilometres or 11 per cent of the total land surface of Peru. The climate is humid and warm with relatively even temperatures throughout the year. There is little rainfall, but the Coast is cut across by rivers flowing down from the Andean mountains. Almost all of the 807,000 hectares of cropland (5.6 per cent of the Coast) are irrigated. The four main crops are cotton, maize, rice and sugar which together account for 70 per cent of the cultivated area. The high soil quality and favourable climate allow year-round cropping as well as some multiple cropping.

The Highlands (Sierra), formed by the Andean mountains, is located at over 1,500 metres above sea level. It comprises around 335,000 square kilometres or 26 per cent of the total land surface of Peru. The climate is dry and cold in winter with rainfall during the summer months. About 6.8 per cent (2,280,000 hectares) of the Highlands' total land area is cultivated. This low percentage is largely explained by the Highlands' difficult ecology – its uneven terrain, its large climatic variations and the low temperatures prevailing at high altitudes. About 90 per cent (14 million hectares) of the country's natural pastureland is found here, mostly located at over 3,000 metres in the *altiplano*. Only 492,000 hectares or about 22 per cent of the cropland are irrigated. The four main crops grown are maize, potatoes, barley and wheat.

The Jungle (Selva) extends eastwards from the Andes reaching the extensive Amazonic area. It is the largest of the three regions, comprising around 806,000 square kilometres or 63 per cent of the total land surface of Peru. The climate is warm with heavy, tropical rainfall throughout the year. Only 0.8 per cent (605,000 hectares) of the Selva's land surface is cultivated as most of the land is either unsuitable for agriculture or of difficult access. The main crops grown are bananas, maize, rice and yucca.

Peru has the lowest amount of arable land *per capita* in South America. Its average of 0.18 hectares is well below the continental average of 0.50 hectares and is also below the European and Asian averages of 0.30 and 0.20 hectares respectively (Alberts, 1978: 18). Within Peru the Highlands is the largest agricultural region accounting for 82 per cent of the agricultural land, 62 per cent of cropland and almost all the natural pastureland. It also contains the bulk of Peru's agricultural population: 67 per cent as compared to 25 per cent on the Coast. However, the Highlands produces only 42 per cent of the country's agricultural produce (including livestock). Soil, climate and terrain are all more favourable for agriculture on the Coast and in economic terms it is this region which is the most important.

II The Agrarian Systems in Historical Perspective

Three main agrarian systems have historically existed in Peru — plantations on the Coast and *haciendas* and peasant communities in the Highlands. Although it is convenient and useful to distinguish these systems conceptually, in reality they did not exist in isolation from each other. *Haciendas* and peasant communities have a long history of conflictual relationship. The Highlands, particularly the peasant communities, also supplied labour to the coastal plantations.

II.I The Coastal Plantations

The uniformity of the terrain, the absence of marked seasonal fluctuations and the necessity of major irrigation works have traditionally favoured large-scale agriculture on the Coast. Plantations occupied about 80 per cent of coastal farmland. Production was largely for the export market and, as this expanded during the second half of the nineteenth century, a major process of land concentration took place. This process was most extreme in the case of the sugar plantations and foreign capital played an important role. It was less extreme on the rice and cotton plantations as these were much smaller in size.

II.1.1 Sugar Land consolidation of sugar holdings became particularly acute from the 1880s onwards. Old plantations and small farms were annexed and consolidated into larger, more efficient production units. Small, independent producers, who had formed a sizeable rural middle class, were dispossessed in the process, often becoming proletarians and working on the same plantations which had usurped their lands (Burga, 1976). Within two or three decades whole valleys became dominated by three or four sugar plantations. The core plantations multiplied many times over in size, comprising several thousand hectares. This process of land concentration was matched by an even more dramatic centralisation of the milling process. The growing, processing and marketing of sugar became dominated by a few vertically integrated, corporate plantations. Sugar production continually expanded from 30 million tons in 1887 to 112 million in 1900 to 339 million tons in 1930. Over the same period, sugar exports accounted for approximately 20 per cent of the total value of exports (Klaren, 1977: 238-9).

Changes in the land tenure structure also affected the labour structure. Until the mid-nineteenth century plantation owners relied upon Negro slaves and later Chinese coolie labour. After 1874 when coolie labour was restricted and as sugar plantations became more profitable, planters began to attract labour through a free wage system. A stable permanent work-force was drawn from nearby villages and included many former small farmers. As local supply did not meet demand, however, the *enganche* system was introduced. This involved recruiting Indian labour from the Highlands to work on a temporary basis, ranging from three months up to two years. The flow of labour was particularly intense during the slack months of peasant agriculture in the Highlands. As a way of enticement, the labour contractor offered a cash advance which tied the worker to the plantation until his debt had been paid. The coercive features of this recruitment system were later muted as many peasants voluntarily came to seek wage employment on the plantations as a means of supplementing the meagre incomes from their small plots (Favre, 1977). The flow of labour from the Highlands became particularly intense after the crisis of the 1930s. Demographic pressures meant that the Highlands could no longer economically support the growing population and some now came to seek permanent employment on the Coast. The *enganche* system thus completely disappeared by the 1950s (Scott, 1976).

The plantations developed many features of a dual labour market structure. Most Highland labour was contracted on a temporary basis as *eventuales* and had to perform the more menial tasks, such as cane cutting and loading. The permanent workers, the *contratados*, undertook the more specialised jobs. On average the permanent worker's wage was 15 to 20 per cent higher than that of the temporary worker although field workers also received a food ration (Klaren, 1977: 245). On many plantations resident workers were given a small garden plot in usufruct to grow foodstuffs.

The increasingly supply of sugar workers, as rural out-migration accelerated in the 1950s, did not affect the rising trend of wages amongst the permanent work-force as these were organised into unions. The workers' bargaining strength became firmly established during the mid-1950s as national political parties supported their claims. The political power of organised labour was thus able to offset the impact of an increased labour supply and even gain substantial wage increases. Seasonal workers, who were not organised, were the main victims. Unionisation thus accentuated the structural differentiation between permanent and temporary workers (Scott, 1978: 132-3). Partly in response to the increased political power of unionised labour, sugar planters mechanised heavily and cut their total work-force by approximately 50 per cent between 1950 and 1969 (Miller, 1964; Horton, 1976: 95).

II.1.2 Rice and Cotton   Unlike sugar, the rice and cotton plantations relied to a large extent on tenants and sharecroppers for their labour. On some estates tenants and sharecroppers cultivated over half the land. The most common system on the rice estates was the *colonato* whereby the tenant or *colono* was given a small parcel of land to grow rice and some other subsistence crops. The estate provided the rice seed and fertiliser and facilitated cash advances to be repaid at harvest time. In return tenants paid a fixed rent in rice and sold most of the remainder to the estate store at below market price. They also had to work a certain number of days on the demesne for less than the going wage (Horton, 1976: 189).

Sharecropping, or *yanaconaje*, was more common on the cotton plantations. A sharecropper, or *yanacona*, paid rent in cotton which was either a fixed percentage of the total harvested crop (20 per cent or more), or less frequently a fixed amount of cotton (Matos Mar and Carbajal, 1974: 107). The plantation provided seeds, fertilisers, insecticides and cash advances. As in the case of a *colono*, a *yanacona* had these loans and interest charges discounted after harvest. The remaining cotton had to be sold to the plantation owner at a cut-price rate (Castro-Pozo, 1947). The plantation's administrative expenses were also charged to the *yanaconas* on a proportional basis. Sometimes the plantation rented draught animals and even machinery to the *yanaconas*. Where *yanaconas* managed a large plot they, themselves, sometimes engaged seasonal wage labour.

The main difference between *colonato* and the *yanaconaje* was that while a *colono* had to work a substantial number of days on the demesne, a *yanacona* rarely had to do so. While both paid kind and/or money rents, the *colono* also had to pay a labour rent. A *colono* was, therefore, a more direct part of the estate's labour force while a *yanacona* was more of a small-scale producer.

Landlords began expelling or proletarianising *yanaconas* from the 1940s onwards as wage-labour-based production became more profitable. This process proceeded at different paces and continued right up until the agrarian reform. It is estimated that in 1936 there were between 80,000 and 100,000 *yanaconas*; in 1961 their number had fallen to 18,000 (Matos Mar, 1976: 46). A study of the coastal valley of Chancay indicates, however, that although diminished in number (from 3,000 in 1942 to under 2,000 in 1945 and only 787 in 1964), *yanaconas* still cultivated a third of the estates' cultivated area in 1964 (ibid.: 111; Mejía, 1979: 37-8). Expulsion or proletarianisation was most intense on the more capitalised estates and was generally strongly resisted. *Yanaconas* on traditional estates could purchase their tenancy, a practice which later spread to modernising estates as owners feared expropriation through agrarian reform (Greaves, 1968).

II.2 The Highland Haciendas

Haciendas originated during the colonial period but fully developed and consolidated towards the turn of the nineteenth century, largely by appropriating land from the indigenous peasant communities. This was done either by outright expropriation or by fraudulent purchase at derisory prices (Chevalier, 1966). Two types of *hacienda* can be distinguished: the vast livestock estates and the not-so-large livestock-crop estates. The livestock estates were mainly located in the central and southern Highlands at altitudes unsuitable for cropping. The smaller mixed farming *haciendas* were scattered throughout the Highlands and located at lower altitudes.

II.2.1 Livestock Haciendas   The expansion of the livestock *haciendas* was closely linked to the growth of the textile industry, especially in Great Britain, and to the rise in the price of wool on the international market. This process of expansion, which had many features of an enclosure process, was more marked in the central Highlands where the most profitable livestock estates were located. The largest landlords in this area were closely associated with the even more profitable mining industry and had greater capital resources than landlords in the south, where the enclosure process remained incomplete. Transport connections to markets were also more favourable in the central region.

At first landlords rented most of the pasture land to independent shepherds or *arrendatorio-pastores* who owned most of the sheep. Later these shepherds became *huacchilleros* (sheep-tenants, a type of *colono*) who looked after the landlord's flock in return for the right to graze their own animals and increasingly received a small wage payment. The main factor behind the proletarianisation of shepherds was the increasing productivity of the landlord's sheep as a result of investments to improve the quality of their stock. As landlords and *huacchilleros* competed for the same pastureland, landlords attempted to dispossess the *huacchilleros* totally and transform them into shepherds paid only with a wage. *Huacchilleros* resisted their further proletarianisation as it meant losing what little entrepreneurial independence they had and because they believed that the wage payment would not compensate for the income lost if they gave up their herds. In the 1950s about half of a typical *hacienda*'s pastureland in the central Highlands was still used by *huacchilleros*. Only on the better administered estates, often belonging to corporate enterprises, did the proportion fall to about a fifth (Martínez-Alier, 1977: 145). As the enclosure process did not proceed thus far in the south, landlords here did not exert such direct control over sheep rearing as in the central region and peasants often retained over half of the total livestock population (Horton, 1976: 116).

II.2.2 Livestock-crop Haciendas   On the livestock-crop or solely crop *haciendas* the process of proletarianisation was not very advanced as landlords relied mainly on *colonos*, or *feudatarios* as they were often called, for labour. These tenants were largely in charge of production as they cultivated most of the *hacienda*'s cropland and owned most of the livestock. On the more traditional *haciendas*, landlords were often absentee owners and retained only a fraction of the estate's land under direct cultivation. This meant that surplus was extracted in the form of product rent or money rent rather than in the form of labour rent. Production methods were primitive. The technology used on the small demesne did not differ substantially from that of the peasants. Productivity was low and little capital accumulation took place.

On some *haciendas* a process of proleterianisation of the *colonos* was, however, beginning to take place. Landlords on these *haciendas* took a greater interest in the running of the estate and the demesne was larger than on traditional *haciendas*. Labour rents were, therefore, more important as the larger demesne required more labour services. In more recent decades the *colonos*' labour services were partly remunerated with a small kind and/or monetary payment which gradually took the form of a small wage.

II.3 The Peasant Communities

The peasant community is a pre-colonial organisation which existed throughout Peru. In the remote past almost all land belonged to the communities but since the Spanish conquest much had been appropriated by landlords. Today peasant communities are confined to the Highlands but they still account for about 40 per cent of the rural population (CENCIRA, 1977: 3).

In the past a community was distinguished by economic links between members as well as by shared social, political, cultural and religious bonds (Matos Mar (ed.), 1976). The peasant community was an association, membership of which conferred certain rights and obligations. Community authorities determined the proportion of land to be worked collectively and distributed individual plots in usufruct to members. Pastureland was always used communally. Members (*comuneros*) collectively worked that part of the land whose proceeds went to finance communal activity and were also called upon to undertake collective tasks such as construction and maintenance work. The reciprocal exchange of labour between members for the cultivation of individual plots was a common practice (Alberti and Mayer (eds), 1974).

This picture began to change with the intrusion of the *hacienda* system into community land and with the penetration of capitalist market relations. The conflict between *hacienda* and community was permanent and often erupted in land invasions by *comuneros* demanding the restitution of their lands. The invaders were usually forcefully evicted and *comuneros* often suffered casualties in the ensuing clashes. Very rarely were *comuneros* successful in obtaining a favourable ruling by the judiciary and in establishing their property rights (Kapsoli, 1977).

The loss of community land to the *hacienda* forced some members to become labour-service tenants on the estates or rent pastureland from the *hacienda*. Some members later sought seasonal wage employment on the Highland estates and, above all, on the coastal plantations. Others migrated for good.

The penetration of capitalist market relations began a process of privatisation of communal cropland. Legislation also speeded up the establishment of private property relations over cropland which *comuneros* had previously only held in usufruct. Natural pastureland was generally exempt from this privatisation process, being too costly to fence and of poor quality (Figueroa, 1978: 25). In any case, communal ownership did not prevent richer *comuneros* from grazing a larger number of privately owned animals than poorer members. Consequently collective work arrangements had over time dwindled to essential construction and maintenance work.

Comuneros also became differentiated in the wake of the development of labour and product markets. Richer *comuneros* contracted some wage labour but they were mainly differentiated from poorer members by their incursion into other economic activities – commerce, transport, small-scale crafts (Sánchez, 1979: 3-5). Poorer *comuneros* had increasingly to sell part of their labour for a wage, eventually forming a semi-proletariat (Campana and Rivera, 1978). The traditional free exchange of labour between *comuneros* was gradually replaced by monetary payments although some forms of reciprocity survived.

At first only the richer *comuneros* sold a significant part of their agricultural production on the capitalist market and bought commodities in return. Later the majority of *comuneros* were incorporated into the product market and monetary exchange supplanted barter (Gonzáles, 1979).

All these processes meant that peasant communities had largely evolved into associations of private peasant producers by the late sixties. Privatisation and differentiation notwithstanding, however, *comuneros* retained their organisation and many non-*comunero* peasants attempted to form legally constituted peasant communities as a way of pursuing land claims more effectively and of obtaining credits and other services from the state.

III The Pre-reform Setting

III.1 Land Tenure Structure

Table 5.1 gives a picture of the extremely high degree of land concen-

Table 5.1: Distribution of Number and Area of Farm Units by Socio-economic Category, 1961

Socio-economic Category[a]	Number of Farm Units (in %)	Area (in %)	Average Size of Holding (in hectares)
Coast			
Sub-family	83.2	10.0	2.9
Family	11.4	4.0	8.4
Multi-family medium	3.7	6.0	39.0
Multi-family large	1.7	80.0	1 126.1
Total Coast	100.0	100.0	23.8
Highlands			
Sub-family	83.4	4.7	1.2
Family	12.5	4.8	8.2
Multi-family medium	2.7	5.0	39.8
Multi-family large	1.3	75.0	1 284.8
Peasant communities[b]	0.1	10.5	1 985.1
Total Highlands	100.0	100.0	21.6
Peru			
Sub-family	84.4	6.1	1.6
Family	11.6	4.7	8.9
Multi-family medium	2.7	5.4	43.3
Multi-family large	1.2	75.2	1 338.1
Peasant communities	0.1	8.6	1 985.1
Total Peru	100.0	100.0	21.8

Notes: a. The size groups are defined by CIDA as follows: sub-family: farms too small to employ a single family (two workers with the typical incomes, markets and levels of technology and capital prevailing in each region); family: farms large enough to employ 2-3.9 people, on the assumption that most of the farm work is carried out by members of the family; multi-family medium: farms large enough to employ from 4 to 12 people; multi-family large: farms large enough to employ more than 12 people; peasant communities: registered land-holding peasant communities.

b. The 1961 census greatly underestimates the amount of natural pastures held by peasant communities thereby magnifying land concentration as most *comuneros* own sub-family farms (CIDA, 1966: 35; Caballero, 1979: 89, 93).

Source: CIDA, *Tenencia de la Tierra y Desarrollo Socio-Económico del Sector Agrícola, Perú* (Washington, DC, 1966), pp. 41, 56.

tration and the large differences in size between types of agricultural enterprise. The category peasant communities refers only to lands held in common ownership, largely natural pastures of poor quality. Sub-family farms are the typical *minifundia*. These, together with the family farms, are roughly equivalent to the term peasant enterprise used in this chapter. The multi-family large farms correspond to the category of *hacienda*, plantation or estate.

The data in this table, however, tend to overestimate the degree of land concentration in the Sierra as it fails to consider the quality of land. Converting different qualities of land into a standardised quality does not alter the picture very much for the Coast, where all agricultural land is irrigated and of similar quality, but it does make a substantial difference in the Sierra.

Table 5.2: Distribution by Number and Area in Simple and Standardised[a] Hectares of Farm Units in the Sierra, 1972[b]

Farm Unit (by size)	No. of Farm Units (in %)	Average Size in Hectares		Area in %	
		Simple	Standard-ised	Simple	Standard-ised
Minifundio (less than 2 ha.)	45.1	0.97	0.58	2.0	13.5
Family (2-10 ha.)	44.2	3.73	1.76	7.9	40.6
Small enterprise (10-50 ha.)	8.5	15.31	4.18	6.9	18.6
Medium enterprise (50-500 ha.)	1.8	111.20	8.97	10.3	8.2
Large enterprise (over 500 ha.)	0.3	2 257.70	71.44	42.0	11.9
Communal[c] (50 and more ha.)	0.1	6 620.00	129.40	30.9	7.2
Total	100.0	19.57	1.92	100.0	100.0

Notes: a. The Highlands average hectare of cultivated irrigated land has been taken as the standard unit.

b. Although the data refer to 1972, they still reflect the pre-agrarian reform situation as large-scale expropriations in the Highlands took place after only 1972.

c. The land exploited individually by *comuneros* is not considered in the communal category but under the respective size category. 'Communal' therefore refers only to land collectively used, largely pastures.

Source: J. M. Caballero, 'La Economía Agraria de la Sierra Peruana en los Albores de la Reforma Agraria', typescript (Instituto de Estudios Peruanos, Lima, 1979), p. 93.

Table 5.2[1] unequivocally shows that when land is converted into standardised hectares, land concentration in the Sierra is much less pronounced as a large part of the estates' land consisted of natural pastures. Thus large enterprises which made up 0.3 per cent of farm units held 42 per cent of the land but only 11.9 per cent of the standardised land. Likewise, at the other extreme, *minifundios* which formed 45.1 per cent of farm units held 2.0 per cent of the land but 13.5 per cent of the standardised land.

The revised figures presented in Table 5.2 are particularly relevant as they show that medium and large enterprises only controlled 20.1 per cent of the standardised land. Thus the scarcity of land for redistribution in the Sierra was to pose a problem for any agrarian reform process. Little awareness existed of this when the 1969 agrarian reform law was passed. The prevailing image was the one propagated by the massive CIDA study which indicated that 'large multi-family' farms owned three-quarters of the land in the Sierra as shown in Table 5.1.

Apart from the low level of standardised land held by the estates, landlords only owned 16.0 per cent of the Highlands' livestock in 1972, expressed in standardised sheep units,[2] as compared to the peasants' 60.2 per cent (Caballero, 1979: 98).[3]

The peasant units of production were highly fragmented. At one time this had the advantage of spreading risk as plots were located at different altitudes and ecological levels.[4] However, fragmentation through inheritance had become excessive and uneconomical in recent decades. By 1972 an average *minifundio* was fragmented into 4.3 plots of 0.22 hectares each and a typical farm between 2 and 10 hectares was likely to be scattered in 6.1 plots (ibid.: 93).

III.2 Population and Labour Structure

Before the agrarian reform, Peru was still largely a rural country, but one characterised by a rapid process of urbanisation. Of the total population, 74.6 per cent were rural in 1940, 67.2 per cent in 1961 and 52.5 per cent in 1972. Rural population stagnated in the 1960s, growing at only 0.6 per cent yearly while the urban population expanded at ten times this rate at 6.4 per cent (Figueroa, 1976: 15). Rural out-migration became particularly intense in the same decade, contributing to the rapid growth of the urban population. The Sierra, however, remained a rural region *par excellence*, accounting for 45 per cent of the total population but 80 per cent of the total rural population (Caballero, 1979: 106).

The active agricultural population followed a similar pattern. In 1940 62.5 per cent of the total active population was employed in agriculture; the proportion was 50.5 per cent in 1961 and 42.5 per

cent in 1972 (Maletta, 1980: 4). The male active agricultural population practically stagnated, growing at an average annual rate of 1.11
per cent between 1940 and 1961 and at only 0.33 per cent between
1961 and 1972 (ibid.: 2). In most coastal areas the active agricultural
population declined slightly between 1961 and 1972 as a result of
mechanisation. The total active agricultural population in the country
has remained around 1.5 million since 1961.

Important changes took place, however, in the composition of the
active agricultural population. The self-employed who made up half the
total in 1961 rose to two-thirds in 1972; the proportion of wage labourers, by contrast, decreased from under a third to just over a fifth in
the same period. The remainder were family workers. In absolute terms
the number of self-employed increased by 24 per cent while that of
wage workers fell by 31 per cent and that of family workers by 12 per
cent (ibid.: 12). The majority of the self-employed are smallholders
whose numbers swelled through colonisation of the Selva or by some
tenants on estates becoming *minifundistas* (small proprietors).

The regional occupational structure immediately prior to the agrarian reform is set out in Table 5.3. This shows that the large majority
of agricultural families had access to land either as proprietors (76.9
per cent) or as tenants (14.0 per cent). Wage labourers constituted only
9.1 per cent of agricultural families.[5] In the Highlands, 69 per cent
of agricultural families were *minifundistas* and only 5 per cent were
wage labourers. On the Coast, wage labourers were more important,
constituting 20 per cent of agricultural families, but *minifundistas*
(57 per cent) still constituted the most significant group.

Within the category of wage labour, seasonal wage labour tended
to increase in importance over time *vis-à-vis* permanent labour. This
is explained by the landlords' tendency to switch from permanent
to seasonal wage labour in their efforts to reduce costs, especially
where permanent workers were organised as on the Coast. An important section of landless seasonal wage labourers thus emerged on the
Coast, forming 'permanent *eventuales*' with characteristics of a sub-
proletariat. The growing incidence of seasonal labour was also due to
the increased number of smallholders in the Sierra who were seeking
temporary wage labour to ensure a subsistence income.

As over 90 per cent of the *colonos* and wage labourers were employed
by large estates, Table 5.3 also permits a rough view of the estates'
labour structure. In 1961 the large estates absorbed 28 per cent of the
active agricultural population, including owners and administrative
personnel (CIDA, 1966: 52). By 1969 this percentage had fallen to

Table 5.3: *Socio-economic Status of Agricultural Families circa 1969*

Occupational Category	Coast (in thousands)	Highlands	Selva	Total
Estate owners and managers	15 (5.0)	30 (3.8)	4 (4.0)	49 (4.0)
Family farmers	40 (13.3)	40 (5.0)	7 (7.0)	87 (7.3)
Sub-family farmers	170 (56.7)	555 (69.4)	62 (62.0)	787 (65.6)
Wage labourers	60 (20.0)	40 (5.0)	9 (9.0)	109 (9.1)
Colonos	15 (5.0)	135 (16.8)	18 (18.0)	168 (14.0)
Total	300 (100.0)	800 (100.0)	100 (100.0)	1 200 (100.0)

Note: Figures in parentheses are percentages.

Source: D. E. Horton, 'Haciendas and Co-operatives: A Study of Estate Organisation, Land Reform and New Reform Enterprises in Peru', unpublished PhD dissertation, Cornell University, Ithaca, 1976, p. 109.

around 25 per cent. On coastal estates, two-thirds of the work-force were wage workers, a sixth were tenants and a sixth were owners and employers. On Highland *haciendas*, approximately a fifth of the work-force were wage workers, two-thirds were tenants and the remainder were owners and employers. Wage labour thus predominated on coastal estates and *colonato* predominated on Highland *haciendas*.

III.3 The Crisis of the Agrarian Economy

III.3.1 Production   During the 1960s the agricultural sector stagnated. The inability of production to keep pace with population growth and the consequent failure to generate more productive employment and to increase living standards reflected a generalised crisis of the rural sector. The underlying causes of the crisis lay in the land tenure structure and, in particular, the Highland *hacienda* system. A precipitating factor was the state's economic policy, especially import substitution industrialisation which discriminated against agriculture.

The agricultural sector grew at an average yearly rate of 3.7 per cent between 1945 and 1961. The population growth rate for the same period was 2.2 per cent per annum. This positive *per capita* rate of growth turned negative between 1962 and 1969, when agricultural production grew by only 1.1 per cent yearly and population at 3 per cent per annum (Hopkins, 1979: 34).

Agriculture was certainly the least dynamic economic sector; its contribution to the gross national product fell from 35 per cent in 1950 to 15 per cent in 1969 (Alberts, 1978: 7). Agricultural production was particularly stagnant in the Sierra. The four main crops produced in the Highlands grew at a yearly average rate of 0.67 per cent between 1964 and 1972 and livestock production increased by 0.65 per cent yearly (Caballero, 1979: 198, 201). The value added per worker in agriculture as a whole in 1971 was only a fifth of that of other commodity producing sectors and a quarter of that service sector (Horton, 1976: 85). When agricultural regions are compared the poor performance of the Sierra once again comes to the fore. In 1972 the value added per worker in the Sierra was US$385; on the Coast $1,730 and in the Selva $923 (Caballero 1979: 197).

Although these regional differences can largely be explained by uneven rates of technological development, any attempt to blame the Sierra's poor economic performance on these grounds alone is inadequate. As pointed out at the beginning of the chapter, the difficult ecological conditions in the Highlands exercise an important constraint on increasing productivity regardless of land tenure systems or state policies.

The differing and often low levels of technological development can be illustrated by analysing the use of fertilisers, certified seeds and machinery. In 1961 only about a third of crop farms in the Highlands, accounting for 27 per cent of the total cultivated area, used fertilisers (Roquez, 1978: 71). Only 7.2 per cent of farms used chemical fertilisers or *island guano* (Maletta and Foronda, 1979: 99). The purchase of seeds, generally indicating improved quality, was also uncommon on the Sierra crop farms. With respect to the four main crops, the percentage of crop farms purchasing seeds was: potatoes (23 per cent); wheat (8 per cent); maize (10 per cent); and barley (8 per cent) (Caballero, 1979: 219).

As for machinery, in 1961 only 5.3 per cent of farms over 5 hectares in the country used tractors and these were mainly rented. This figure increased to 8.6 per cent in 1972. The yearly increase in the number of tractors over the same period was only 1.3 per cent (Maletta and Foronda, 1979: 116). Most of the tractors were in use on the Coast. In 1962, 80 per cent of tractors were found on the Coast and only 17 per cent in the Sierra (Roquez, 1978: 69). Given the nature of the terrain, tractors can be used on only a quarter of the cultivated area in the Sierra. Ox-drawn ploughs (79 per cent of farms) and foot-ploughs (52 per cent of farms) were commonly used; only 4 per cent of farms used a tractor.

The *latifundia-minifundia* land tenure structure generated both inequality and inefficiency (Hunt, 1974; Griffin, 1976). Landlords owned too much land and employed too little labour while peasants had too little land and used too much labour. Large estates left a high proportion of their land idle and cultivated the rest less intensively than the peasants. Peasant farms were small and often fragmented into numerous parcels. Furthermore, the peasants' poverty forced them to cultivate too intensively which tended to erode the quality of their cropland, just as overgrazing tended to erode the quality of their pastureland (Janvry and Garramón, 1977). Average land and capital productivities were higher on peasant farms than on estates (see Table 5.4), but the opposite held for average labour productivity.

State economic policy only tended to aggravate regional and economic inequalities. In the decade before the agrarian reform, over three-quarters of public credit (and almost all private credit) went to large landowners. Furthermore 75 per cent of state credit was allocated to the Coast (Haudry, 1978). Similarly, public agricultural investments were monopolised by a few large irrigation projects on the Coast and

Table 5.4: Value of Capital per Hectare and Productivities of Land and Capital by Tenancy System, 1962[a]

Tenancy System	Capital[b]	Gross Income	Net Income[c]	Gross Income Capital Co-effi- cient (%)	Net Income Capital Co-effi- cient (%)
	(soles per used hectare[d])				
Coast					
Multi-family large	36 909	8 460	819	22.9	2.2
Multi-family medium	38 691	7 045	2 279	18.2	5.9
Family	27 381	9 460	2 852	34.6	10.4
Sub-family	29 043	11 711	4 245	40.3	14.6
Highlands					
Multi-family large	916	116	33	12.6	3.6
Multi-family medium	6 099	1 869	365	30.3	9.3
Family	16 990	3 569	935	20.8	5.5
Sub-family	13 425	5 162	2 367	38.5	17.6
Comunero	10 996	6 180	3 337	55.6	30.3
Colono	2 539	2 321	481	90.9	18.9

Notes: a. The data are obtained from a non-representative survey of 65 farms on the Coast and 121 farms in the Highlands.

b. Capital has been defined as the value of land (which automatically took account of quality), livestock, machinery, plantations, buildings and others.

c. Net income is the difference between gross income and expenditure. The latter includes a 6 per cent imputed rent value on the land plus an estimated value of the non-remunerated operator's labour.

d. 1 US dollar = 43.5 soles.

Source: CIDA, *Tenencia de la Tierra y Desarrollo Socio-Económico del Sector Agrícola, Peru* (Unión Panamericana, Washington, DC, 1966), pp. 72, 86.

largely benefited the estates. The cultivators most discriminated against by these policies were the small Highland farmers (Figueroa, 1976; 196).

Agriculture's poor economic performance was not helped by the adoption of an import substitution industrialisation strategy from 1959. This strategy discriminated against agriculture, depriving it of important investment funds and exacerbating the declining trend in production (Alberts, 1978). The foreign exchange policy, which over-valued the local currency, had the effect of making food imports cheaper and agricultural exports less profitable. Agricultural imports rose from 9 per cent of domestic agricultural production in 1960 to 15 per cent in 1969 (Figueroa, 1976: 58). Value of agricultural imports exceeded that of agricultural exports for the first time in 1967 and Peru ceased to be a major agricultural exporting country (Roquez, 1978; 15). Agriculture, which had contributed 51 per cent of the total value of foreign exchange earnings between 1944 and 1959, only contributed 27 per cent between 1960 and 1967 (Havens, 1976: 23). In addition to the foreign exchange policy, price controls on food products were extended and more strictly enforced. These policies led to a deterioration of agriculture's terms of trade *vis-à-vis* the rest of the economy in the 1960s (Alvarez, 1979: 24).

However, agriculture was not uniformly affected by these policies. Table 5.5 shows how this strategy affected three categories of agricultural commodity by their market destination.[6] This table shows the abrupt decline of exports in the 1960s which up until then had had the highest rate of growth. Agricultural products destined for the urban market fared better than those consumed by the rural population. These changes had regional implications. Around 80 per cent of export crops were grown on the Coast. Since 1963 they increasingly geared production to the domestic urban market which experienced the most rapid expansion (Horton, 1976: 89). It is also likely that some large exporters transferred capital into more profitable industrial ventures (Saldívar, 1978: 10-11). Highland *haciendas* and peasant communities were more severely affected as their production patterns were more rigid and more geared to the rural market. Peasant family incomes suffered as purchasing power of agricultural products failed to keep pace with that of other commodities.

III.3.2 Employment   Rural surplus labour generated by a combination of a high rate of population growth and agriculture's growing inability to absorb labour did not lead to high levels of rural unemployment and underemployment as out-migration provided a safety valve. Practically the whole annual increase in rural population tended to emigrate to the urban sector (Caballero, 1980: 22).

Official estimates put agricultural unemployment at 0.3 per cent in 1969 and underemployment at 66.6 per cent (Maletta, 1978b: 34). The equivalent figures for the whole population were 5.9 per cent and 46.1 per cent. However, Maletta (1978b) suggests, on the basis of a thorough examination of official statistics, that official figures under-estimate the rate of unemployment and over-estimate the rate of underemployment. He bases his judgement on the following observations. First, official figures are based on an under-estimation of the economically active agricultural population since those seeking employment for the first time are excluded by definition. Second, official estimates of seasonal labour requirements in agriculture are under-estimates since labour requirements for livestock farming are left out and presumed technical conditions of production correspond to those existing only on the most efficient farms. Third, official estimates of the underemployed include those whose incomes from agricultural activities fall below a minimum subsistence level. Maletta concludes that the rate of unemployment was probably around 5 per cent while the rate of underemployment was much lower than the official estimate. It would thus appear that unemployment and under-employment in Peru's agriculture, though significant, were not severe. But this was only because unemployment was simply transferred to urban areas through migration.

Table 5.5: Rate of Growth of Agricultural Commodities by Type of Market and their Relative Importance, 1950-69

Agricultural Commodities Destined for	Rate of Growth (annual average)		Share of Total Production (%)	
	1950-9	1960-9	1950	1969
Internal market:				
Urban	3.2	3.6	27.9	52.4
Rural	0.4	1.1	50.1	32.7
Export market:				
Exports	5.7	−0.1	22.0	14.9
Total	3.0	2.0	100.0	100.0

Source: E. Alvarez, 'Politica agraria y estancamiento de la agricultura, 1969-1977', *Primer Seminario sobre Agricultura y Alimentación* (Pontificia Universidad Católica del Peru, Lima, 1979), pp. 4-5.

III.3.3 Income Inequality and Poverty   The structural characteristics of the economy and the pattern of growth produced severe inequalities in the distribution of incomes among the population. The inequalities had three distinct dimensions. First, average urban income was much higher than average rural income. In 1971/72, for example, average urban income was observed to be more than three times the average rural income (Table 5.6). Urban households, constituting 45.6 per cent of all households, accounted for 73.3 per cent of the total income. More strikingly, 20.1 per cent of the households living in metropolitan Lima accounted for 43.7 per cent of the income, while 54.4 per cent of the households, living in rural areas, received only 26.7 per cent of the income. There are reasons to believe that these data on income inequalities overstate, to some extent, the actual inequalities in levels of living; but even when one adjusts the data for differences in costs of living, inequalities remain very considerable (Maletta, 1980: 23).

Second, within the rural sector, there were considerable regional inequalities. Thus in 1971/2, the average rural income in the Coast was almost twice that in the Sierra (see Table 5.6); 74.3 per cent of the rural households lived in the Sierra, but they accounted for 61.8 per cent of the total income generated in the rural sector.

Table 5.6: Average Family Income and Standardised Expenditure by Area, 1971/2

Area of Residence	Average Family Income (soles/ month[a])	Average Standard-Family Expendi- ture	Number of Families (%)	Income (%)
Urban	7 299	–	45.6	73.3
Metropolitan Lima	9 860	7 640	20.1	43.7
Other urban centres	5 281	–	25.5	29.6
Rural	2 230	4 595	54.4	26.7
Coast	3 300	5 837	7.5	5.5
Sierra	1 861	4 050	40.4	16.5
Selva	3 266	6 546	6.5	4.7
Peru	4 500	–	100.0	100.0

Note: a. 1 US dollar = 43.5 soles.
   b. The nominal expenditure has been standardised by taking prices in Lima of the 1,000 calories contained in the average diet for each respective area.

Sources: C. Amat and H. León, *Estructura y Niveles de Ingreso Familiar en el Perú* (Universidad del Pacífico, Lima 1979), p. 32, and H. Maletta, 'El empleo en la agricultura peruana', *I Seminario sobre el Problema del Empleo en el Peru* (Universidad Católica del Peru, Lima, 1980), p. 24.

Third, intra-regional inequalities in the distribution of rural incomes were also considerable. This can be seen from the data presented in Table 5.7.

Low incomes combined with distributional inequity led to wide-spread poverty. Even when the official minimum wage was taken as an indicator of minimum subsistence requirements, the *per capita* expenditure of almost a quarter of rural families fell below this level in 1971/2 (Maletta, 1980: 26-7).[7] It has also been estimated that in 1971/2 members of 54 per cent of rural families failed to satisfy the minimum requirement of calories and those of 43 per cent of rural families failed to satisfy the minimum requirement of protein (Amat and Curonisy, 1979: 73d).

Table 5.7: Percentage Distribution of Rural Families and Incomes by Level of Income, 1971/2

Income level (soles/month[a])	Coast		Highlands		Total Rural Sector	
	Number of Families (%)	Family Income (%)	Number of Families (%)	Family Income (%)	Number of Families (%)	Family Income (%)
Less than 670	5.8	0.6	37.0	6.9	31.2	4.8
671-1,330	10.8	3.4	25.0	13.4	22.6	9.9
1,331-2,000	19.5	9.8	13.1	11.9	14.0	10.6
2,001-2,670	15.9	11.3	7.3	9.3	8.9	9.4
2,671-4,000	23.0	22.6	6.6	12.0	9.2	14.0
4,001-6,000	15.7	23.0	5.9	16.1	7.6	17.1
6,001-9,670	5.9	13.4	3.0	12.0	4.0	13.2
9,671-15,000	2.4	9.1	1.6	10.2	1.8	10.4
15,001 and over	1.0	6,8	0.5	8.2	0.7	10.6
Total	100.0	100.0	100.0	100.0	100.0	100.0

Note: a. 1 US dollar = 43.5 soles.
Source: Amat and León, *Estructura y Niveles de Ingreso Familiar en el Perú*, pp. 36, 155, 156.

IV The Agrarian Reform: Origin, Objectives and Implementation

IV.1 Origin and Objectives

An explanation of the 1969 agrarian reform entails examining the antecedents of the 1968 military *coup* which overthrew the Belaúnde government. From the mid-1950s rural labourers made their presence increasingly felt on the national political scene and social conflicts became more intense in the countryside. In the early 1950s Highland *hacienda* tenants began to unionise and organise strikes (Handelman, 1975). In the second half of the 1950s a wave of strikes hit the sugar plantations on the Coast. More importantly between 1960 and 1963 peasant communities staged a series of land invasions in the central and southern Highlands, challenging the domination of the *hacienda* system from outside ('asedio externo'). At the same time, tenants increased pressure on the *hacienda* system from inside ('asedio interno') complementing the *comuneros'* traditional struggle against the landlords (Valderrama, 1976: 41).

The army seized power in a *coup* in 1962 and attempted to quell peasant insurgency. This short 10-month military government decreed the Law of Foundations of the Agrarian Reform (Decree No. 14238). Although not designed to have any practical application, this decree was the first legislation on agrarian reform and marked a legal recognition of the existing land tenure problem (Caballero, 1975: 45). Shortly afterwards in early 1963, the military administration decreed the first agrarian reform law proper (No. 14444) to deal with the rural areas most in conflict. This was an emergency measure brought about by a major peasant insurrection in La Convención Valley in the southern Highlands (Fioravanti, 1974). The estates in this valley were expropriated and the land distributed in private property to the *arrendire* tenants. Landlords retained the central enterprise (demesne).

Belaúnde, elected into office in 1963 on a reformist platform, was faced with a mobilised and militant peasantry. His promises of agrarian reform in his election programme were later set out in the 1964 agrarian reform law (No. 15037). However congressional opposition severely limited funds for its implementation (Pease, 1977: 61). Eventually only some minor expropriations were carried out in areas of intense social conflict and the government had to rely on the military to repress the peasant movement and liquidate the incipient guerrilla struggle (Valderrama, 1976: 42). Only 11,760 families benefited, receiving 380,000 hectares mainly on an individual basis (Petras and Laporte, 1971: 112). This represented an insignificant fraction of the total rural population and land area.

The military seized power again in late 1968 with the intention of carrying out major changes in economy, society and polity. A new concept of 'national security' came to form part of military thinking, whereby the solution of internal conflicts was seen as a prerequisite for the defence of national sovereignty (Pásara, 1978: 147).

By mid-1969 a new agrarian reform decree-law (No. 17716) came into effect. The spirit of this law was not markedly different from its 1964 predecessor. A report cited by Harding (n.d.: 121) states one of the aims of the new reform as 'the elimination of the traditional systems of excessively large and small landholdings, and the concomitant promotion of small and medium-sized economically viable holdings, and group operation for larger units where economies of scale would suffer from sub-division'. Where it differed from the previous law, being more radical in conception, was in making all commercial estates over a certain size liable for expropriation irrespective of whether they were efficient or not or whether they formed agro-industrial complexes such as the sugar plantations or not (ibid.: 121-2).

The underlying objectives of the agrarian reform are not always explicitly stated in legislation and sometimes come to light only with the unfolding of the process. The agrarian reform law and related legislation also undergo change over time as a result of group pressures. At the economic level, the major aims of the reform seem to have been to increase the agricultural marketable surplus for the rapidly growing urban centres as well as expanding the internal market for incipient local industry. The military government not only continued the import substitution industrialisation strategy begun in 1959 but assigned a new urgency and a greater directing role to the state in pursuing this development strategy.

At the political level, perhaps the more important for the military, the agrarian reform attempted to incorporate the peasantry into the political system under the tutelage of the state, widening the social base and thereby legitimising its rule (Valderrama, 1978: 97). This aim formed part of the military government's wider objectives to create a strong, autonomous national state, free from oligarchic control as in the past.

IV.2 Administrative Framework and the Process of Implementation

A vast and complex administrative machine was developed to implement the reform programme. Three sets of measures can be distinguished: those dealing with the expropriation process, those dealing with the organisation and functioning of the reformed sector and finally those relating to the organisation of the peasantry.

IV.2.1 Expropriation   Decree-law No. 17716 set maximum limits to the amount of land which could be retained by individual owners. Land above these limits was liable to expropriation. On the Coast the limits were 150 hectares for irrigated and 300 hectares for non-irrigated farmland. In the Sierra they were between 15 and 55 hectares for irrigated farmland (depending on the province), or a landholding sufficient to sustain 5,000 head of sheep for pastureland. These maximum retainable holdings could be adjusted upwards if specified efficiency conditions were met. Land owned by corporations and agro-industrial complexes was subject to expropriation. Apart from size and type of ownership almost 30 other conditions are cited as causes for expropriation. Among these were inefficient land use, indirect farm management, illegal labour conditions, existence of social conflicts, acts of sabotage, the needs of the reform process in the locality, etc.

The military government first expropriated the twelve enormous sugar estates on the Coast. Within 24 hours of issuing Decree No. 17716 the sugar estates were militarily occupied and taken over. Unlike the Belaúnde reform, the military government thus started by expropriating the most profitable and capital-intensive agricultural enterprises, showing their determination and ability to liquidate the rural oligarchy. Expropriations in the Sierra came later but followed the same pattern of starting with the more economically significant *haciendas*.

At first expropriations were carried out on an individual enterprise level. Later they were implemented on a regional basis following directives set out in the *Planes Integrales de Desarrollo — PIDs (Integral Development Plans)* and the *Proyectos Integrales de Asentamiento Rural — PIARs (Integral Rural Settlement Projects)*.

Expropriated landowners were entitled to compensation. The amount of this was calculated by basing land values on self-assessed values as registered in the 1968 returns of the newly established land tax. Other fixed assets were paid at book value and livestock at market value. Forms of compensation differed. The maximum cash payment was $100,000 for efficiently managed farms, $50,000 for inefficient owner-operated land and $25,000 for indirectly cultivated or idle land. The remainder was to be paid in bonds. Up to $1 million cash was paid for expropriated capital equipment with the balance in bonds. The terms of payment for bonds were as follows: 20 years and 6 per cent interest for efficiently managed farms, 25 years and 5 per cent interest for inefficient estates and 30 years and 4 per cent interest for idle lands. Bonds were non-negotiable but could be redeemed at nominal value if the proceeds, accompanied by the equivalent amount of cash, were invested in industrial enterprises specified by regulations (Horton, 1976: 219-20). However, the amount of compensation landowners actually received was undercut by the fact that book values were well below real market values and by the high levels of inflation in the Peruvian economy. Few landowners found the government's option of tying the redemption of bonds to industrial investment to be profitable and this only occurred in a few instances (Caballero, 1976: 37-40).

The law contained a series of norms allowing landowners to carry out their own voluntary reform in areas not declared agrarian reform zones. Many landowners took advantage of this opportunity and subdivided their estates into units below the maximum retainable. These subdivisions were often false, the farm being divided between relatives whilst continuing to be farmed as a unit. As a result of peasant pressure, the government modified the conditions for private parcellisation of estate lands five months later. Decree No. 18003 practically eliminated such practices (Harding, n.d.: 122; Saldívar, 1978: 29-33).

An innovation of the 1969 agrarian reform law was the creation of the Fuero Privativo Agrario. This new body was set up as the traditional judiciary was seen to be too closely linked to oligarchic interests to favour the expropriation process. This judicial body, whose members were nominated by the President, was in charge of resolving conflicts related to the application of the agrarian reform and to settle land disputes. It was a decentralised public organ operating through the *juzgados de tierras* (land inspectorates) and *tribunales agrarios* (agrarian tribunals) which existed throughout the country. It certainly speeded up judicial resolutions over land conflicts and these no longer tended to favour landlords as in the past. It also gave peasants access to a defence lawyer free of charge for the first time (Pásara, 1978: 57-74).

The financing and control of all public agencies dealing with agriculture were centralised within the Ministry of Agriculture and a comprehensive planning process initiated. The Land Reform and Settlement Administration (Dirección General de Reforma Agraria y Asentamiento Rural) within the Ministry of Agriculture had a high priority and the budgets of the Ministry and Land Reform Agency were substantially increased (Horton, 1976: 222-3). The land reform agency operated through twelve Agrarian Zones each of which had three administrative subdivisions: Land Reform, Natural Resources, and Production and Commercialisation. The directors of the zonal offices had broad authority over local decisions but had to co-ordinate their actions with policies laid down in Lima. Within the 12 agrarian zones, agricultural policies were implemented via 53 agrarian offices and 205 agrarian agencies each of which was responsible for a specific geographical area (ibid.: 224-6).

IV.2.2 Reform Sector   The military plans for the reformed sector were aimed at developing co-operative forms of organisation at enterprise and regional levels. Expropriated land was adjudicated to groups of peasants as well as to individuals. The land allocated to groups of peasants was organised into Cooperativas Agrarias de Producción – CAPs (Agrarian Production Co-operatives), Sociedades Agrarias de Interés Social – SAISs (Agrarian Societies of Social Interest), Cooperativas Comunales (Communal Co-operatives) or Grupos Campesinos (Peasant Groups).

The CAP (Agrarian Production Co-operative) is an individual production unit in which ownership and usufruct of all productive assets are collective. No individual production is permissible. The members (*socios*), who all work on the collective enterprise, participate in the CAP's management through democratically elected bodies. The decision-making institutions are the general assembly, the administrative council and the supervisory council. The general assembly is the supreme authority in the management of co-operative affairs. One member of the administrative council is elected the co-operative's president. The administrative council receives advice from various specialised committees and gives instructions to the management office. The manager is often a hired professional who is in charge of the daily running of the enterprise via the production units. Profits are only distributed after a series of obligatory deductions for reserve, investment, social security, education and development funds. The remainder is either distributed equally among members or according to the number of days worked (ibid.: 20-2). An overwhelming majority of coastal estates were organised into CAPs, but only a few Highland *haciendas*.

The SAIS (Agrarian Society of Social Interest) is a co-operative enterprise like the CAP but membership extends beyond former estate workers. The SAIS is legally owned and managed by a service co-operative (*cooperativa de servicios*) made up of the workers from the expropriated estates and by a number of neighbouring peasant communities designated by the reform agency as land reform beneficiaries. *Comuneros* from affiliated peasant communities do not work on the production units but communally receive part of the profits for infrastructural investments. Furthermore the development division should provide technical assistance and credit for the exclusive benefit of affiliated communities. The governing bodies of the SAIS correspond to those of the CAP. The service co-operative and each affiliated peasant community are equally represented in the Assembly of Delegates. As each SAIS has several community members and only one service co-operative, the communities in principle control the management of the enterprise and receive most of the surplus (ibid.: 22-5). The rights of members to farm individual plots and to pasture animals on SAIS land continue where they existed before (Eckstein *et al.*, 1978:

28). The SAIS represents an attempt to extend agrarian reform benefits beyond the tenants and workers of the former *hacienda* to the neighbouring peasant communities. They were, therefore, only established in the Highlands and were frequently very large, grouping together several former *haciendas*. They were intended to be transitional units, eventually becoming CAPs.

The Communal Co-operatives were formed by distributing land to peasant communities with the requirement that this be collectively exploited. The internal organisation of this co-operative paralleled that of the CAP, the only difference being that members must be *comuneros*.

Adjudications to peasant groups were conceived as transitional, precooperative rulings made as a step to the eventual formation of CAPs. Unlike the case of peasant communities, peasant groups were explicitly organised for the purpose of distributing land to groups of *minifundistas* and former tenants. Although land was adjudicated to the group collectively, it was generally farmed individually, despite the government's preference for group farming (Caballero, 1976: 79).

The agrarian reform programme also contemplated the co-operative organisation of the non-reform sector by restructuring peasant communities and by regrouping smallholders. Smallholders were to be regrouped and organised through the Cooperativas Agrarias de Servicios – CAS (Agrarian Service Co-operatives) and the Cooperativas Agrarias de Integración Parcelaria – CAIP (Agrarian Parcel Integration Co-operatives). In the CAS peasants would organise co-operatively for the provision of inputs, credits, marketing and other services but would retain individual ownership of their plots and would carry on production individuallly. In the CAIP, members would give their own land in usufruct or in ownership to the co-operative and these lands would be collectively farmed. However the restructuring of communities along co-operative lines met with fierce resistance as it meant replacing traditional communal authorities and collectivising land ownership and land use. It, therefore, did not proceed very far (Horton, 1976: 79).

Although co-operative modes of adjudication predominated, some expropriated land was adjudicated on an individual basis giving beneficiaries private property titles. This generally occurred where tenants exercised strong pressure for the individual redistribution of land and where the central enterprise was small.

By late 1972 the government began to draw up integral Rural Settlement Projects (PIARs) with a view to fusing several farms into larger production units. Within each PIAR both estate land and all agricultural land were made liable to be affected by agrarian reform measures, including parcel concentration and integration (ibid.: 228-9). The basic units created by the agrarian reform process were seen in turn as the building blocks for a larger-scale regional integration, enabling a broader rural reconstruction and eventually allowing agriculture to be planned at the national level (Smith, n.d.: 101). The PIAR were to be constituted by CAPs, SAISs, peasant groups, restructured peasant communities, CASs and CAIPs. These base-level units were to be linked to a regional Central de Cooperativas (Central Co-operative) which was expected to fulfil several social and economic functions. The primary purpose of the Central Co-operative was to reduce socio-economic differences between members resulting from unequal access to productive resources. For this purpose it was planned that each member should give between 45 and 70 per cent of its economic surplus to the Central Co-operative for compensation and investment funds (Matos Mar and Mejía, 1980: 56). The organisation of a large Central Co-operative would also allow members to reap economies of scale when purchasing inputs and commercialising agricultural production. Through the Central Co-operatives it was also hoped to facilitate implementation of agricultural policy, to reduce the Ministry of Agriculture's financial and administrative burdens and to stimulate economic diversification in the rural areas via the regional Integral Development Plans (PIDs).

IV.2.3 Peasant Organisation   Before the agrarian reform the main existing peasant organisations were the peasant communities and the rural unions. About 1,300 communities and 425 unions were officially registered in 1970 but their real numbers are estimated at 2,500 and 1,500 respectively (CENCIRA, 1977: 3; Pásara, 1978: 149). Some of the unions were affiliated to the Confederación Nacional Campesina del Perú — CCP (National Peasant Confederation of Peru) founded in 1947.

In the face of both increasingly militant and independent peasant action, and landlord opposition to the agrarian reform, the government in one decree (No. 19400) withdrew legal recognition from rural unions and abolished the landlords' organisation Sociedad Nacional Agraria (Valderrama, 1976: 89). The government aimed to replace these class organisations in the rural sector with a single corporate body — the Confederación Nacional Agraria — CNA (National Agrarian Confederation) which grouped together producers and labourers in a co-operative structure. Behind these aims lay the thinking that class conflict and thereby the need for class organisation had become redundant. The agrarian reform, by expropriating the large estates and setting up collective enterprises in agriculture, was seen as eliminating the class divide as all were now producers. This new organisational ideology was promoted by the Sistema Nacional de Apoyo a la Movilización Social – SINAMOS (National Support System for Social Mobilisation) which was in charge of setting up the CNA. SINAMOS was formed in mid-1971 with the purpose of spreading government doctrine of cooperative organisation, self-management and social solidarity (Pease, 1977: 109).

By forming the CNA the government hoped to gain the peasantry's political support whilst at the same time establishing corporatist control over peasant actions. This official institutional network of peasant representation included four levels of association. The base-level organisations were the peasant communities and the Asociaciones Agrarias (Agrarian Associations). The latter included various enterprises of the reformed sector, landless peasants and even those landowners not affected by the land reform. These base units were grouped together into second-level organisations, the Ligas Agrarias (Agrarian Leagues) formed at Province or valley level. The third level – the Federaciones Agrarias (Agrarian Federations) – was formed at Departmental level (the country's largest regional administrative unit). These finally went to make up the CNA (Valderrama, 1976: 264). SINAMOS started to promote the formation of Ligas in the second half of 1972 and two years later the foundation congress of the CNA was held. At that time it had approximately half a million peasant members, grouped together in about 2,500 base-level organisations, around 50 Ligas and 20 Federations (Pásara, 1978: 156-7).

The government's plans for peasant organisation were based on controlling and restricting peasant participation within clearly defined parameters. At the economic level, peasants could participate in the economic decision-making processes from enterprise level up to sectoral planning level through the co-operative structures mentioned in the previous section. At the political level, they could participate through the CNA. Peasant action outside these structures was made difficult by the illegal status of the CCP (the peasant national union) and by restrictions of political party activity. Nevertheless, as will be seen below, the government was not wholly successful in encompassing peasant action within these structures and the CCP continued as an autonomous and rival body.

V The Agrarian Reform: An Evaluation of Achievements

V.1 Land Redistribution and Beneficiaries
The military government planned a massive land redistribution, expro-

priating around 15,000 farms (9.7 million hectares) and benefiting around 365,000 families. Approximately 41 per cent of the nation's agricultural land was thus to be expropriated benefiting almost 39 per cent of the peasant families. In standardised terms (i.e. hectares of homogeneous quality) about 44 per cent of the total land would in effect be expropriated (Caballero, 1976: 10-11). This vast programme, planned to be completed by 1975/6, fell somewhat behind schedule, especially with regard to adjudications. But by 1979, ten years after the promulgation of the law, 15,826 farms (10.5 million hectares) had been expropriated, though only 7.7 million hectares had been adjudicated benefiting 337,662 families (Matos Mar and Mejía, 1980: 64).

Table 5.8 shows the progress in implementation of the agrarian reform until 1979. The co-operativist character of the agrarian reform clearly predominated and little land had been adjudicated in favour of individuals. Most of the land (62.3 per cent) was distributed to fully or partly collectivised large-scale farm units, i.e. CAPS and SAISs, which accounted for 45.2 per cent of family beneficiaries. Table 5.8 also reflects the enormous average size of the CAPs and SAISs which individually incorporated, on average, 13 expropriated estates. For example the CAP Tupac Amaru was formed by bringing together 105 adjoining farms (Martínez, 1977: 24). This consolidation of land, particularly in the Highlands, was aimed at obtaining economies of scale.

A more accurate picture of the reform is gained by distinguishing types of land – both by quality and degree of centralisation of management – and types of beneficiary. Table 5.9 shows that in terms of standardised hectares the collectives received an estimated 76.6 per cent of total expropriated land while peasant communities received only 6.8 per cent, peasant groups 10.3 per cent and individual peasants 6.2 per cent. However, the overwhelming predominance of collective, co-operative and group farming is only apparent since, as Table 5.9 clearly shows, a large part of this continues to be farmed on an individual basis contrary to government intentions. Although only 6.2 per cent of total expropriated land (expressed in standardised units) was distributed to private individuals, 48 per cent is actually farmed individually, constituting the decentralised land referred to in the table.

This high incidence of individual farming, in the face of strong government pressure to the contrary, reveals the enduring influence of the pre-reform agrarian structure. Tenant-labour relations continued to be widespread in the Highland *haciendas*, whilst coastal estates had largely centralised management and a proletarian work-force. It is thus not surprising to find that most Highland collective enterprise land is in fact farmed individually, i.e. 24.4 per cent of total standardised land, compared with 7.5 per cent farmed collectively (see Table 5.9). The opposite holds for the coastal collectives where only 0.3 per cent of total standardised land is managed individually and 44.4 per cent is managed collectively. In the Highlands, moreover, much more estate land was distributed to peasant groups (9.3 per cent), who cultivate it individually, than on the Coast (1.0 per cent). Some Highland estates were almost totally farmed by tenants, who strongly resisted collectivisation.

Table 5.8: Progress of the Agrarian Reform, up to 1979

Type of Adjudication	Adjudi-catory Units	Adjudicated Land		Family Beneficiaries	
		Hectares	%	Number	%
CAP	604	2 557 366	29.7	108 726	29.0
SAIS	60	2 805 048	32.6	60 954	16.2
Peasant groups	834	1 685 382	19.6	45 561	12.1
Peasant communities	448	889 364	10.4	117 710	31.4
Independent peasants	42 295	662 093	7.7	42 295	11.3
Total	44 241	8 599 253	100.0	375 246	100.0

Source: J. Matos Mar and J. M. Mejía, *Reforma Agraria: Logros y Contradicciones 1969-79* (Instituto de Estudios Peruanos, Lima, 1980), p. 67.

To assess the impact of the agrarian reform on the rural population, it is useful to distinguish between types of beneficiaries: full and marginal. Full beneficiaries are those peasants who are either members (including full-time workers) of the collective enterprises or who received usufruct or property rights over a legally defined *unidad agrícola familiar* (agricultural family unit) which supposedly secured a minimum living standard. Marginal beneficiaries sometimes hardly benefited at all from the agrarian reform. One such group is formed by members of the peasant communities affiliated to an SAIS. Although entitled in principle to a share of the SAIS's economic surplus, little or no surplus has been redistributed in practice as most SAISs have yet to produce large distributable profits and this situation is unlikely to change in the future. It also has to be borne in mind that member peasant communities have no access to SAIS land. Thus, a third of the members of Highland collectives are marginal beneficiaries. A second large group of marginal beneficiaries is formed by *comuneros* in Highland peasant communities as these received only a small share of the redistributed land.

Table 5.9: Estimated Land Distribution by Type of Adjudication Unit and by Region (in standardised hectares)

Type of Adjudication	Centralised Land	Decentralised Land	Total Land
		(in %)	
Coast			
Collective enterprise[a]	44.4[b]	0.3	44.7
Peasant community	–	0.0	0.0
Peasant group	–	1.0	1.0
Individual enterprise	–	2.5	2.5
Sub-total Coast	44.4	3.8	48.2
Highland			
Collective enterprise[a]	7.5	24.4	31.9
Peasant community	–	6.8	6.8
Peasant group	–	9.3	9.3
Individual enterprise	–	3.7	3.7
Sub-total Highland	7.5	44.2	51.7
Coast and Highland			
Collective enterprise[a]	51.9	24.7	76.6
Peasant community	–	6.8	6.8
Peasant group	–	10.3	10.3
Individual enterprise	–	6.2	6.2
Total	51.9	48.0	99.9

Note: a. Collective enterprise refers to CAP and SAIS.

b. The data are rough estimates based on the projected goals on completion of adjudication around the early 1980s.

Source: J. M. Caballero, *Reforma y Reestructuración Agraria en el Perú*, Publicaciones CISEPA, no. 34 (Universidad Católica del Perú, Lima, 1976), p. 23.

A comparison of Tables 5.9 and 5.10 indicates regional inequalities in the land distributed to beneficiaries. Although three-quarters of the beneficiaries are in the Highlands, they received only half the expropriated standardised land, while the reverse holds true for the Coast.

A more precise and complete illustration of the unequal distribution of the benefits of the agrarian reform is presented in Table 5.11.

Table 5.10: Estimated number of Beneficiaries by Type of Adjudication and by Region

Type of Adjudication	Full Beneficiaries[a]	Marginal Beneficiaries[b]	Total Beneficiaries
		(in %)	
Coast			
Collective enterprise[c]	20.3[d]	1.6	21.9
Peasant community	0.0	0.1	0.1
Peasant group	0.4	–	0.4
Individual enterprise	1.9	0.6	2.5
Sub-total Coast	22.6	2.3	24.9
Highland			
Collective enterprise[c]	28.7	13.3	42.0
Peasant community	6.5	9.8	16.3
Peasant group	9.1	2.2	11.3
Individual enterprise	3.5	2.0	5.5
Sub-total Highland	47.8	27.3	75.1
Coast and Highland			
Collective enterprise[c]	49.0	14.9	63.9
Peasant community	6.5	9.9	16.4
Peasant group	9.5	2.2	11.7
Individual enterprise	5.4	2.6	8.0
Total	70.4	29.6	100.0

Notes: a. Full beneficiaries are those peasants who are either members of the collective enterprises or are full-time workers on the collective enterprises or received usufruct or property rights over a legally defined *unidad agrícola familiar* (agricultural family unit).

b. Marginal beneficiaries are those peasants who either belong to a peasant community which is affiliated to an SAIS or have obtained individual or communal access to land which is smaller than a *unidad agrícola familiar* for each beneficiary.

c. Collective enterprise refers to CAP and SAIS.

d. See note b, Table 5.9.

Source: Caballero, *Reforma y Reestructuración Agraria*, pp. 8, 9, 24.

The extremes are represented by the Highland peasant communities and the coastal agro-industrial sugar plantations. Capital value per family beneficiary was evidently much higher on coastal collectives (CAPs) than those in the Highlands (CAPs and SAISs). But the latter still have higher capital values per family beneficiary than all non-collectives (peasant groups, peasant communities, individuals).

The relatively privileged position of reform beneficiaries *vis-à-vis*

Table 5.11: Land Distributed, Value of Adjudications[a] and Bene-ficiaries by Type of Adjudication, 31 December 1975

Type of Adjudication	Land	Value	Family Beneficiaries	Value per Family
		(in %)		thousand of soles[b]
SAIS (all in Highlands)	43	14	25	23
CAP	33	81	41	80
Agro-industrial				
(all on Coast)	(2)	(38)	(12)	135
Other coastal	(10)	(37)	(17)	88
Highland	(19)	(5)	(11)	18
Jungle	(2)	(1)	(1)	33
Peasant groups	16	4	11	13
Peasant communities	7	1	19	2
Individual	1	0	4	7
Total	100	100	100	40

Note: a. Value of adjudications includes land, cattle, buildings, machinery and equipment.

b. 1 US dollar = 43.5 soles.

Source: H. Martínez, *La Reforma Agraria en el Perú*, 1er Seminario Problem-ática Agraria Peruana (Ayacucho, 1977), pp. 20-1.

non-beneficiaries is shown by the fact that beneficiaries who constitute almost 30 per cent of the nation's rural families have access to almost 50 per cent of the nation's cropland. *Comuneros* and *minifundistas* have largely been excluded from the land redistribution. *Comuneros*, who still comprise 29 per cent of rural families, only possess 19 per cent of the cropland (Horton, 1976: 273).

V.2 Performance of the Reform Enterprises

This section examines CAPs and SAISs, as the remaining types of enterprise set up after the agrarian reform are of little significance and are difficult to study for lack of data.

V.2.1 Coastal Production Co-operatives (CAPs)

Two types of CAPs were set up on the Coast: twelve enormous sugar co-operatives, and others, cotton, rice and mixed farming co-operatives. We shall consider each type separately.

Sugar co-operatives. These CAPs cover a sixth of the expropriated land on the Coast and slightly less than a third of coastal beneficiaries (Caballero, 1976: 23-4). However, they account for 38 per cent of the capital value of all reform enterprises (see Table 5.11). The government keeps a close watch over their functioning and information is more readily available than for the other reform enterprises.

The agrarian reform did not disrupt production and productivity levels in the sugar industry. The sugar agro-industrial complexes were expropriated immediately following the agrarian reform law to prevent massive decapitalisation. From 1969 to 1977 sugar production increased by about 40 per cent, productivity, expressed in metric tons of sugar-cane per hectare, by 15 per cent and the cultivated area by 24 per cent (Gonzáles and Munaiz, 1979: 18, 20, 22). Much of this increase, however, occurred between 1969 and 1975, after which frequent droughts disrupted production of sugar-cane.

Incomes of workers increased sharply in the immediate aftermath of the formation of the CAPs, but stagnated thereafter and may even have fallen in recent years. Average real personal incomes rose by 62 per cent from 1968 to 1972 for both workers and employees of sugar CAPs. Real incomes *per capita* for permanent workers increased by 78 per cent whilst those of employees and temporary workers (*eventuales*) increased by only 27 per cent and 20 per cent respectively during the same period (Roca, 1975: 53). Thus income differentials between white-collar and blue-collar *socios* declined but those between *socios* and *non-socios* increased. In 1968, the ratio of the average daily wage of a permanent worker to that of an *eventuale* was 2.7; by 1972 this had risen to 4. Differences in wage levels between co-operatives continued, those with the highest average levels of productivity paying the highest average wage for *socios* (ibid.: 61).

Total employment expressed in days worked increased by 1.3 per cent between 1968 and 1972. However, employment of *eventuales* grew by 43 per cent whilst that of *socios* fell by 4 per cent during the same years (ibid.: 44, 46). The *eventuales'* contribution to the total number of days worked thus rose from 11 per cent in 1968 to 16 per cent in 1972. With respect to the total number of workers (*socios* and *eventuales*) employed, this increased by 21 per cent between 1970 and 1978 but decreased thereafter (Gonzáles and Munaiz, 1979: 24). Labour productivity expressed in metric tons of sugar per worker remained stagnant between 1968 and 1977 (ibid.: 22).

The sugar co-operatives made substantial profits, reaching their maximum level in 1975 when sugar prices boomed (Ríos, 1979: 10). Since 1976 the industry has suffered losses because of the steep fall in sugar prices in the international market, mismanagement, droughts and inappropriate government intervention. In the good years the government creamed off a large part of the surplus, thus constraining investment. Between 1971 and 1974 government revenues from the tax on profits amounted to US$65.3 million or 51 per cent of the co-operatives' total gross profits. Prior to 1969 this tax had never exceeded 35 per cent (Scott, 1978: 326). Furthermore government interference in pricing proved detrimental to the co-operatives as domestic prices in particular were fixed far below the equilibrium market price (Gonzáles and Munaiz, 1979: 13).

Cotton and rice co-operatives. Information on these co-operatives tends to be incomplete and fragmentary in nature, making generalisation difficult and tentative. A 1973 study of six cotton CAPs distributed throughout the Coast shows that production and investment records were reasonably favourable. Employment increased moderately and real wages rose significantly (Horton, 1976: Appendices). A regional study of the Chancay Valley indicates that cotton workers' real wages were higher by about 10 per cent in 1973-6 than in the pre-reform period (Matos Mar and Mejĭa, 1980: 95). Other regional studies of the Piura area — a major cotton-producing area — revealed a slight increase in output per hectare following the reform (Rubin, 1977: 36; Revesz, 1979: 12). However, the cultivated cotton area in Piura has fallen in recent years because of drought and the government's unfavourable price policy enforced through its complete control of cotton marketing (Revesz, 1979: 12-14). Over half the labourers were *eventuales* in the Piura cotton co-operatives in 1974/5 (Rubin, 1977: 35-6).

With respect to the rice co-operatives, the best available information comes from a study of the Jaquetepeque Valley — one of the most important rice valleys containing over 15 per cent of the nation's cultivated rice area (CENCIRA, 1976: 110). The Valley's 22 CAPs account for two-thirds of the rice area in the region, the remainder being cultivated by small- and medium-sized farms (ibid.: 110-12). Income differentials between *socios* accentuated as former tenants maintained their plots but no longer paid rent. In 1974/5 these tenants constituted 21 per cent of the *socios* and controlled 12.1 per cent of the CAPs' cultivated rice area (ibid.: 108, 122). Real wages have risen significantly since the formation of the co-operatives but wage differentials between co-operatives remained considerable, particularly for *socios*. Wage differentials between permanent and temporary workers within a co-operative persisted as well (Scott and Manrique,

1973: 13). The average daily wage of a *socio* was 28 per cent higher than that of an *eventuale* in 1974/5. This difference was magnified by the different ways in which *socios* and *eventuales* obtained their daily wage. *Eventuales* continued to be paid by *tarea* (workload) and to earn their daily wage they undertook two *tareas* – the equivalent of a full day's work. *Socios*, meanwhile, who were also paid by *tarea* before the reform, now received a fixed daily wage which only required carrying out one *tarea* – half a day's work. In the 22 CAPs under examination, *eventuales* provided 63 per cent of the total number of days worked in 1974/5 and *socios* the remainder. But in terms of numbers, *eventuales* formed only 43 per cent of the total workforce (CENCIRA, 1976: 184-5).

Contradictions: eventuales, socios and the state. Two tendencies seem to have been at work within the co-operatives leading to an erosion of their co-operative character. Both are connected with the existence of divisions within the work-force between members and non-members. In general, the *eventuales*, who form a quarter of the coastal co-operative's labour force (Caballero, 1978: 62), work more hours per day, do the most irksome tasks and receive considerably lower wages than the *socios*. Their generally higher productivity and lower wages create a surplus which is appropriated by the co-operative. Yet they have no say in the running of the co-operative nor do they receive a share of profits or other benefits which the *socios* receive.

The first major tendency at work in the sugar co-operatives is for members (*socios*) to shift to administrative, technical and supervisory positions substituting *eventuale* labour for their own. Greater reliance on low-paid *eventuale* labour increases profits whilst restricting the number of *socios* increases each individual *socios'* share of distributable profit. There is a tendency, therefore, for *socios* to degenerate into quasi-capitalist employers of cheap and unorganised labour (ibid.: 60). This explains the *socios'* strong opposition to incorporating additional members and their eagerness to contract *eventuales*. The only counterbalancing force to these tendencies is the *socios'* fear that too much dependence on *eventuales* might lead to greater pressure for their incorporation as members.

The second tendency at work, particularly in the cotton and rice co-operatives, involves the pressure exerted by tenant members to appropriate more co-operative land and water for private use. If this process of *asedio interno* (or internal encroachment) goes unchecked, the collective will eventually be reduced to a collection of individual farms operated by former tenants who privately appropriate the surplus generated by employing cheap, *eventuale* labour. Whilst centralised production is secure on the sugar co-operatives, on cotton and rice co-operatives the predominance of tenant-beneficiaries threatens to decentralise production totally.

In both cases, the *eventuales'* labour contributes to the surplus appropriated by the *socios*. However in the sugar co-operatives the surplus is produced collectively and *socios* assume the position of collective capitalists, so to speak. In the cotton and rice estates, the *socios* have steadily been developing into individual capitalists.

These conflicts between the *socios* and the *eventuales* relate to wider conflicts between the *socios* and the state. On the sugar co-operatives, the *socios'* main aim is to maximise personal disposable income in the form of wages and other benefits. They have little interest in increasing the non-wage component of value-added as this mainly goes to paying agrarian debts and other funds whose benefit to individual members is unclear. The size of these deductions also means that often little is left over for redistribution. The state, however, has a clear interest in increasing the share of the surplus *vis-à-vis* wages to ensure that taxes and debts get paid and investment takes place. Furthermore, the state has an interest in increasing the size of the surplus itself by raising work intensity, a process which the *socios* naturally resist.

On the rice and cotton estates the main tug of war between the *socios* and the state centres around the question of land allocation. Tenants exert constant pressure to increase the proportion of land individually worked and thereby to reduce the size of the collective favoured by the state.

In a bid to save the co-operative model, the state sought to limit the autonomy of the co-operatives from early on. Decree-law 18299 promulgated in 1970 imposed a series of mandatory accumulation provisions together with measures designed to secure the payment of taxes, debts and loans provided by the State Bank (García-Sayán, 1977: 170-2). These provisions were institutionalised by the Sistema de Asesoramiento y Fiscalización de las CAPs – SAF-CAP (System of Advice and Control of the CAPs) which exercises considerable control over the functioning of the co-operatives, particularly relating to land use, employment, wages and managerial nominations. Furthermore the state fixes the prices of sugar, cotton and rice and completely controls their marketing through a state monopoly of marketing enterprises.

As the conflicts outlined above have increasingly surfaced, there has been a creeping process of growing state intervention in the affairs of the co-operatives. This process has not been uniform, the degree of state intervention varying directly with the economic importance of the enterprise. Far more control is exercised over the sugar CAPs than the rice or cotton CAPs.

In 1972 the state forced some CAPs to incorporate some permanent *eventuales* as *socios*. Although legislation gave these workers membership rights, they had been ignored until the government exercised pressure (Scott, 1978: 333). On some co-operatives a compromise was reached whereby a few *eventuales* were incorporated as *rentados*, giving them job stability and a fixed wage throughout the year. But they did not receive the full range of benefits accruing to the socios, nor were their wages as high (Raviñes, 1979: 12).

As co-operatives began to experience severe financial problems, state control over their functioning was stepped up. In 1976, Decree-law 21585 ruled that all reform enterprises had to have the Ministry of Agriculture's authorisation to raise wages or change work conditions (Pásara, 1978: 136). As a result of the sugar co-operatives' mounting financial crises, Decree 21815 of early 1977 abolished what little autonomy they had enjoyed, replacing elected representatives with government functionaries (Scott, 1978: 321). On the cotton and rice co-operatives, the former tenants' private appropriation of collective resources threatened their financial viability. Bankruptcy was only postponed by the greater use and exploitation of *eventuales* and by favourable international prices for their crops for a few years. When export prices fell steeply in 1975, the State Bank put pressure on members to increase co-operative work by reintroducing payment by *tarea* (workload) for the *socios* (Rubin, 1977: 45).

In some cases the state tried to eliminate or reduce the tenancy system to forestall a co-operative's disintegration. Whether these measures have been successful or not is difficult to tell. It is likely that the expansion of the tenancy system has been checked particularly on those co-operatives which are in financial difficulties (Scott and Manrique, 1973: 7). For example in one CAP where tenant members had increased their share of total rice production from 13.5 per cent in 1972 to 64.4 per cent in 1975, the Ministry of Agriculture eliminated the tenancy system when the co-operative underwent a financial crisis in 1977 rather than permit its disintegration (Raviñes, 1979: 16, 20). Tenant members had the choice of giving up their tenancy and remaining *socios* or keeping their plots and thereby losing all the benefits of co-operative membership. One hundred and twenty-three out of 144 tenant members decided to give up their tenancy and the few who decided against were not, surprisingly, the larger tenants (ibid.: 16).

The increased level of state intervention in the CAPs raises the question of whether the CAPs can best be characterised as state capitalist or self-managed enterprises. Since their establishment the CAPs have formed a type of hybrid enterprise, characterised by a changing balance between state control and workers' self-management (Scott, 1978: 321). Growing state involvement in the CAPs has pushed the balance in favour of state capitalism. This has gone furthest in the case of the sugar co-operatives. The fact that some sugar complexes were militarily occupied in order to break a strike by beneficiaries shows that workers' self-management had virtually ceased to exist. The sugar CAPs can, therefore, best be described as state enterprises, their expropriation leading in the end to the state supplanting former landlords (Matos Mar and Mejía, 1980: 78).

V.2.2 Highland CAPs and SAISs Data on the performance of the Highland reformed enterprises are fragmentary and inadequate. Generalisation is also made difficult by the wide variety of situations encountered. One Ministry of Agriculture official even proclaimed that each of the reform enterprises in the Highlands had a different problem. Some general features can nevertheless be identified.

The enduring influence of the pre-reform situation on the functioning of the reform enterprises manifests itself in production patterns, technology, labour structure and articulation of conflicts. The performance of a particular SAIS or CAP is conditioned by the pre-existing degree of capitalisation and of centralisation of its production process, for these had implications for the degree of proletarianisation of labour and the level of technological development. In general, *haciendas* which had more developed capitalist relations of production tend to perform better as co-operative enterprises after expropriation than those which were relying largely on pre-capitalist relations of production.

The two types of enterprises — the livestock and livestock-crop enterprises — continued to be distinct after expropriation. SAISs tended to be set up on the former and CAPs on the latter (Martínez, 1977: 78). The livestock units tend to perform better than livestock-crop enterprises. The former, it may be noted, were better managed and employed more wage labour than the latter before expropriation. The technology involved in sheep rearing is also more amenable to changes designed to realise economies of scale than that involved in crop cultivation (Eckstein *et al.*, 1978: 58). Post-reform performance is also influenced by the degree of decapitalisation landlords carried out immediately before expropriation. The more capitalised livestock estates were expropriated first with a view to preventing a stripping of their capital assets. Subsequent state policies have tended to favour the more profitable units in terms of technical and financial support thereby widening the initial differences.

SAISs. A study of five sheep-SAISs in the Southern Highlands in 1973 found that production of the central enterprise had increased in three of them since expropriation and remained unchanged in two (Eckstein *et al.*, 1978: 59). Shepherds (*huacchilleros*) who owned between a sixth and a third of the SAISs' total flock obtained larger increases in production than the central enterprise (ibid.: 58-9). Although pasture fees for the peasants' sheep (*huaccha*) were increased, their flock did not diminish. Methods of production were intensified on only a few enterprises but employment rose substantially on most. All units were profitable and investment increased on all (ibid.: 59). A brief reassessment of these units in 1976 found that the number of individually owned sheep had continued to grow and that profits and investment had slumped. Land invasions by associated peasant communities, disappointed with the prolonged inability of the enterprises to earn substantial profits for redistribution, had occurred (ibid.: 92).

A more recent case study of a SAIS in the Southern Highlands confirms this tendency for private farming to grow. In this case the former *huacchilleros'* livestock had increased from 26,000 sheep units in 1974 to 39,000 in 1975 (Auroi, 1980: 67). To counter this tendency, the management introduced a pasture right fee which succeeded in reducing the peasants' livestock to 29,000 sheep in 1978. The relative importance of *huaccha* livestock thus fell from 31 per cent to 27 per cent between 1974 and 1978 and the central enterprise's higher-quality livestock expanded by 34 per cent in the same period (ibid.:52, 67).

The same case study showed that wages in real terms had increased substantially between 1974 and 1977, falling in 1978 but still remaining above 1974 level. Wages of the 43 wage labourers increased more than those of the 93 shepherds and the 15 overseers (ibid.: 61). The payment of the agrarian debt absorbed between 8 and 10 per cent of the SAIS's gross income and that part left for distribution to associated peasant communities was so derisory that it provoked one community to withdraw in 1978 and to stage a land invasion a year later (ibid.: 63, 65-6).

Turning now to the Central Highlands, a total of seven SAISs existed here and all were dedicated to sheep rearing. The four most important of these totalled 637,531 hectares, of which 120,000 were unusable. Only 3.4 per cent of the usable land was cultivated, the remainder being natural pasture (Caycho, 1977: 41). Of the 8,647 families who were beneficiaries, 86.1 per cent were *comuneros*, 7.2 per cent *feudatorios* (former *huacchillero* tenants) and 6.7 per cent permanent wage labourers. Former tenants retained both their tenancy (*parcela*), though this was small as most production income derived from sheep rearing, and their pasture rights (*hierbaje*). In one SAIS the average size of a *parcela* was 0.3 hectares (Valcarcel, 1978: 15) and was probably similar in the others. Another SAIS, Tupac Amaru I, had already elim-inated tenancies before expropriation.

With respect to the labour composition, both former tenants and permanent wage workers — jointly totalling 1,204 families on the four SAISs — belonged to the service co-operative. However only 595 worked permanently on the central enterprise as the other preferred to work on their own enterprises. As in the Southern Highland case studies, the administration of the largest SAIS in the Central Highlands (Cahuide) tried to reduce the shepherd's livestock by offering in this case monthly bonus payments for each sheep unit reduced and by setting a maximum number of sheep units each category of worker could pasture on the SAIS's land (ibid.: 9). Despite these measures the *huaccha* livestock still grew by 4 per cent between 1971 and 1977 but a potentially explosive rate of growth was prevented. As the central enterprise's livestock expanded faster (14 per cent), the peasants' livestock fell from 13.7 per cent of the total (in sheep units) in 1971 to 11.2 per cent in 1977 (ibid.: 10).

The growth of the private peasant enterprise may be indirectly reflected in the employment of seasonal wage labour, which made up 35 per cent of the work-force on the collective enterprise and generally received half the wage of permanent workers (Caycho, 1977: 97).

Data on the evolution of real wages, available for only one SAIS, indicate that wages almost doubled during the two years following expropriation and then remained stagnant for two years up until 1974. Even so, the daily wage was still four times higher than the legal mini-mum (ibid.: 96). Production increased substantially between 1971 and 1972 in the two SAISs for which information is available (ibid.: 102). Profits rose significantly on all four SAISs during 1972 to 1973 but fell between 1973 and 1974 except in Tupac Amaru I where the rise continued. A large proportion of profits was reinvested and the remain-der distributed to associated peasant communities on a collective basis (ibid.: 114). My own calculations for 1972 show that on a *per capita* basis this would mean each *comunero* receiving between 1.6 per cent and 6.3 per cent of a permanent worker's average annual wage (ibid.: 41, 119).

CAPs. Although CAPs were planned as fully collectivised units of production, individually cultivated plots and individually owned livestock make up a third to a half of total hectares cultivated and animals raised (Eckstein, 1978: 51). As CAPs are partly crop enterprises they tend to be smaller in size than SAISs. The ecological conditions of the Highlands make cropping particularly suitable on small peasant plots, so government efforts to centralise production have been rather unsuccessful here. Strong peasant pressure for private cultivation was reinforced by ecological considerations. Large-scale enterprises may in fact create diseconomies of scale and become unmanageable, as occurred in some instances (Martinez, 1977). The state, like the landlords before it and for the same reasons, failed to proletarianise the work-force fully. Highland ecology simply does not make it profitable for state or private capital to commit important resources into developing the technological conditions of production for crop cultivation. Despite the major change in formal property relations then, the structural processes of production remained the same. Bearing this in mind, it is hardly surprising that in 1975/6, 67 per cent of the land of 28 CAPs in the Southern Highlands was under peasant cultivation. In late 1976, 38 per cent of the land and 28 per cent of the livestock on nine CAPs and seven SAISs in another region of the Southern Sierra were in peasant hands (Martinez, 1977: 74, 57). Finally, on 20 CAPs in the Northern Highlands, 14 per cent of the land was worked privately by co-operative members and 57 per cent of livestock, largely cattle (expressed in standardised units) was owned individually (ibid.: 57, 75).

A 1973 study of seven livestock-crop CAPs shows that on three production increased moderately since expropriation and on three it fell. However, production did increase substantially on the members' land as *socios* spent more time working on their plots (Eckstein *et al.*, 1978: 53). Employment grew slightly in only one CAP and fell in the remainder, reflecting in part the increased time devoted to individual production. Wages, in spite of significant rises, still did not match the income peasants derived from working longer on their own plots. Profits rose in three co-operatives, but four suffered losses; productive investment increased in four but decreased in three (ibid.: 54-5). These discouraging results were found to be even worse when the enterprises were reassessed in 1976. *Socios* had increased their individual proportion of co-operative cropland and further reduced their labour time on the central enterprise. Virtually no CAP remained profitable and decapitalisation through lack of investment was common (ibid.: 91-2). This negative picture of falling labour productivity and decapitalisation is confirmed by Martínez in his extensive evaluation of Highland reform enterprises (1977: 117-18).

Contradictions: internal and external encroachment. Reference was made in the historical background to two major conflicts between landlords and peasants in the Highlands. First, tenants (the internal peasant enterprises) resisted proletarianisation consequent upon modernisation efforts by landlords. Second, landlords and peasants were locked in a struggle as the peasant communities (the external peasant enterprises) attempted to encroach on the *haciendas* and regain their traditionally held land. As will be seen below, the agrarian reform rather than resolving these conflicts gave them a different expression. The conflict between peasants and landlords over proletarianisation was now expressed between beneficiaries and the state. Whilst before the reform this conflict was settled to the landlords' advantage, after the reform peasants could exert greater pressure and internal encroachment of the peasant enterprise on collective land increased. The traditional conflict between peasant communities and the *hacienda* system also came to be expressed as one between the peasant communities and the state as the latter guarded the territorial integrity of the reformed enterprises. Furthermore, a secondary conflict arose between different categories of members.

Taking the conflict over proletarianisation first, the greater ability of peasants to resist this process after the reform often led to the collective's economic crisis. The fact that former tenants were more successful in encroaching on the central enterprise's land, that they initially received large wage increases and that they no longer paid labour rents — all contributed to the collective's financial difficulties. These were most acute where decentralised (i.e. tenant) production and pre-capitalist relations predominated before expropriation.

As the surplus is small, beneficiaries prefer to work on their individual plots and maintain private pasture rights. The small size of the surplus does not permit wage increases large enough to entice beneficiaries to give up their tenancies and pasture rights and develop the collective enterprise (Martínez, 1977: 54-6). This neglect of the collective then helps to keep the surplus low. This vicious circle or low-level equilibrium trap can only be broken by the state committing major capital resources into developing the productivity of the central enterprises. As this has not occurred the process of internal encroachment is gaining the upper hand and effectively subverting the collective by a *de facto* individual appropriation of its resources.

Beneficiaries themselves are not over-concerned about the financial collapse of the central enterprise as they hope this might accelerate its disintegration and result in the private adjudication of the land. To safeguard at least part of the co-operative from further internal encroachment, the government has retreated in some cases by allowing former tenants to set up an agrarian service co-operative (CAS) which belongs to the CAP or SAIS (Martínez, 1977: 50). Strictly speaking this is illegal as beneficiaries do not privately own the plots of land.

On collectives with a degree of central management, the management has strictly limited means at its disposal to correct the decline in profitability. Legally, rents cannot be charged as the rental of land is prohibited, nor can members be forced to work on the collective. Attempts to introduce a pasturage fee have generally foundered through strong opposition from former *huacchilleros*. In some situations, the fee has not been high enough to reduce the *huaccha* livestock. Minimum wage laws have to be respected; social benefits, which landlords were never obliged to pay, have to be paid. A forcible elimination of usufruct rights is too explosive a measure to be seriously contemplated. One way of overcoming some members' unwillingness to work on the central enterprise is by employing more non-beneficiary wage labour as has occurred in some instances. But many CAPs or SAISs lack the financial means to do this and there is the risk of creating new problems if contracted wage labourers press for membership rights. Conflicts already exist between beneficiaries with usufruct rights and wage labourers, for the surplus which the wage labourers generate is seen as going to pay for the land (via the agrarian debt payments) whose fruits are appropriated privately by former tenants (Arce and Mejía, 1975).

The likely long-run outcome of these internal contradictions is either a complete dissolution of the collective through a process of parcellisation or a separation of the peasant enterprises from the central enterprise through an adjudication of land to former tenants and a transformation of the central enterprise into a state farm operated with wage labour.

Turning now to the second historical conflict between peasant communities and the *hacienda* system, a large part of the Peruvian agrarian reform's originality lies in the SAIS model which was set up to solve just this conflict. The SAIS model favoured large-scale agriculture by maintaining the original estate or merging several estates into an even larger unit. At the same time, local peasant communities were incorporated into the SAIS as associate members which in fact gave them a majority presence in the top decision-making body and a major share of the distributable surplus generated by the SAIS's production units. However, profits were generally meagre or nil and *comuneros* revealed their discontent by staging several invasions of 'their' enterprise, taking over lands for community use. As has been seen, profits of the SAIS's central enterprise were kept down by tenants' gradual and persistent appropriation of central resources. Thus, in some cases, external encroachment was precipitated by internal encroachment bringing *comuneros* into indirect conflict with tenants and wage workers.[8] Where profits were reasonable, little remained for redistribution to peasant communities after deducting agrarian debt payments and investments in the central enterprise. In any case this was not distributed individually but consumed in public works which the state should have paid for anyway, so *comuneros* perceived little economic advantage.

To deter land invasions, the SAIS management sometimes made concessions which strictly speaking were illegal. Where *comuneros* refused to perform unpaid work for improving SAIS infrastructure as they considered the redistributed surplus to be too small, the SAIS eventually paid an individual wage under the guise of a food quota (Valcarcel, 1978: 18-19). Similarly, where the external encroachment went too far the SAIS indirectly rented some pastureland to the community and discounted this 'rental' from future SAIS profits owed to the community (ibid.: 19-20). It is difficult to know how common these concessions were but whatever the case may be, they only postponed rather than resolved the fundamental problem. Land hunger, together with the peasant communities' fierce desire to recuperate land which had once belonged to them, could not be contained by future promises of surplus redistribution or other tactics. From 1977 a new wave of land invasions swept the Highlands, threatening the very survival of the SAIS. The land invaders were often violently dislodged but in some cases the government initiated a process of *redimensionamiento*, i.e. restructuring the reform enterprise by relocating boundaries. This often meant adjudicating some of the less productive land to the most militant peasant communities in order to defuse the conflict. The inability of the agrarian reform to solve the peasant communities' land problem is best illustrated by the case of the CAP Tupac Amaru II in the Southern Highlands. (This CAP is actually misnamed as it functions as a

SAIS). This large CAP, established as a government show-piece, is today
being liquidated through a series of community land invasions (Lovón
et al., 1979: 1). When constituted in 1971, 26 peasant communities
were affiliated to it. Successive invasions began in late 1976 and continued until late 1979 when, after successive *redimensionamientos*,
what remained was finally taken over as well (García-Sayán, 1980:
45). In this case then the external encroachment by peasant communities succeeded in dismembering the entire CAP, appropriating land
and other capital resources (Lovón *et al.*, 1979: 15-18).

VI Agrarian Reform in Peru: Concluding Observations

Peru's agrarian reform remains to date one of the boldest experiments
of its kind undertaken in Latin America. It effectively destroyed the
hitherto existing oligarchical order and created a new institutional
framework which, in spite of its weaknesses, was eminently suitable for
pursuing simultaneous growth and equity. This being said, the question
must be asked: to what extent have the reforms actually succeeded in
bringing about desirable changes in the basic problem areas of Peru's
agriculture, viz. production, income distribution and poverty?

In the post-reform period, agricultural production continued to grow
at the same low rates as in the 1960s. Between 1970 and 1976 agricultural production grew at an annual average rate of 1.8 per cent
(Alvarez, 1978: 8) which was well below the population growth rate of
2.5 per cent (CELADE, 1979: 8). Meanwhile the country's gross
domestic product grew at 5.1 per cent yearly during the same period
(Actualidad Económica, 1980: 10). Agriculture's performance appears
particularly dismal in view of government plans for a 4.2 per cent
growth rate per year from 1971 to 1975 (Matos Mar and Mejía, 1980:
56). In 1977 agricultural production did not grow at all and in 1978
it fell by 3 per cent (Actualidad Económica, 1980: 10). These negative
results, however, were due to the severe economic recession in Peru in
those years compounded by a drought in 1978.

Trends in pre-reform production patterns also continued. Between
1970 and 1976 agricultural products destined for urban consumers
remained more dynamic, increasing by 4.5 per cent yearly, whilst those
destined for rural consumers fell by 1.1 per cent yearly and those destined for the export markets decreased by 1.3 per cent yearly (Alvarez,
1979: 5). Thus in 1976 agricultural commodities produced for the
urban market represented 65.1 per cent of total agricultural production

whilst those for the rural and export markets contributed 27 per cent and 7.9 per cent respectively (ibid.: 4). As the Highlands produce primarily for the rural market and the Coast primarily for the urban market, regional inequalities are likely to have been accentuated. So far then, the agrarian reforms have had very little impact on the trend and structure of agricultural production.

With respect to employment, the national development plan for 1971-5 stated that 307,800 additional jobs would be created in agriculture (Matos Mar and Mejía, 1980: 57). However, official estimates show that only 171,900 new agricultural workplaces were created between 1969 and 1978 (Choy, 1980: 21, 25). Nevertheless, the 0.9 per cent yearly rate of employment growth achieved between 1969 and 1978 is almost three times the annual rate of male agricultural employment growth from 1961 to 1972 (Maletta, 1980: 2). Thus the agrarian reform did have some positive effect on employment although far less than anticipated. Owing to agriculture's still low rate of absorption of labour, the high rate of rural outmigration continued (Matos Mar and Mejía, 1980: 102).

The redistributive capacity of the agrarian reform was quite limited. Approximately only 1 to 2 per cent of the national income has been redistributed to about a third of peasant families who form less than a sixth of the country's families (Webb and Figueroa, 1975: 132-3). Furthermore, the redistributive effect of the agrarian reform was very uneven. According to available estimates, the real income per family increased by 80 per cent between 1969 and 1973 on the sugar co-operatives and by 45 per cent on the other coastal co-operatives. In the Highlands incomes of beneficiaries on the expropriated livestock-crop *haciendas* increased by 69 per cent and on the livestock ones by 17 per cent (Horton, 1976: 324). Incomes of other beneficiaries who received land whether in private or collective property also increased substantially. It is likely that the incomes of private and peasant group beneficiaries increased more than those of peasant community beneficiaries who received land as land-labour ratios are more favourable in the first group. The incomes of *comuneros* who belonged to peasant communities affiliated to SAISs rose only marginally as profits were generally low (ibid.: 325-6).

The uneven nature of the redistribution becomes even more evident when the income increases gained by each peasant group are related to the group's relative importance within the total peasant population. Thus, to take two extremes, sugar workers — a minority group — received the highest incomes before the reform and obtained the largest increases after it whilst *comuneros* – the largest and poorest group before the reform – were largely excluded and obtained the smallest increases after it (Figueroa, 1976: 172). The agrarian reform, therefore, did not significantly alter income inequalities within the peasant population (Caballero, 1978b: 41-3). The beneficiaries have also suffered, in economic terms, in recent years as some reform enterprises have experienced acute financial problems. In short, the agrarian reform has been unable to ameliorate rural poverty significantly, as it left the traditionally poor majority unaffected (Matos Mar and Mejía, 1980: 98).

General economic policies of the state must bear a major responsibility for the poor performance of the post-reform agrarian economy. As already noted, the military government continued the 'import substitution industrialisation strategy' which discriminated against the agricultural sector. The foreign exchange policy favoured agricultural imports even more than before and penalised agricultural exports. Price controls on agricultural products were further extended and more strictly enforced. Furthermore food subsidies were introduced in 1973 to prevent food prices from rising. Between 1973 and 1976, 87 per cent of these agricultural subsidies went to imported food products creating unfair competition for local producers (Alvarez, 1979: 29). In 1973 the subsidy on wheat and meat imports alone represented around 5 per cent of the agricultural sector's income (Figueroa, 1979: 21).

The government's cheap food policy not only negatively affected rural incomes but also, by reducing agriculture's profitability, indirectly affected investment, production and employment in this sector. Attempts to correct these deleterious consequences, by increasing real agricultural credits – mostly State credits – by 5.4 per cent yearly from 1970 to 1978 and by raising public agricultural investments, were insufficient to compensate for the damage done by landlords' decapitalisation and the inadequate investment effort of the reform sector (Figueroa, 1976: 64; Maletta and Foronda, 1979: 106).

The agrarian reform by itself could not overcome the negative effects of these economic policies. Furthermore, the type of agrarian reform the government implemented did not take full advantage of the benefits a reform programme can generate for rural development. The military government's statist-co-operativist type of reform was generally unsuccessful in significantly increasing production, creating more employment and achieving higher incomes for the majority of peasants and reducing conflicts. Within the context of a market economic system and given the characteristics of the pre-existent agrarian system, it is likely that a redistributive type of reform which assigned land to peasant farmers and peasant communities would have been more successful. State farms could be restricted to agro-industrial, capital-intensive estates, such as the sugar plantations, where economies of scale were clearly important. Giving land to tenants and peasant communities would also have removed a major source of rural conflict. As it was, the government's model of large-scale and co-operativist agriculture for the reform enterprises met with substantial opposition from sectors of the peasantry, preventing its full implementation.

The co-operativist model had the best chance of success on the sugar plantations as large-scale agriculture and wage labour predated the reform. Upon expropriation the state took over large, centrally managed and capital-intensive enterprises. The sugar workers saw little advantage in dismantling the estate into small plots as economies of scale had also benefited them in the form of higher wages, so there was little pressure for privatisation. However, conflict between wage workers and the state came to focus on the question of control of the new co-operatives. From the start, the state loaded the dice against the workers by giving technical and managerial staff over-representation on decision-making bodies. When workers' demands for wage increases, backed up by strike action, threatened to eat too far into the surplus, the sugar co-operatives, being too economically important to allow the government to sit back, were subjected to military intervention. Although the sugar workers did not resist the co-operativist economic organisation upheld by the military, their wage struggle pierced the ideological trappings of the model by forcing the government to abandon any pretence of these being self-managed economic enterprises. Paradoxically, the coastal sugar plantations which had the most favourable economic conditions for the model's success have experienced the greatest degree of state intervention. The model, in this case, was secured by force.

In the Highlands (and non-sugar coastal co-operatives) the greater presence of tenant labour and the existence of peasant communities meant that the model met with strong opposition from its inception. The peasants' opposition to the co-operativist scheme, expressed by both internal and external encroachment of the central enterprise, struck at the very heart of the model making for a much more conflictual situation in the Highlands (Pásara, 1978: 88, 95-9). Land seizures here questioned the basis of adjudication, as illustrated by the large-scale invasions in Andahuaylas in mid-1974 when over 40,000 peasants seized 78 *haciendas*, demanding their expropriation and individual adjudication (García-Sayán, 1980: 44). The ability of peasants to resist the collective model is shown by the fact that the CAPs, planned as fully collectivised enterprises, were in fact only partly so. Nor was the government able to transform the other associated enterprises – SAIS, land adjudicated to peasant communities and peasant groups – into fully collectivised CAPs as intended. Few peasant communities have been restructured because of the *comuneros'* hostility to the prospect of losing their autonomy by becoming integrated into the state co-operative model (Caballero, 1978b: 38). Few agrarian service co-operatives (CAS) and no agrarian parcel integration co-operatives (CAIP) have been established among smallholders because of their opposition to joint farming (Horton, 1976: 27). Only a third of base-level associative enterprises have joined regional central co-operatives, given the fear that they may become instruments of government control and that the surplus appropriated would be channelled to a central fund (ibid.: 29). This hindered government attempts at regional planning. Only 42 regional central co-operatives, linked to 87 integrated rural settlement projects (PIARs), have so far been established so that the aim of their serving as instruments of income compensation could not be achieved (Caballero, 1978b: 38; Matos Mar and Mejía, 1980: 74). Integrated development projects (PIDs) have never been formed at all.

Although peasant resistance was strong, the government chose not to wage an all-out battle to secure its model in the Highlands. The terrain is difficult, the number of units involved overwhelming, and their economic importance not so great as the coastal sugar plantations. Rather the government has responded to peasant pressure with a number of concessions such as restructuring boundaries (*redimension-amiento*) to deal with external encroachment and permitting the development of individual peasant enterprises in response to internal encroachment. Unlike the coastal sugar workers, the Highland peasantry do enjoy a greater degree of self-management but of an individual nature.

The ability of the peasantry to resist the agrarian reform model in the Highlands and to organise and stage strikes on the coastal sugar estates points to the government's failure to fulfil its political objective of co-opting the peasantry via the agrarian reform. This is not to deny that the state came to exercise a greater degree of control and influence in the countryside. However, even the government's own organisations set up to organise and channel the peasant movement eventually escaped their control. Initially the government had some success in organising the peasantry – through SINAMOS – into the CNA, a parallel national peasant organisation. At one point CNA had twice as many members as the pre-existing and autonomous CCP (Matos Mar and Mejía, 1980: 116-17). But in 1978 the CNA was dissolved as its actions had increasingly questioned the role of the state in the country-side (ibid: 18). As a final stab in the back of government policy, CNA eventually came to co-ordinate some actions with its one-time rival, the CCP.[9]

Indeed, in retrospect, it would seem that this failure to rally the peasants and workers around its central goals has proved to be one of the basic weaknesses of the whole reform programme. In the absence of a genuine mass participation, the collective co-operative model had to be imposed from above and was perceived as a constraint by the very people it was supposed to benefit. This important limitation notwith-standing, the reforms have undoubtedly dismantled the old order in the countryside and have gone some way towards creating an environment in which growth and equity are compatible objectives. But there is no denying that they have so far largely failed either to stimulate produc-tion or eradicate rural poverty.[10] There are tendencies, too, which threaten to undermine some of the positive achievements of the reforms.

Notes

1. Tables 5.1 and 5.2 are not strictly comparable as categories differ. The 'sub-family' category in Table 5.1 approximates the '*minifundio*' category in Table 5.2. The 'multi-family large' category in Table 5.1 approximates the 'large enterprise' and part of the 'medium enterprise' category in Table 5.2. See CIDA, 1966: 40.

2. A sheep unit is a standardised measurement of livestock. Different types of livestock are converted into a homogeneous unit of measurement.

3. Landlords, for this purpose, are defined as proprietors of farms over 50 hectares, excluding peasant communities. Peasants are those who farm production units below 10 hectares, including *comuneros*.

4. For an analysis of verticality and its significance for the economy of Andean societies, see Murra, 1975: 59-116.

5. The data under-estimate wage labour as the majority of sub-family farmers also worked as part-time wage labourers (Amat and León, 1979: 48).

6. For details of this classification, see Hopkins, 1979, and Alvarez, 1979. The partial data have to be taken as rough approximations and only to indicate trends.

7. The minimum wage varies according to region but in 1971-2 the average was around 1,000 soles per month, i.e. US$280 yearly. Many authors agree that the minimum wage does not ensure a minimum standard of living.

8. A member of the service co-operative articulates this conflict when complaining about the transfer of the surplus to the peasant communities: 'In the old days we had *one* landlord who appropriated and enjoyed our work. Today we

have *many* landlords.' A *comunero*, for his part, replies: 'The *SAIS* wants to give us money for our pastureland ... but we don't want money. We want pastures and space for our animals' (Caycho, 1977: 130-1).
9. Some conflicts persist between the CNA and the CCP. The former largely represents the interests of members of the reform enterprises while the latter is more closely associated with the demands of peasant communities and those excluded from the agrarian reform (Pásara, 1978: 166).
10. It must be noted, however, that this failure reflects the weaknesses of the agrarian reform programme only partially. It was, in large part, attributable to the inappropriate policies pursued by the government in the spheres of industry and trade.

References

Actualidad Económica (1980) 'Información Estadística', *Actualidad Económica*, *3* (27)
Alberti, G. and Mayer, E. (eds.) (1974) *Reciprocidad e Intercambio en los Andes Peruanos*, Instituto de Estudios Peruanos, Lima
Alberts, T. (1978) 'The Underdevelopment of Agriculture in Peru, 1950-75', *Research Policy Program University of Lund*, Discussion Paper no. 121
Alvarez, E. (1974) 'La Agricultura Alimenticia Peruana, 1960-70', *Tesis de Bachiller en Ciencias Sociales con Mención en Economica*, Universidad Católica del Perú, Lima
— (1979) 'Política Agraria y Estancamiento de la Agricultura, 1969-77', *I Seminario sobre Agricultura y Alimentación*, Universidad Católica del Perú, Lima
Amat, C. and Curonisy, D. (1979) 'El Consumo de Alimentos en el Perú y sus Efectos Nutricionales 1972', *I Seminario sobre Agricultura y Alimentación*, Universidad Católica del Perú, Lima
— and León (1979) *Estructura y Niveles de Ingreso Familar en el Perú*, Universidad del Pacífico, Lima
Arce, E. and Mejia, J. (1975) 'Algunas Consecuencias del Proceso de Reforma Agraria Sobre el Campesinado de la Provincia de Cajamarca', mimeo., Instituto de Estudios Peruanos, Lima
Auroi, C. (1980) 'Contradictions et Conflits dans la Reforme Agraire Peruvienne', *Itineraires Notes et Travaux*, 7, Institut Universitaire d'études du developpement, Geneva
Barraclough, S. and Collarte, J.C. (eds.) (1972) *El Hombre y la Tierra en América Latina*, Editorial Universitaria, Santiago
Burga, M. (1976) *De la Encomienda a la Hacienda Capitalista*, Instituto de Estudios Peruanos, Lima
Caballero, J. M. (1975) 'Aspectos Financieros en las Reformas Agrarias: Elementos Teóricos y Experiencias Históricas en el Perú', *Serie Documentos de Trabajo CISEPA*, 25, Universidad Católica del Perú, Lima
— (1976) 'Reforma y Reestructuración Agraria en el Perú', *Serie Documentos de Trabajo CISEPA*, 34, Universidad Católica del Perú, Lima
— (1978a) 'Los Eventuales en las Co-operativas Costeñas Peruanas', *Economica*, *I* (2)
— (1978b) 'La Reforma Agraria y Más Allá: Del Fracaso del Modelo Agrario del Régimen Militar', *Critica Andina*, 2
— (1979) '*La Economía Agraria de la Sierra Peruana en los Albores de la Reforma Agraria*', typescript, Instituto de Estudios Peruanos, Lima
— (1980) *Agricultura, Reforma Agraria y Pobreza Campesina*, Instituto de Estudios Peruanos, Lima

Campaña, P. and Rivera, R. (1978) 'El Proceso de Descampesinización en la Sierra Central del Perú', *Estudios Rurales Latinoamericanos, 1* (2)

Castro-Pozo, H. (1947), *El Yanaconaje en las Haciendas Piuranas*, Compañia de Impresiones y Publicidad, Lima

Caycho, H. (1977) *Las SAIS de la Sierra Central*, ESAN, Lima

CELADE (1979) 'América Latina: Población Urbana y Rural', *Boletín Demográfico, 12* (24)

CENCIRA (1976) *Los Eventuales y los Mercados de Trabajo en la Agricultura*, CENCIRA, Lima

— (1977) *Comunidades Campesinas*, CENCIRA, Lima

Chevalier, F. (1966) 'L'Expansion de la grande Propriété dans la Haut-Pérou au XXe siècle', *Annales ESC, 21* (4)

Choy, E. (1980) 'Cuadros Básicos', *I Seminario sobre el Problema del Empleo en el Perú*, Universidad Católica del Perú, Lima

CIDA (1966) *Tenencia de la Tierra y Desarrollo Socio-Económico del Sector Agrícola: Perú*, Unión Panamericana, Washington, DC

Eckstein, S. *et al.* (1978) *Land Reform in Latin America: Bolivia, Chile, Mexico, Peru and Venezuela*, World Bank Staff Working Paper no. 275, Washington, DC

Favre, H. (1977) 'The Dynamics of Indian Peasant Society and Migration to Coastal Plantations in Central Peru' in K. Duncan and I. Rutledge (eds.), *Land and Labour in Latin America*, Cambridge University Press, Cambridge

Figueroa, A. (1976) *Estudio por Países sobre el Empleo Rural: Perú*, ILO, Geneva

— (1978) 'La Economía de las Comunidades Campesinas: El Case de la Sierra sur del Perú', *Serie Documentos de Trabajo CISEPA*, 36, Universidad Católica del Perú, Lima

— (1979) 'Política de Precios Agropecuarios e Ingresos Rurales en el Perú', *Serie Documentos de Trabajo CISEPA*, 45, Universidad Católica del Perú, Lima

Fioravanti, E. (1974) *Latifundio y Sindicalismo Agrario en el Perú*, Instituto de Estudios Peruanos, Lima

García-Sayán, D. (1977) 'La Reforma Agraria Hoy' in H. Pease *et al.*, *Estado y Política Agraria*, DESCO, Lima

— (1980) 'Luces y Sombras de las Tomas de Tierras', *Que Hacer*, 3

Gonzáles, A. and Munaiz, J.A. (1979), 'Crisis en la Industria Azucarera', *IV Seminario Nacional Problemática Agraria Peruana*, Universidad Nacional Técnica de Cajamarca

Gonzáles, E. (1979), 'La economía de la Familia Comunera', *Serie Documentos de Trabajo*, 39, Universidad Católica del Perú, Lima

Greaves, T (1968), 'The Dying Chalán', unpublished PhD dissertation, Cornell University, Ithaca

Griffin, K. (1976) *Land Concentration and Rural Poverty*, Macmillan, London

Handelman, H. (1975) *Struggle in the Andes*, University of Texas Press, Austin

Harding, C. (n.d.) 'Agrarian Reform and Agrarian Struggles in Peru' in R. Miller, *et al.* (eds), *Social and Economic Change in Modern Peru*, Monograph Series, no. 6, Centre for Latin-American Studies, University of Liverpool

Haudry, R. (1978) 'El Crédito Agropecuario en el Perú 1966-76', Tesis de Bachiller en Economía, Universidad Católica del Perú, Lima

Havens, E. (1976) 'Hacia un análisis de la estructura agraria peruana', *Departamento de Ciencias Sociales*, Universidad Católica del Perú, Lima

Hopkins, R. (1979) 'La Producción Agropecuaria en el Perú, 1944-1969', *Serie Documentos de Trabajo CISEPA*, 42, Universidad Católica del Perú, Lima

Horton, D. (1976) 'Haciendas and Co-operatives: A Study of Estate Organisation, Land Reform and New Reform Enterprises in Peru', unpublished PhD dissertation, Cornell University, Ithaca

238   *Agrarian Reform in Peru*

Hunt, S. (1974) 'The Economics of Haciendas and Plantations in Latin America', *Research Papers in Economic Development*, Woodrow Wilson School, Princeton

Janvry, A. de and Garramón, C. (1977) 'The Dynamics of Rural Poverty in Latin America', *The Journal of Peasant Studies*, 4 (3)

Kapsoli, W. (1977) *Los Movimientos Campesinos en el Perú 1879-1965*, Delva Editores, Lima

Klaren, P. (1977) 'The Social and Economic Consequences of Modernisation in the Peruvian Sugar Industry, 1870-1930' in K. Duncan and I. Rutledge (eds.), *Land and Labour in Latin America*, Cambridge University Press, Cambridge

Lovón, G. *et al.* (1979) 'Reforma Agraria y Comunidades Campesinas: El Caso de Anta', *IV Seminario Nacional Problemática Agraria Peruana*, Universidad Nacional Técnica de Cajamarca

Maletta, H. (1978a) 'Perú. ¿Pais campesino?' *Análisis*, 6

— (1978b) 'El subempleo en el Perú', *Apuntes*, 4 (8)

— (1978c) 'La Absorción de Mano de Obra en el Sector Agropecuario', *Seminario Empleo y Población en el Perú*, Asociación Multidisciplinaria de Investigación y Docencia en Población, Lima

— (1980) 'Volúmen y evolución del empleo agricola', *I Seminario sobre el Problema del Empleo en el Perú*, Universidad Católica del Perú, Lima

— and Foronda, J. (1979) 'Acumulación de Capital en la Agricultura Peruana', *IV Seminario Nacional Problemática Agraria Peruana*, Universidad Nacional Técnica de Cajamarca

Martínez, H. (1977) 'La Reforma Agraria en el Perú: Las Empresas Asociativas Andinas', *II Seminario Problematica Agraria Peruana*, Universidad Nacional de San Cristóbal de Huamanga

Martínez-Alier, J. (1977) 'Relations of Production in Andean Haciendas: Peru' in K. Duncan and I. Rutledge (eds.), *Land and Labour in Latin America*, Cambridge University Press, Cambridge

Matos Mar, J. (1976) *Yanaconaje y Reforma Agraria en el Perú*, Instituto de Estudios Peruanos, Lima

— (ed.) (1976) *Hacienda, Comunidad y Campesinado en el Perú*, Instituto de Estudios Peruanos, Lima

—— and Carbajal, J. (1974) *Erasmo, Yanacón del Valle de Chancay*, Instituto de Estudios Peruanos, Lima

— and Mejía, J.M. (1980) *Reforma Agraria: Logros y Contradicciones, 1969-79*, Instituto de Estudios Peruanos, Lima

Mejía, J. M. (1979) 'Los Eventuales de Valle de Chancay', mimeo., Instituto de Estudios Peruanos, Lima

Miller, S. (1964) 'The Hacienda and Plantation in Northern Peru', PhD dissertation, Columbia University, New York

Montoya, R. *et al.* (1974) 'La SAIS Cahuide y sus Contradicciones', mimeo., Universidad Nacional Mayor de San Marcos, Lima

Murra, J. (1975) *Formaciones Económicas y Políticas del Mundo Andino*, Instituto de Estudios Peruanos, Lima

Pásara, L. (1978) *Reforma Agraria: Derecho y Conflicto*, Instituto de Estudios Peruanos, Lima

Pease, H. (1977) 'La Reforma Agraria Peruana en la Crisis del Estado oligárquico' in H. Pease *et al.*, *Estado y Politica Agraria*, DESCO, Lima

Peek, P. (1978) 'Agrarian Change and Rural Emigration in Latin America', World Employment Programme Research Working Paper, ILO, Geneva

Petras, J. and LaPorte, R. (1971) *Perú: Transformación Revolucionaria o Modernización?*, Amorrortu Editores, Buenos Aires

Ravines, C. (1979) 'Estructura Agraria de la CAP "Ucupe" ', *Realidades* (Chiclayo), 3

Revesz, B. (1979) 'Movilizaciones Campesinas en Piura y Comercialización Estatal del Algodón', *IV Seminario Nacional Problemática Agraria Peruana*, Universidad Nacional Técnica de Cajamarca

Ríos, J. R. (1979) 'La Crisis Agraria en la Industria Azucarera', *IV Seminario Nacional Problemática Agraria Peruana*, Universidad Nacional Técnica de Cajamarca

Roca, S. (1975) *Las Cooperativas Azucareras del Perú*, ESAN, Lima

Roquez, G. (1978) 'La Agricultura Peruana: Estadisticas Agrarias 1950-1968', *Serie Ensayos Generales*, 1, Taller de Estudios Andinos, Universidad Nacional Agraria

Rubin, E. (1977) *Las CAPs de Piura y sus Contradicciones*, CIPCA, Piura

—— (1978) *¿Qué Piensa el Campesino de la Reforma Agraria? Caso Piura*, CIPCA, Piura

Saldívar, R. (1978) *Elementos para un Enfoque General de la Reforma Agraria Peruana*, n.p., Lima

Sánchez, R. (1979) 'Capitalismo y Persistencia del Campesinado Parcelario: El Caso de la Sierra Central', *IV Seminario Nacional Problemática Agraria Peruana*, Universidad Nacional Técnica de Cajamarca

Scott, C. (1976) 'Peasants, Proletarianisation and the Articulation of Modes of Production: The Case of Sugar Cane Cutters in Northern Peru, 1940-69', *Journal of Peasant Studies*, *3* (3)

—— (1978) 'Machetes, Machines and Agrarian Reform: The Political Economy of Technical Choice in the Peruvian Sugar Industry, 1954-74', *Monographs in Development Studies*, 4, University of East Anglia

—— and Manrique, M. (1973) 'El Problema de los Eventuales en la Agricultura de la Costa Norte', *Serie Materiales de Trabajo del Taller de Investigación Rural*, 4, Universidad Católica del Perú, Lima

Smith, C. (n.d.) 'Agrarian Reform and Regional Development in Peru' in R. Miller *et al.* (eds.), *Social and Economic Change in Modern Peru*, Monograph Series no. 6, Centre for Latin-American Studies, University of Liverpool

Valcarcel, M. (1978) 'Economía campesina, economía empresarial y reforma agraria', *Serie Documentos de Trabajo*, Taller de Coyuntura Agraria, Universidad Nacional Agraria

Valderrama, M. (1976) *7 Años de Reforma Agraria Peruana 1969-1976*, Fondo Editorial Universidad Católica de Perú, Lima

Webb, R. and Figueroa, A. (1975) *Distribución del Ingreso en el Perú*, Instituto de Estudios Peruanos, Lima

6 AGRARIAN REFORM AND STRUCTURAL CHANGE IN CHILE, 1965-79

Leonardo Castillo and David Lehmann

I Introduction

In 1965 Peter Dorner, an American agricultural economist who was working in Chile, wrote an unprecedented 'Open Letter' to the Chilean landlords, in which he argued as follows:

> if the present income generated by the agricultural sector is distributed in such a way that a substantially larger share goes to the workers, then these workers will spend more on consumer goods of a kind that Chile can produce internally. The landlord, it is reasoned, would spend much more in luxury consumption and imports or possibly send dollars abroad. The traditional pattern of the large rural properties has to be broken. Labourers will then have higher incomes and the power of the landlords in other activities will be reduced. Greater consumption expenditures on internally produced goods would encourage investment in Chilean industries. Investment in Chilean industries would in turn create more jobs and increase the demand for food as well as for internal products. The greater income and opportunity for decision making would then make possible the increased education of rural people. Given credit and education, these workers would gradually become better entrepreneurs and agricultural production would rise. In this way the economic development which everyone seems to want would come about.[1]

Seen from the vantage point of 1980, the letter seems naïvely optimistic, but it expresses well the climate of opinion of the time both in Chile and in the 'aid community'. Dorner's mistakes were to assume, first, that industrial investment follows smooth curves and responds to demand emanating from low-income sectors fairly rapidly, and, secondly, that the problem of reorganising agricultural production could be solved by magic. Indeed, this latter problem received little mention in his letter. Yet these were the two major problems which bedevilled the agrarian reforms of both the Frei government (1964-70)

240

and the Allende (or UP – Unidad Popular) government (1970-3). In each case the government initiated at the beginning of its period a rapid rise in lower-class incomes: rural wage earners in Frei's case, and rural and urban wage earners in Allende's. In each case the expansion of demand first generated an expansion of industrial production, especially of finished goods such as textiles, as industrialists utilised unused capacity and increased their overall profits while prices remained fairly stable, and in each case this was followed by a crisis (more pronounced under Allende than under Frei) in which industrial production reached a ceiling while demand continued to expand, thus initiating once again the old inflationary cycle. Food prices were held down during both periods by direct controls and subsided imports. This, among other factors, produced balance of payments pressures which in their turn fed back into price inflation.[2] The Frei government adopted a 'crawling peg' devaluation strategy, but by the end of its period inflation was nevertheless returning to the 30 per cent mark. The Allende government, faced with similar difficulties on a magnified scale (shortfalls in domestic food production and rapidly expanding demand) could not hold inflation down at all.

An OAS report described Frei's policy succinctly:

an intelligent administration of price controls, a more efficient use of productive capacity, the rise of copper prices on the world market, and two good harvests, allowed a reduction of unit, but not of total, profits at the same time as a considerable rise in real wages, a substantial reduction of the rate of inflation, and a significant increase in the national product. But this threefold result could not repeat itself indefinitely because the margins of unused capacity grow narrower as they are used up, profits have a limited elasticity and good harvests are followed by years of low agricultural yields.[3]

In 1965, the rural minimum wage was raised to the level of its urban equivalent; it was raised by 67 per cent as against an increase in the official cost-of-living index of 38.4 per cent. Moreover, due to the availability of unused capacity, the Frei government was able to expand the economy and to reduce the rate of inflation at the same time. But once the point of full capacity utilisation was reached, it had to decide whether to pursue structural change by taking over in part the task of industrial investment itself and by imposing fairly strict planning and perhaps even a rationing system, or whether to have recourse to more 'orthodox' techniques of deflation through credit restrictions and the reduction of the rate of growth. The Frei regime chose the latter option.

The first year of the Unidad Popular was, in these respects, very similar. Having come to power in November 1970, the government awarded wage increases in January 1971 with the intention of favouring the lower strata of the wage earners in particular.[4] By this time, the economy was back into stagflation, and inflation was running at over 30 per cent while 40 per cent of industrial capacity lay idle. During 1971 GNP rose by 8 per cent while industrial production rose by 15 per cent and the share of wages in national income rose from an average of 51.7 per cent in 1960-9 to 59 per cent. These trends could not be sustained, however, and soon the Allende regime was faced with the same choices as its predecessor. This is not the place to go deeply into the macro-economic problems of the Allende period; suffice it to say that where Frei had retreated to orthodox counter-inflationary policies the UP government found itself unable to impose either orthodox stabilisation or a more radical response in the form of rationing. Demand continued to spiral, exercising pressure on the balance of payments and frustrating the government's attempts to control prices by administrative means. The early spurt of growth soon gave way to stagnation, even decline.

Under neither of these governments did industry or agriculture respond in the manner predicted by Dorner. Industrialists, of course, lost confidence and ceased to invest under Allende. In agriculture, the new organisational forms established in the state sector and the agrarian reform co-operatives inevitably took time to settle down. In any case, the assumption that agricultural production could be rapidly reorganised and could readily respond to increases in demand proved to be hopelessly optimistic.

II The Urban Character of Chilean Society

When the 'lessons' of Chile are searched, it must be recalled that this is a highly urbanised society in which any programme of agrarian reform is bound to interact very significantly with patterns of urban food demand in particular. Furthermore, it is a highly dependent economy. Food imports accounted for 11.7 per cent of total imports in 1965-70, 15.7 per cent in 1971, 33.1 per cent in 1972, and (in spite of a compression of internal demand) 9.9 per cent in 1977; copper accounted for some 75 per cent of exports until the military government created conditions for the development of 'non-traditional exports' and the price of the metal continued downwards, whereupon its share declined steadily to 50 per cent in 1978. The relationships between urban food subsidies, the price of copper, the balance of payments and domestic price index constituted a delicately balanced network, and the Chilean economy had for some time lurched violently from crisis to crisis – from demand-led inflations (reinforced by structural bottlenecks) to severe state-induced deflations. The other special feature was the remarkable degree of organisation of the workers. True, the rural workers only began to organise themselves on a substantial scale after 1964, and the industrial unions were fragmented in small plant units[5] (with the exception of the copper miners); but compared with their counterparts in most other developing countries, the Chilean workers were remarkably well organised.

Chilean politics, and the Chilean economy, were thus driven above all by forces based in the cities of Santiago and Concepión, and in the mines. This had been the case for a long time. Even before the nitrate boom in the last quarter of the nineteenth century, and the later development of copper, the country had experienced a process of industrialisation,[6] and when nitrate extraction speeded up in the northern desert after the War of the Pacific in 1879 a large-scale movement of labour occurred from the central agricultural zone to the mines. Already in 1920 scarcely more than half the population lived in rural areas, and in 1940 agriculture's share of national income was already below 15 per cent (see Table 6.1) – hardly the stereotype of an underdeveloped country. Since then, however, neither agriculture nor manufacture nor mining absorbed much, if any, of the growth in the labour force: the share of manufacturing, mining and construction in the labour force actually declined from 28.7 per cent in 1952 to 24.5 per cent in 1970, and manufacturing itself accounted for only 15.9 per cent in 1970. The burgeoning 'services' sector had to absorb almost the whole of the increase in the labour force and unemployment, both rural and urban, had become a major social problem.[7]

III The Problem of Slow Growth in Agriculture and the Various Interpretations Thereof

Agricultural production – and especially the production of wheat – had long been experiencing a slow rate of growth. From 1939 to 1964 the sector grew only by 53.1 per cent or by 1.72 per cent per year.

Having once been a net provider of foreign exchange, the sector became a net burden: the balance of trade in agricultural products turned negative in 1942 and this deficit increased steadily until it reached $124 million in 1963.[8] *Per capita* wheat production, after rising from 155 kg. per annum in 1909-10 to 180 kg. in 1921-5, declined steadily to 150 kg. in the early sixties.[9] Wheat has for long been a staple in Chile – unlike the more northerly Andean countries where it has only recently become an important component of the popular urban diet.

Several theories were elaborated to account for this poor agricultural performance in a country endowed with fertile land and a favourable climate, at least in the Central Valley (stretching from Aconcagua in the north to Ñuble in the south). One thesis which was eventually to dominate Chilean politics between 1964 and 1973 started out from an analysis of inflation which it explained largely in terms of the slow growth of agricultural production.[10] But the analysis went further, and explained that slow growth in terms of the landowners' unwilling-

Table 6.1: The Agricultural Sector in the National Economy

	Share of Income	Share of Employment	Share of Male Labour Force	Rural Share of Population
1940	14.9	37.3	43.5	47.5
1950	13.2	32.2	n.a	41.3
1960	11.3	30.7	34.0	34.7
1970	9.2	23.2	n.a	24.0

Note: All figures are percentages.
Sources: First two columns: Markos Mamalakis, *Growth and Structure of the Chilean Economy* (Yale University Press, New Haven and London, 1976), p. 129. Third column: Jeannine Swift, *Agrarian Reform in Chile* (Heath, Lexington, Mass., 1971), p. 12. Fourth column: Patricio Mueller and Carol Rahilly, 'Characteristics of the Labour Force in Chile' in Juna J. Buttari (ed.) *Employment and Labour Force in Latin America* (ECIEL and OAS, 1979), vol. II. The authors show an overall drop in the rural population between 1940 and 1970 from 2.4 million to 2.1 million.

ness to invest,[11] and this, in its turn, was explained by the agrarian structure itself: the unequal distribution of land which arose from the history and functioning of the institutions of tenure.[12]

The Central Valley is suitable for intensive production of fruit, vegetables and grapes for wine, yet it was largely given over to wheat production and cattle pasture. This was clearly brought out in the CIDA report which also showed that family units (i.e. those with enough land to maintain a family) produced 30 per cent more per arable hectare than medium multi-family units, employing up to 12 permanent workers, and 50 per cent more than those with more than 12 workers. Sub-family units, defined as not capable of fully employing a family's work-force, produced three times as much per arable hectare as even the family units.[13] The overall inverse relationship between farm size and intensity of land use stood out clearly from the data as a whole.

The conclusion from these various considerations was that agricultural stagnation (and economic stagnation in general), as well as the inflationary process, could be reversed by a redistribution of land in Chile. The precise form or mechanism of redistribution, and above all the precise or even the broad outlines of the institutions of production to follow upon such a reform were, significantly, not considered in detail by CIDA (whose authors may have considered such matters beyond their brief). Agrarian reform would transfer land from those who used it extensively to those who used it intensively. This would, at the same time, expand the internal market for manufactures through raising the incomes of the new property holders. Growth of production and market expansion would fuel economic growth in general and reduce inflation. This was known as the structuralist position.

There were other views, of course. One of these, sustained by the school of thought whose ideas eventually came 'to power' in 1973, explained the sluggish performance of agriculture in neo-classical terms. According to this thesis, the sector suffered from severe discrimination by the strategy of import substituting industrialisation followed in Chile ever since the creation of the Development Corporation (CORFO) in the 1930s. This strategy maintained an artificially high rate of exchange which discouraged agricultural exports and encouraged agricultural imports, thus protecting domestic industry at the expense of the agricultural sector. Government policy frequently fixed low prices for urban food products and sustained them with cheap or subsidised imported food, thus doubly punishing the sector with high prices for (locally manufactured) inputs and low prices for its wheat and cattle products.[14]

One difficulty with this theory was that it emphasised the negative effect on farmers of 'highly' priced domestically manufactured inputs as if farmers depended on these. In fact, however, a high, probably overwhelming, proportion of manufactured inputs in Chilean agriculture was imported, and thus subsidised by the overvaluation of the currency. The theory also overlooked the fact that a further subsidy was provided by cheap and fairly easy credit. The one point on which this theory was on firm ground was that concerning low food prices, which were undoubtedly held down by governments. However, the extent to which the large farmers were in fact responsive to prices was a further subject of controversy. Floto[15] concludes from various studies on the subject that while the cropping pattern was responsive to changes in relative prices, it is not clear that the total cropped acreage responded at all to price changes. The implication is that a policy of improving farm prices need not necessarily induce large farmers to increase their overall production.

For the proponents of the neo-classical view, agrarian reform was inconsistent with the capitalist system, while the structuralists saw in it one key to achieving a seemingly elusive capitalist development. A third school agreed with the structuralist analysis, but was critical of reform for the very reason which impelled structuralists to support it. Its effects, according to the third school, would be only superficial and they might, by partial or illusory improvements, broaden the support for a discredited system among the masses.[16]

For politicians, agrarian reform presented possible solutions to two problems: one was the hope — all else having failed — of putting an end to the wild gyrations of the Chilean economy. Runaway inflation was interspersed by periods of severe deflation as governments came under conflicting external and internal pressures. It seemed that as soon as a process of growth was initiated with the various subsidies to imports that it entailed, the balance of payments came under pressure and the state held out until forced into massive devaluation-cum-deflationary measures, which in their turn laid the basis for a new but short-lived spurt of growth. One factor in this vicious circle was the unresponsiveness of food production to increases in demand, and the consequent pressures on the balance of payments as food was imported at preferential exchange rates. Perhaps an agrarian reform would reduce this vulnerability. Furthermore, a redistribution of land would hurt only the interests of a numerically small landed elite but might attract the political support of the rural proletariat and the peasantry who were beginning to free themselves from the grasp of the landlords in the early sixties.[17]

But much depended on the extent to which one could assume that structural changes in the countryside would not lead to immediate and serious dislocations in other sectors.[18] For if agrarian reform led to too much of a shortfall in production in the short term, or if its beneficiaries increased their consumption much more quickly than they increased their production, then dangerous instabilities might develop, with renewed inflationary pressures and social conflicts jeopardising a reform programme which would inevitably raise popular expectations.

IV Tendencies in the Agrarian Structure
on the Eve of Agrarian Reform

A comparison of the 1955 and 1965 agricultural censuses shows that a process of structural change was under way in the organisation of production in Chile's agriculture. The 'classic' *hacienda* had been a decentralised enterprise with a large resident peasant population, known generally as *inquilinos*, who would work every day, including Sundays, from dawn to dusk in return for a small money wage and access to a piece of *hacienda* land. But between 1955 and 1965, on farms over 50 hectares in the ten provinces of the Central Valley, total land allocated to permanent employees declined by 14.9 per cent, reaching an average of only 4.6 per cent of total arable land in 1965.[19] The number of workers with such rights had also declined by 5.2 per cent in the period; in 1965, they constituted only 31.9 per cent of the permanent employees on estates of over 50 hectares and a quarter of these, consisting of office and supervisory personnel, were not strictly members of the rural proletariat.[20] Taking the agricultural sector in the country as a whole, Billaz and Maffei estimated that in 1965-7 out of 691,900 members of the agricultural labour force, permanently employed wage earners constituted 28.1 per cent, landless workers 21.5 per cent, unpaid family workers 26.7 per cent, and smallholders (with less than enough land to satisfy their family requirements) 23.7 per cent.[21] If double-counting in the 1965 Census is allowed for[22] then the total would decline to 637,000 and the proportion of permanent employees would rise to 30.5 per cent, but even so it is clear that the figure of the *inquilino* no longer occupied the (quantitatively) dominant position which it did in, say, McBride's account written in the 1930s.[23] Moreover, the larger estates were expanding the landlord enterprise at the expense of the peasant enterprise within their boundaries, which in itself would be evidence of some intensification of their agricultural activity. This also implied that the estate labour force was becoming proletarianised in the sense that an increasing percentage of employees of the large estates was losing access to land, and that the numbers of both landless workers without secure employment and of smallholders dependent on temporary wage labour were probably increasing. These changes had profound implications for the functioning of post-reform productive institutions in agriculture, to which we shall return, but for the moment it is important to note them as a sign that the internal structure of the large estates was changing, and so was the social composition of the landed elite.

There are indications too that the large estates may have been modernising rapidly during this period. Certainly, Frei's election to power was followed by a spurt in investment in agriculture. The research of Wayne Ringlien[24] shows a marked process of investment by large farmers during the early years of the Frei government, when the annual rate of growth of agricultural production rose from 1.6 per cent (the rate for 1956-64) to 2.8 per cent (between 1965 and 1970).[25] Most, if not all, of this improvement must have been due to the performance of the sectors which remained unaffected by Frei's agrarian reform, since the reformed sector only accounted for some 30 per cent of arable land even by the end of the period. These farmers presumably increased their production in response to cheap and rapidly expanding credit for all farmers, and perhaps to the government's proclaimed policy of expropriating only those properties which were undercultivated. However, much of the credit for investment was for semi-industrial activities, such as pig-breeding and battery hens, and these enterprises were also largely immune to expropriation. At worst, those expropriated could expect to be reimbursed in full for the value of the capital stock and buildings. Thus while nascent agricultural capitalism was showing some dynamism, wheat production declined at an annual rate of 0.2 per cent between 1966 and 1970.[26]

It is often claimed that in the mid-twentieth century, the landed, industrial and financial elite were in effect one class, united by family ties. Research on bankers and high executives in industry has shown that 42 per cent of the former and 31 per cent of the latter were either landowners themselves or counted landowners among their close relatives.[27] This implied that the power of the landed elite stretched across different sectors of economic and political activity, and that the content of that power was oligarchic in the strict sense that it relied on ties of personal dependency. The emergence of an important stratum of 'yeoman' farmers, on the other hand, meant the rise of farmers who had few non-agricultural sources of income at their disposal. The owners of the largest estates were at the same time politicians, lawyers, diplomats and so on, and could do without their estates, but the emerging modern farmers were largely dependent on agriculture. The same was true of the farmers of the southern Frontier and Lake regions where the great estate, as a specific type of production enterprise, had never existed at all.[28]

On the eve of Frei's reforms, then, Chilean agrarian society was not characterised only by a confrontation of landlords and peasants, although this certainly was the most striking feature. There were 'old' and 'new' landlords, 'old' tenant labourers (or *inquilinos*) and 'new' permanent workers with no access to land, landless workers with neither secure employment nor land, and a mass of land-poor small-holders, or *minifundistas*. Whereas in the nineteenth century the rural poor had been stratified largely in a hierarchy of relations of personal dependency reaching from the landlords down through a chain of rich, middle and poor peasants, they were now gradually being brought under the more impersonal dominance of markets. The situation of labour shortage, which had originally obliged landlords to give land to workers in order to obtain their services, was also giving way to a situation of labour surplus. The transformation was evidently incomplete; one-third of workers on the estates still enjoyed access to land, and they constituted something of an elite, with living standards superior to those of the majority of *minifundistas* outside the estates, the workers who had secure employment but no access to land, and temporary land-less labourers. These divisions were to influence the nascent rural labour movement in the sixties. The independent smallholders themselves were differentiated by wealth in the Central Valley in particular, and also regionally. Their problem was evidently very different from that of the partially and completely proletarianised workers, and to conciliate all these claims by distributing land was to prove more difficult than any-one had foreseen in the halcyon days of reformism.

V Land Reform under the Christian Democratic Government: 1964-70

When the Frei government came to power and committed itself to carrying out agrarian reform it was faced with three basic choices. It might have decided to redistribute parcels of land from the great estates primarily to the landless. Such a policy was feasible, but it seemed risky, given the potentially powerful pressures from the urban classes in an economy heavily dependent on food imports. It was likely to involve a massive dislocation of technical relations of production in the countryside as the use of estate land would have changed from exten-sive wheat and cattle production (in the Central Valley of the country) to crops which would be cultivated more intensively, but were not the staple diet of the urban population. It might have involved massive administrative complications and social conflicts since the bureaucracy

would have been under an obligation to allocate parcels to more claimants than it could satisfy: the permanent but landless workers on
estates, the land-poor *minifundistas* outside the estates, and the landless
without permanent employment.

Alternatively, the government might have chosen to direct its attention to the *minifundistas* in areas of small holding, and could seek to
distribute land to them in a bid to make them viable production units.
The difficulty here was that it would have led to endemic conflicts with
those workers already employed on the estates who took it for granted
that any distribution would be in their favour. Furthermore, relations
between the two groups were not likely to be free of conflict, since it
was the unorganised landless or land-poor from small-holding areas who
provided seasonal cheap labour to the estates and thus undercut the
wages of the permanently employed.

Finally, the government could seek to transform the estates into
some form of co-operative and thus avoid as far as possible a disruption
of the technical conditions of production, while redistributing the fruits
of that production. This was the solution adopted in 1966, and was
justified on grounds of preserving the 'economies of scale' of the
haciendas — an argument which stood in evident contradiction to the
standard criticism which claimed that the *haciendas'* monopoly of land
prevented them from using it efficiently, yet which also underlined a
fear of a sudden fall in agricultural production.

The restructuring of agricultural production, moreover, was not
exclusively a problem of reorganising relations of production 'on the
ground'. It required changes in the economic environment in which
agricultural production was carried out: price relations, the distribution
of labour between the different types and sizes of holding, the allocation of resources within that sector, and between agriculture and
industry, as well as foreign trade policy. This range of issues, more
macro-economic in its implications, was not really confronted by the
Christian Democratic government. Neither the agricultural economy
nor the structural changes which the government was carrying out were
integrated into macro-economic decision-making. Thus, during the
Christian Democratic government, while the urban and industrial
sectors were not subjected to structural reforms or collectivist institutions, agriculture was being restructured through the promotion of co-
operative farming which by 1970 accounted for some 20 per cent of
arable land.[29] Similarly, the political discourse directed at the
campesinos, calling for solidarity and egalitarianism among them, sat
uneasily alongside the scarcely modified individualistic assumptions

which prevailed in other sectors. It was as if all the reforming zeal of the Christian Democratic Party, which came to power in 1964 proclaiming a 'Revolution in Liberty', was concentrated on the 568 farms and 19,000 families which by 1969 belonged to the new co-operatives (*asentamientos*). Yet there were some 637,000 workers in Chilean agriculture in 1965.[30]

The Agrarian Reform Law (No. 16,640), passed after an almost interminable parliamentary debate in 1967, allowed the government to expropriate any rural property of over 80 standardised hectares. (A hectare of very productive land in the Maipo Valley near Santiago was taken as a standard hectare.) The law established conditions of repayment according to the degree of undercultivation but left much to the discretion of officials in this matter. Various other provisions are mentioned below in their appropriate context.

The Christian Democratic government initiated its experiments in new forms of agricultural organisation in the northern Valley of Choapa, one of the very few areas of Chile where there was a long-standing rural union movement. As a result it had difficulty in imposing a new model. At first, the beneficiaries were asked to enter into a sharecropping relationship with the Agrarian Reform Corporation (CORA), but this provoked violent reactions, and so CORA, after negotiating with the unions in Choapa, adopted a system known as the *asentamiento*, which was later consecrated in the 1967 Agrarian Reform Law. The idea was that beneficiaries should form a self-managed company jointly with CORA as a transitional arrangement while their new enterprises became independently viable. These *asentamientos*, technically known as Sociedades Agrícolas de Reforma Agraria, were to be managed by a Council of elected members, plus an official of CORA who would have veto power on decisions. The Council included, as its most prominent members, a President and a Secretary, and one of these generally had to devote almost all his time to managerial activities and to dealings with CORA concerning credit, production plans and the like.

The Agrarian Reform Law stipulated that compensation was to be paid promptly and in full on machinery, buildings and animals when an estate was expropriated, but compensation for the land was much less favourable: 10 per cent in cash and the remainder in bonds which were likely to become rapidly worthless as a result of inflation.[31] The result was that farms were handed over without any capital except the land, and *asentados* had to borrow large sums of money – albeit at low interest rates – in order to have the minimum implements to work the land, and they also needed to borrow annual credits for consumption.[32]

The *asentados* were allowed about 1.5 hectares in private plots in the Central Valley, and more in the southern province of Cautín, apart from generous (and uncontrolled) pasture rights, all of which meant a substantial improvement in the economic position of the workers compared with their previous position in the *haciendas*. But they now needed a regular money income while awaiting the year-end distribution of profits from their collective enterprises. Yet if they were to be members of a collective enterprise such payments could not be called wages: rather they had to be thought of as 'advances' on eventual profits. These advances were financed by cheap loans from CORA,[33] but the problem was how the workers could be persuaded to work together in pursuit of collective profit in the expectation that they would then benefit from the redistribution thereof. Further problems arose with respect to employment: the co-operatives seemed to embody a built-in incentive to keep the number of members (full-time workers) participating in such profits to a minimum, relying for the remaining work on temporary, or even permanent, hired labour. Of course, this latter incentive could hardly be expected to operate if no profits were expected at all, and thus the real attraction of membership lay in the prospect of assured full-time employment and access to a private parcel. But then if the number of full members expanded, the importance of private parcels also expanded, thus undermining the co-operative character of the enterprise.

These problems were not insurmountable and need not have been inimical either to the development or to the fair distribution of the fruits of reform. But the objectives of the programme were themselves ambiguous. The government proclaimed its faith in co-operative production, but the system it established encouraged private production. Indeed, the 1967 Reform Law envisaged an ultimately individualist model, providing for collective organisation only during a transitional period and subsequently only if the beneficiaries so desired.[34] Under the *asentamiento* system private plots co-existed with and were in effect subsidised by the collective, and the land held collectively remained undercultivated now as on the old *haciendas*, since the incentives to work on one's private plot were far greater. The logic of incentives is simply stated. Since the 'advance' paid monthly to *asentados* was not technically a wage, it was based on the number of days worked by each full member of an *asentamiento*, and not on the productivity of the workers, either individually or as a group. The result was that distrust developed when people saw that a laggard earned the same as another who worked conscientiously. It was therefore hardly surprising that *asentados* devoted their most strenuous efforts to private plots and did not pursue the intensification of collective production. Nevertheless, the collective unit was obviously essential to the maintenance of the private plot, since it was the source of monthly money payments, as well as inputs and other help, and it would therefore be wrong to claim that the logic of this behaviour meant that the *campesinos* 'wanted' to subdivide the land. What they presumably 'wanted' was the maximum available income and equal returns to equal work.

The *Diagnóstico*[35] claims, on the basis of CORA's monitoring system (*Registro de Control de Avance*), that in 1972, 13 per cent of land under annual crops on reformed estates was in individual hands, and the annual cultivation plans (*planes de explotación*) showed that each beneficiary possessed an average of 6.9 privately owned animals. Field work in Colchagua in 1972 by Lehmann revealed that in 10 out of 13 farms (unfortunately not randomly sampled) private parcels constituted 10 per cent of irrigated land; when we add pasutre rights to this the result would obviously yield a far higher figure than the 8.3 per cent given for the province of Colchagua as a whole by CORA's *Registro*. On this basis, we may legitimately suspect that the national figure of 13 per cent was also an underestimate.

The importance of private parcels, however, varied in accordance with the special regional characteristics of the pre-reform situation. The Central Valley provinces, from Coquimbo in the north to Ñuble in the south, accounted for the lion's share of production and population, and it was here that the process of reduction of labour tenants to pure proletarian status had gone furthest on the *haciendas*. Further south, especially in Cautín, we come to the areas where the Mapuche Indians are more numerous. The Mapuche were smallholders, and opportunities for wage employment in the areas were few. During the nineteenth century, the State created reserves (*reducciones*) for the Mapuche. This measure radically reduced their access to land. They also lost land subsequently through innumerable illegal unsurpations. The Mapuche took advantage of the reform to carve out private parcels within the reformed units, and also to reserve a high proportion of the collective production for distribution among the membership. In his field work,[36] Lehmann studied six reformed units in Cautín, and, taking both privatisation of land and the distribution of collective production into account, he found that on average 20 per cent (varying from 9.2 per cent to 35 per cent) of the arable land on these units was being used in one way or another for individual enterprises.[37]

Table 6.2: Evolution of Agricultural Minimum Wage

	Wage (escudos)	Annual Increase (%)	Increase in Official Cost of Living Index (%)
1962	0.905		
1963	1.28	41.7	27.7
1964	1.955	52.7	45.4
1965	3.264	67.0	38.4
1966	4.104	25.7	25.9
1967	4.800	14.5	17.0
1968	5.851	21.8	21.9
1969	7.500	28.2	27.9

Note: The figures up to and including 1964 refer to the average legally established minimum agricultural daily wage for the nine Central Valley provinces. From 1965 onwards the legal agricultural minimum wage was uniform throughout the country, and was also placed on the same level as the industrial minimum wage. The cost-of-living index, although it has its limitations, serves as a rough index to show the increase in real agricultural wages. These are only the legal minimum wages; those who were affected by special collective wage agreements experienced an even higher growth of wages, and by 1969 these were some 33 per cent of the agricultural wage labourers.
Source: Banco Central, *Boletín Mensual* (1965 and 1969).

VI Unidad Popular Tries to Reform the Reform: 1970-3

The economic and political programme of Unidad Popular has been documented in many places and will not be repeated here.[38] We have already noted the reactivation of the economy in 1970/1. However, the government did not succeed in restructuring the economic system and was unwilling to impose a draconian deflationary policy, and an economic crisis was soon on the horizon. The retail price index grew by 34.9 per cent in 1970, 53.4 per cent in 1971, 145.6 per cent in 1972, and beyond that subsequently.[39] Chilean society became a whirlpool of political forces, as hitherto dormant social groups (the seasonally employed in agriculture, the Mapuche Indians, urban poor, urban women) mobilised in support of new demands, and the pressure on resources of all kinds became very strong. In these circumstances, the problem of agrarian reform took on an altogether new importance, for the demand for food was enormous and the state was using its dwindling foreign exchange resources in importing it and selling it to the populace at heavily subsidised prices. Food imports, which had cost an average of 11.7 per cent of export earnings in 1965-70, rose to 15.7 per cent in 1971 and 33.1 per cent in 1972, due partly to rising demand, and partly to the increase in import prices (which rose by 8 per cent in 1971 and 41 per cent in 1972). Between 1971 and 1973 wheat production declined at an annual rate of 13.6 per cent.[40]

At the same time the new political situation — with a government in power whose declared aim was gradually to install socialism in the country — transformed the relatively, though not completely, peaceful relationship which had developed between the reformed and the private sectors of agriculture during the previous government's tenure of office. Large private farmers did not have the confidence to invest, and crop output fell by 9.5 per cent in 1971/2. Wheat — an essential staple — grew by 12.3 per cent in 1970/1 and then fell by 19 per cent.[41] So now the government had to advance rapidly — for political and economic reasons — in the sphere of agrarian reform, in order to establish a new order and to set production on a steady path again. But to achieve this it had to decide on the shape of the new system. The tendencies which looked quite harmless under Frei — underemployment on the reformed units, apparent concentration of effort by beneficiaries on their private plots — now became matters of serious concern.

No group within the UP coalition proposed a policy of subdivision of the estates. Certain factions of the Socialist Party[42] proposed wholesale collective agriculture with little space for private plots, while the Communist Party, less eager for conflict with the beneficiaries or with the Christian Democrats, supported a continuation of the *asentamiento* system. In practice the latter view came to prevail. Some factions proposed the creation of Centros de Reforma Agraria (CERA) which would admit a far wider membership, impose more egalitarianism and permit less private parcellisation. In practice, those expropriated farms whose workers wished to show their political support for the UP government called themselves CERA while others remained or became *asentamientos*, but all functioned in much the same way. By 1972 the majority of expropriated estates were being known officially as *Comités Campesinos* because, although they were like *asentamientos* in practice, the bureaucracy did not want to formalise this. A further category consisted of 'intervened' farms, where conflict between workers and their employers had led to a strike or land invasion, and the government had taken over the management — but not the ownership — of the farms. These farms were in practice managed as if they were reformed units but they had no legal status as such.

Apart from the quickening pace of expropriation the major change in the internal operation of the 'reformed units' which occurred under the UP government was the broadening of the criteria for admission to full *asentado* status. Consequently, the number of beneficiaries increased far more than the arable land expropriated. Not only a breadwinner, but also his adult sons, could obtain both secure employment and access to a plot of land on all reformed units, while on CERAs the members' wives, daughters and sons of over 16 years of age were supposed to have voting rights.

At the same time the government implemented policies which had long been advocated as part and parcel of an agrarian reform. There was a redistribution of credit in favour of smallholders,[43] and the state also invervened in or took over some marketing networks, though the latter was a reaction less to the problems of agriculture than to the pressing need to get supplies directly to the urban consumer at low, official prices. The expropriation process advanced with great rapidity, so that by mid-1972 there was hardly anything left for the government to expropriate under the 1967 Law. Frei's administration had expropriated 1,412 farms in six years; within two years Allende's government had expropriated 3,278.[44] But whereas under Allende there were more than twice as many beneficiaries, there was only 50 per cent more irrigated land available for expropriation (see Table 6.3). The pressure

Table 6.3: Area Covered by Expropriations

	Irrigated Hectares	Total Hectares
1965	37 887.9	447 271.0
1966	57 877.3	511 155.9
1967	51 649.5	239 845.0
1968	44 689.5	650 559.5
1969	56 778.1	1 022 668.7
1970	41 493.7	1 701 319.3
1971	177 481.4	2 026 720.7
1972 (30 June)	193 748.5	2 741 250.2
Total	661 831.0	8 860 310.0

Sources: 1965-70, CORA lists; 1970-2 and Total, *Diagnóstico*, Tables III-2, III-3.

for land remained, and the law could not be changed since the government had no majority in the Congress. Had the government had the time or the political space it might still have succeeded in creating effective employment opportunities for all beneficiaries on the undercultivated estates it was taking over, but it was carried along by the pressure of events, and had no quick solution to the problems. Agricultural production was clearly being gravely affected by the disruptions which accompanied great political crises and, by its nature, agriculture resists quick solutions.

VII Employment and Distributional Consequences of the Two Reforms (1965-73)

The Agrarian Reform Law stipulated that the number of beneficiaries on each expropriated estate should be governed by the number of 'family farms' of 8 standard hectares into which the estate could eventually be divided after the transitional period of the *asentamiento* was over. Since full *asentados* were considered to be the potential definitive beneficiaries, who would obtain their own titles to the land (individually or as members of an independent co-operative), this provision might have restricted their number. However, the number of *asentados* was actually higher than the number of permanent workers on the pre-existing estates (and in addition, a substantial number of wage labourers with rights neither to a private plot, nor to a vote in the assembly, nor to a definitive title were employed by the *asentamientos*). Lehmann found that six farms in the Central Valley province of Colchagua for which data were available contained 103 permanent

Table 6.4: Agricultural Credit by Major Lending Institutions, 1965-76 (in millions of escudos at December 1969 prices)

	1967[c]	1970[c]	1973[c]	1976
State Bank[a]	803.5	1 015.1	1 022.9	934.2
CORA and INDAP[a]	146.7	323.6	146.6	100.8
Other[b]	540.0	506.2	197.4	535.4
Total	1 490.2	1 844.9	1 366.7	1 570.4

Notes: a. CORA and INDAP lend to small farmers. In 1971, 1972 and 1973 the State Bank took over the administration of credit to reformed units.

b. Includes CORFO, Central Bank and the private sector, all of which lend to large and medium-sized farmers.

c. Three-year averages.

Source: World Bank, *Chile: An Economy in Transition* (Washington, DC, 1980), p. 448.

Table 6.5: Absolute Numbers of Permanent and Temporary Workers on
Sample of 105 Farms, 1965/6 and 1970/1

| | Permanent Workers | | % | Temporary Workers | | % |
	1965/6	1970/1	Change	1965/6	1970/1	Change
Unexpropriated	1 281	1 312	2.5	2 371	1 540	−35.0
Completely expropriated	1 210	1 489	23.1	971	1 700	85.4
Partially expropriated	784	1 051	34.1	1 145	787	−31.3
Subdivided	933	828	−11.3	1 432	2 416	68.7
Total	4 208	4 680	11.2	5 865	6 443	9.9

Note: The last line should not be taken to represent general trends in the agricultural sector, because not all types of farms are included here.
Source: R. Cortázar and R. Downey, 'Effectos redistributivos de la reforma
agraria' in Estudios de Planificación (CEPLAN, Santiago, 1976), no. 53. There is a
table in D. Stanfield et al., 'The Impact of Agrarian Reform on Chile's Large Farm
Sector' (Development Economics Department, World Bank, June 1975), which
draws, for reasons not given, on 85 of the 105 farms sampled, and gives data
which, while showing similar trends, still differ substantially from those in
Cortázar and Downey. I have used this latter because (a) they present data for the
whole sample, and (b) there seems to be a mistake in the Stanfield paper since
their totals for hired outside workers do not add up.

workers at the time of expropriation, yet in 1972 they had 201 full
members; nine farms in Cautín had 165 permanent workers at the time
of expropriation, and in 1972 had 359 permanent workers. If these
data represent anything at all, they certainly contradict claims that the
reforms reduced employment. Full asentados had rights in land, and
since many of these had no such rights previously, the process of proletarianisation of the rural labour force was evidently reversed. It was
being reversed both in the sense that the decline in the number of
permanently employed was being halted, and also in the sense that
among these permanent employees, access to land also ceased to
decline. René Cortázar and Ramón Downey,[45] on the basis of a sample
of estates studied in 1965/6 and again in 1970/1 (see Table 6.7),
demonstrate that the expropriated estates had experienced a 60 per
cent increase in days worked by permanent employees (full members),
although total labour days (worked by all types of employees) rose by
only 47 per cent. This survey was conducted before the expropriations
carried out in the 1970-2 period had taken their effect; figures would
probably have been even higher if the study had been carried out in,
say, 1972/3.[46]

The hiring of seasonal labour is probably unavoidable when crops vary widely in terms of seasonal labour needs. But recent work on Peruvian co-operatives suggest that there may be a built-in incentive in a co-operative to minimise the number of workers with full participation in profits and hence to rely increasingly on hired seasonal labour; for this tendency was observed even in sugar production which, in Peru, is not heavily seasonal in labour requirements.[47] If the *asentamientos* had operated as intended, such a growing importance of hired labour might indeed have been an unintended consequence, as the *asentados* pursued greater and greater profits for distribution to fewer and fewer people. But, in the case of Chile, since the profits tended in so many cases to be eaten up in debt repayments, such motivation could hardly have existed, and hence there was no effective resistance among the *asentados* to the expansion of the number of full members.

In the early years of the Chilean agrarian reform, a study[48] was carried out on 17 *asentamientos* which purported to show that on average 41 per cent of days worked on them were supplied by hired labourers — but the dispersion of the observations was very wide and this was an early period of the reform. The CORA budget for 1968 showed that 29 per cent of total payments for wages and advances was for hired labour and since the usual daily wage was 20 per cent lower than the usual daily advance to the *asentados*, hired labour must have accounted for about 33 per cent of the days worked.

The UP government tried to reduce what it perceived as inequities by allowing at least those people who were being hired regularly from outside to become full members of the *asentamientos*. In this way, it tried to combat the development of new forms of inequality but probably encouraged underemployment in the process. The most significant new inequality was, in all probability, that which divided the *asentados* and their regular employees from the mass of smallholders and temporary labourers, and neither government could do much to reduce this.

Since the reformed sector accounted for only 10-15 per cent of the agricultural labour force, one must obviously look at other sectors of the agrarian economy in order to understand the full impact of the reforms. Under the 1967 Agrarian Reform Law, an expropriated landowner could keep a 'reserve' of approximately 40 standardised hectares. The Christian Democrats allowed owners to select the location of their reserve, and the owners understandably tended to keep the best land. During the UP regime, when the government was less lenient, owners increasingly appealed to the special Agrarian Tribunals established under the Agrarian Reform Law and often obtained favourable judge-

Table 6.6: Number of Paid Man-days per Farm by Type of Labour, 1965/6 and 1970/1: Sample of 105 Farms of More than 100 Hectares in 1965/6

Farm Status	Days Worked per Permanent Worker in the Agricultural Year		Days Worked per Outside Worker	
	1965/6	1970/1	1965/6	1970/1
Unexpropriated	231	275	35	74
Completely expropriated	223	290	36	10
Partially expropriated	236	292	55	33
Subdivided by owners	194	273	44	24
Total	219	283	41	33

Source: Cortázar and Downey, 'Effectos redistributivos de la reforma agraria'.

ments with respect to the location of their reserves and similar matters.[49] Naturally, these landowners proceeded to concentrate their machinery and other capital resources on the reserves, producing a higher capital-labour ratio and more intensive cultivation than either on the former estates or on the *asentamientos*. The emergence of this type of medium-sized commercial farm was further stimulated when the Christian Democrats put a law through Congress in April 1966 (Law No. 16,465) which prohibited any transfer of landholdings unless permission was obtained from CORA. The object of this law was to prevent pre-emptive subdivisions intended to evade expropriation. At the time, this was thought to have been an effective measure, but a

Table 6.7: Changes in the Use of Different Types of Labour on a Sample of 105 Farms, 1965/6-1970/1: Per Cent Changes in Man-days

	Permanent Workers	Temporary Workers	Total
Not expropriated	22.0	37.3	25.3
Totally expropriated	60.0	−48.5	47.7
Partially expropriated	65.9	−58.7	33.9
Subdivided	24.9	−7.9	16.0
Total	42.0	−11.3	31.1

Source: Calculated on the basis of the figures in Tables 6.5 and 6.6.

survey, carried out in 1970/1 by the Land Tenure Centre of the University of Wisconsin, showed that a large number of landowners had in fact obtained CORA's permission to subdivide. In a sample of 105 farms, those subdivided covered 27 per cent of the holdings, 12.6 per per cent of the total area and 20 per cent of the irrigated area.[50]

The 'medium-sized commercial farms' thus prospered during the Frei regime: they either were unaffected by the reforms or emerged in response to the reforms. Credit was abundantly available for their owners[51] (see State Bank credits, Table 6.4) and expropriation even gave them the opportunity to rid themselves of excess resident labour. The Wisconsin sample shows that by 1970/1 the reserves and sub-divided holdings were more mechanised than both the farms which had remained intact and the *asentamientos*.[52] They were, moreover, mech-anising more rapidly than others.

The process of mechanisation in Chilean agriculture during this period seems to have been accompanied by a substantial increase in agricultural employment on all types of large units (Table 6.7). How-ever, Tables 6.5 and 6.6 show that the actual number of permanent workers increased significantly only on the *asentamientos* and retained reserves, but days worked per year by permanent workers increased on farms of all categories. The increase in the latter was particularly pro-nounced on farms which subdivided in order to escape appropriation. Outside the reformed sector, therefore, permanent workers were having to spend more time working for their landlords and had correspond-ingly less time for themselves on their plots. This related to the fact that in the course of reorganisation, the retained reserves and sub-divided farms were able to reduce access to land among their permanent workers.

Temporary workers, on the other hand, did not benefit in terms of increased employment opportunities except when they were able to become full members of the *asentamientos*. This is clearly suggested by the data in Tables 6.5, 6.6 and 6.7. Between 1965/6 and 1970/1, the total number of temporary workers employed by large farms increased, but the total number of days worked by them actually declined so that the average number of days worked by individual workers declined sub-stantially. However, some interesting differences in trends in employ-ment of temporary workers on different types of farms can be observed. On both unexpropriated farms and retained reserves (in the case of partially expropriated farms), the number of temporary workers employed declined. But while on the unexpropriated farms the average number of days worked by individual workers increased, it declined quite drastically on the retained reserves. There could be two possible reasons for this difference. First, unlike the retained reserves on partially expropriated farms, the unexpropriated farms were unable to reduce access to land among their permanent workers. Thus while the retained reserves chose to expand their permanent work-force (many of the permanent workers no longer had rights to private plots), the unexpropriated farms could only increase their reliance on temporary workers. Second, the unexpropriated farms remained vulnerable to expropriation and were much less inclined to invest in labour-saving machinery than the retained reserves. On wholly expropriated and subdivided farms, the number of temporary workers employed increased quite dramatically, but the number of days worked by individual workers declined sharply. The reasons for these tendencies are, unfortunately, not very clear.

In sum, as far as the reformed units themselves were concerned, there was a reversal of the proletarianising trend among permanent workers; they retained or even increased their access to land and benefited from increased employment opportunities. As far as unexpropriated large estates were concerned, there was an increase in overall employment and this again mainly benefited the permanent workers. On subdivided farms and on the reserves there was only a small increase in the demand for labour, and thus employment opportunities, and this increase also benefited only the permanent workers. Conditions of employment of those temporary workers who were unable to find permanent employment on the *asentamientos* worsened as a consequence of the reforms. This picture, of course, relates to the period 1965/6-1970/1, but there are no strong reasons to believe that the trend was different during the ten later years.

Changes in distribution of wealth and income in the countryside are more difficult to trace in view of a paucity of data. There is little doubt that, but for the agrarian reform, the actual beneficiaries would have continued to be affected by the process of proletarianisation which had been under way for some time in Chilean agriculture. However, the beneficiaries constituted only a minority: under Frei full beneficiaries numbered 20,000 and they reached a maximum of 75,000 by the time the process was brought to a halt in 1973. This latter figure represents just over one-third of the total permanent employees on large farms in 1965, not to mention other sections of the rural poor.

The distribution of income in the countryside may nevertheless have changed significantly. Incomes of a large number of *hacienda* owners were reduced substantially (this is suggested by the changes in the structure of landholding; see Table 6.8) and there evidently was a redistribution in favour of the permanent workers through increases in the rural minimum wage and in employment. It must be noted, however, that the observed changes in the pattern of employment imply a worsening of the distribution of income among the workers themselves. As noted above, it was mainly the permanent workers who benefited from the increased wages and employment opportunities, and thus the income differentials between the permanent and temporary workers must have increased. Independent smallholders benefited from the increased availability of credit on easier terms, but they probably lost in terms of off-farm employment opportunities since they could only work as temporary workers. Income differentials between the permanent workers and independent smallholders may thus also have increased.

On the whole, the distributional consequences of the reforms over the period 1965-73 can perhaps be summed up as follows. The major losers were the *hacienda* owners and the major beneficiaries were the permanent workers. The temporary workers and smallholders (these are not, of course, mutually exclusive categories) were unlikely to have benefited much. Thus, although the overall distribution of income in rural areas improved, the distribution among rural workers may well have worsened. These observations are admittedly rather general and somewhat speculative; unfortunately they must remain so until further evidence becomes available.

VIII Post-1973 Reversals and the Emerging Agrarian Structure

Since its assumption of power in 1973, the military government has taken steps to dismantle the *asentamientos* and has either returned land to the former owners or distributed it to individual beneficiaries. An exact calculation of the scale of the operation is extremely difficult since the 'reformed units' in existence in September 1973 included, apart from expropriated farms formally 'on the books' with CORA, the 'intervened' farms whose management had been taken over by the government but which had not been formally expropriated. There were also farms for which the expropriation process had not been formally completed. With these reservations in mind we can reproduce the data in the official records as of 30 June 1979. These are the records of the 'Oficina de Normalización Agraria', which has replaced CORA, dissolved on 1 January 1979.

Table 6.8: Distribution of Land by Size Stratum in Chile, 1965, 1973 and 1977

Size (standardised irrigated hectares)	No. of Holdings			Amount of Land			Per Cent Holdings			Per Cent of Land		
	1965	1973	1977	1965	1973	1977	1965	1973	1977	1965	1973	1977
Less than 5	189 539	190 000	190 000	199 796	200 000	200 000	81.4	79.3	70.1	9.7	9.67	9.6
5-20	26 877	27 000	58 266	263 377	270 000	557 458	11.5	11.2	21.5	12.7	13.05	26.8
20-40	6 959	8 000	8 000	195 015	240 000	240 000	3.0	3.4	2.9	9.4	11.60	11.5
40-60	2 989	6 000	8 160	146 063	300 000	408 700	1.3	2.5	3.0	7.1	14.50	19.6
60-80	1 715	2 909	2 909	118 553	217 833	217 800	0.7	1.2	1.1	5.7	10.53	10.5
Over 80	4 876	—	1 512	1 144 994	—	127 000	2.1	—	0.6	55.4	—	6.1
Sub-total		233 909			1 227 833			97.6			59.43	
Reformed sector		5 655	2 343[a]		839 983	331 621[a]		2.4	0.9		40.57	15.9
Total	232 955	239 564	271 190	2 067 798	2 067 816	2 082 579						

Note: a. The 'reformed sector' in 1977 consisted of land still in CORA hands due to be 'assigned' to beneficiaries or sold off to private purchasers.

Source: Rodrigo Alvayay, 'Las Cambios Estructurales y la Reforma Agraria Chilena', *Boletin del Grupo de Estudios Agro-Regionales*, no. 4 (Santiago, July-Oct. 1979).

The total amount of land expropriated between 1965 and 1973 was 895,752.1 standard hectares in 5,809 holdings. The government has revoked expropriation decrees referring to 1,641 holdings covering 142,716.5 standard hectares, and it has proceeded to the partial restitution of a further 109,810.9 hectares on 2,172 holdings. This leaves 643,224.7 standard hectares on 4,168 holdings still in the hands of the state or of beneficiaries of the agrarian reform. Thus 28.2 per cent of the land in the reformed sector had been restored in one way or another, by the state, to its former owners. However, if we assume that the land on the 'intervened' farms is not included in these figures, and if we assume also that, in accordance with its proclamation, the government has returned all that land to the former owners, then we can see that the proportion could be higher. We do not know the relative importance of those holdings, but they might well have constituted around 6 per cent of total expropriated holdings prior to 1973.[53]

In accordance with its ideas about the ideal organisation of agricultural production, the new government decided to allocate the remaining land to individuals, including the original beneficiaries, in 'family units'. However, there was not enough land to distribute in such units to even all the 75,000 beneficiaries, and a selection process was therefore initiated. In the first months, numerous people were expelled from the reformed units, while these were taken into direct administration by the government, and the elected officers were dismissed. People who were accused of engaging in violent actions such as farm seizures were removed. Given the atmosphere of fear which prevailed for a very long time after the military took power, it is not difficult to imagine that this was an occasion for settling of accounts in numerous cases both among the peasantry and between *campesinos* and landlords. The points system introduced by the government as part of the selection process itself did not recognise any managerial role that people had played, as members of *asentamiento* councils for example; indeed such a record, like participation in unions, tended to count against them. By contrast, outsiders holding university degrees in agronomy scored high points. Small, and some perhaps not so small, traders also gained access to land by various means.

After the initial removal of an unknown number of beneficiaries through the application of the points system, the government proceeded to demarcate the new family units, and allocate them as private property. Under the Frei and Allende regimes CORA had remained the formal owner of almost all the expropriated land, although some were assigned in individual or co-operative tenure.[54] Initially, the average amount of land assigned per beneficiary was 6.75 standard hectares. By June 1979, when the military had more or less completed the process of land assignment, they had assigned 371,262 standard hectares to 36,746 individuals, yielding 10.1 standard hectares for each. A further 95,865 standard hectares were assigned to co-operatives, but 33,590 of these were in the infertile extreme north of the country. There is also evidence to show that the present government has reassigned the land which had already been assigned in purportedly definitive ownership, to co-operatives.[55]

Of the 75,000 original beneficiaries, 36,746, or 51 per cent, have not received any land. This figure might be said to overstate the loss of access to land if we take into account the fact that, especially during the UP government, members of families apart from the breadwinner obtained full membership in reformed units; but these people had by now become breadwinners (for they were mostly adolescent men at the time).

The World Bank report published in 1980 provides the results of a sample survey conducted by ICIRA, a government agency, in December 1976.[56] The survey covered 181 *asentamientos* on which there had formerly been 4,123 *asentados*. Of these, 2,731 (66.3 per cent) were assigned land in individual tenure, while the remainder had to find work elsewhere; 664 (16.1 per cent) could not be traced and their situation was described as 'unknown'. The survey covered 3,178 assignees on these estates, and 187 (5.9 per cent) of these were not former *asentados* at all.

The question remains whether the new beneficiaries actually have the means to operate efficiently the land they now own. The ICIRA study found that individual farmers had 45.6 per cent of their land under natural pasture, which is a puzzlingly high amount for farmers owning an average of 10 standard hectares.[57] The study also states that 68 per cent of the beneficiaries obtained all the credit they sought, but one suspects that this includes advance payments by traders purchasing their crops for the agencies which were created to lend to small farmers in Chile – CORA, INDAP and the State Bank – and which have had their lending capacity drastically reduced since 1973, in both absolute and proportional terms (Table 6.4).

There is other evidence to support the view that many of these beneficiaries are finding it difficult even to pay for the land, let alone cultivate it. Until April 1980 it was not legally permitted for the post-1973 beneficiaries to sell their land but in a Decree (No. 3,262) published on 24 April 1980, the Ministry of Agriculture laid down norms governing the sale of parcels assigned in the period since 1973.[58] The Ministry stated at the time that, out of 37,000 assignees, some 15 per cent had made technically illegal sales of their parcels. On his return visit to the Santa Cruz area in late 1979 Lehmann found that the sale of parcels was common among beneficiaries, and that the leasing of land under sharecropping arrangements was even more so. Sharecropping arrangements, however, tended to be forms of labour exchange among *campesinos*, rather than landlord-tenant relations with connotations of economic and personal dependency which the latter imply. In four case studies conducted in the area of Linares in 1977/8, Gomez and others found that 52 per cent of the area under cultivation in parcels assigned since 1973 was under sharecropping of this kind, in which no money changed hands.[59] The same study also shows a clear difference in cropping patterns between the reallocated parcels and the land which had remained with or had been returned to the former owners: the parcels tended to devote more land to subsistence crops than the larger units and had far fewer heads of cattle.

In the 1965-73 period, and especially during 1970-3, the beneficiaries of land reform and to some extent the independent smallholders as well, enjoyed various forms of subsidy from the state.[60] Credit was lent at interest rates far below current rates of inflation, and especially in the case of the *asentados*, the state was tolerant of bad debtors. Under the present government, interest rates have been well above the rate of inflation, and the small farmers received little subsidy. Besides, one effect of the post-1973 economic policies has been to reduce demand for basic foodstuffs drastically. The theory which held that the sluggish pace of agrarian change, and the sluggish growth of agricultural production, were the consequences of over-valued exchange rates which discriminated against exports while subsidising imports, and had nothing to do with the agrarian strucure itself, is now dictating policies. Total agricultural production, however, remains very depressed, largely because the production of wheat has displayed a fluctuating but declining trend in production and yields ever since 1970/1.[61] In late 1979 the wheat producers were protesting bitterly, claiming that importers were dumping wheat on the market in connivance with flour-mill owners, and thus undercutting them. Milk producers have also complained bitterly.[62] The price of fertilisers, which had been freed from state control, stands on average 50 per cent above 1970 prices in real terms.

The general impression obtained by an observer of Chilean agriculture is that the only farmers who can really do well are those producing for export. The prices of new export crops, especially fruits, have risen faster than those of other products, and credit is available for such production. Agricultural and fishery exports increased fourfold in value between 1974 and 1977, while crop exports tripled.[63] To produce successfully for export requires standardised production and packaging, and this in turn means capital investment. The possibilities for small producers to participate in that market, in the absence of well-organised marketing and processing co-operatives, are thus virtually ruled out. On the other hand, it is not necessary to have a vast estate in order to produce for these markets; a high concentration of capital on farms smaller in size than the old estates is more efficient.

These very sketchy indications of competitive conditions in Chilean agriculture can be added to the pattern of changes in tenure to produce a picture of an emerging agrarian structure in which an impoverished small-scale production sector co-exists not with large *latifundia* but with medium-scale farms which use capital intensively and produce for export or for the upper-income groups.

Table 6.8 shows the changes in the pattern of land distribution over the period 1965-77. One point which emerges clearly is that the category of properties over 80 standardised hectares has more or less disappeared: where once it accounted for 55.4 per cent of the land, it now only accounts for 6.1 per cent. However, it is also noticeable that the land lost by this category went not to the smallest category, but to the intermediate ones. It should be added, however, that the 1973 and 1977 figures are largely based on estimates, and we await a new Agricultural Census to obtain a more accurate picture. But these tables unambiguously indicate a fundamental change in the class structure of Chilean agrarian producers, a change which was present in embryo under the Frei and Allende regimes, and reached its culmination under the military regime. That change can be summarised as the emergence of the middle farmer as the principal actor in Chile's rural economy. The military regime certainly reversed the reform process initiated in 1966. The net result, however, was not a return to the *status quo* but a transformation from a system dominated by huge *haciendas* based on large contingents of tied labour to a system dominated by medium-sized capital-intensive farm units based on wage labour.

It should be noted, however, that the medium-sized farms constituted the most dynamic element in Chilean agriculture even during the period 1965-73. They escaped expropriation since only the farms of over 80 hectares could be expropriated. Their numbers increased as 'retained reserves' and subdivided farms appeared on the scene. During the Frei regime they benefited from increased credit and technical assistance; during the early years of the Allende regime they benefited from an expanded internal market. And then, under the military, it is the old and new farms in this category which have been in the best position to take advantage of opportunities to export fruits and vegetables.[64] They have been the real beneficiaries of the tortuous reform process which began in 1965.

Notes

1. P. Dorner, 'An Open Letter to Chilean Landowners', *La Nación*, 21 June 1965. (*La Nación* was the official government newspaper.)

2. A. Foxley and O. Munoz, 'Income Redistribution, Economic Growth and Social Structure: The Case of Chile', *Bulletin of the Oxford Institute of Economics and Statistics*, vol. 36, no. 1 (Feb. 1974).

3. From 'Problemas y Perspectivas del Desarrollo Económico y Social de América Latina', document presented at the meeting in Trinidad of the Interamerican Economic and Social Council, June 1969. This section reproduced in *Panorama Económica* (Santiago), no. 246 (July 1969).

4. An intention largely frustrated by collective bargaining, which preserved differentials.

5. A. Angell, *Politics and the Labour Movement in Chile* (Oxford University Press, London, 1972).

6. G. Palma, 'Growth and Structure of Chilean Manufacturing Industry 1830-1935', unpublished DPhil thesis, University of Oxford, 1979.

7. L. Castillo, 'The Oligarchic State in Chile: Stability and Decline in a Period of Social Change', PhD thesis, University of Cambridge, 1980.

8. J. Swift, *Agrarian Reform in Chile: An Economic Study* (Heath, Lexington, Mass., 1971), p. 111.

9. E. Floto, 'Chile: The Secular Food Crisis', *Food Policy* (May 1979).

10. N. Kaldor, 'Economic Problems of Chile' in *Essays on Economic Policy* (Duckworth, London, 1964), vol. III.

11. This received some empirical support in M. Sternberg, 'Chilean Land Tenure and Land Reform', PhD thesis, University of California, Berkeley, 1962.

12. This view was developed at length in the report of the Inter-American Committee for Agricultural Development (CIDA), *Chile: Tenencia de la Tierra y Desarrollo Socio-económico del Sector Agrícola* (Santiago, 1966) (henceforth 'CIDA report').

13. CIDA report, p. 150. It should be pointed out that, in applying its definitions to the Central Valley of Chile, CIDA takes a sub-family farm to be under 5 hectares, which is generous: 5 irrigated hectares can support a family of five persons or even more in the region.

14. M. Mamalakis and C. Reynolds, *Essays on the Chilean Economy* (Richard Irwin and Co., Homewood, Illinois, 1965), p. 118.

15. Floto, 'Chile: The Secular Food Crisis'.

16. A. Martinez and S. Aranda, 'La Agricultura' in A. Pinto *et al.*, *Chile Hoy* (México, 1969), Siglo XXI.

17. This was certainly a major motivation for the Christian Democratic Party: their candidate was elected in 1964 with the votes of centre and right combined, and the right offered its support with no strings attached — that is, demanding

neither concessions on policy nor government posts. Yet the party knew that this arrangement might well not repeat itself – as indeed it did not – and that it badly needed a stronger mass base than it enjoyed if it was to win elections in the future.

18. An issue analysed at length in D. Lehmann, 'Political Incorporation versus Political Stability: The Case of the Chilean Agrarian Reform', *Journal of Development Studies* (July 1970).

19. D. Lehmann, 'Land Reform in Chile', unpublished DPhil thesis, University of Oxford, 1974, p. 351.

20. Ibid., p. 350.

21. R. Billaz and E. Maffei, 'La Reforma Agraria Chilena y el Camino al Socialismo', *Cuadernos de las Realidad Nacional*, no. 11 (1972).

22. A. Corvalán has offered these corrections to allow for the double-counting of permanently employed workers with rights to land, who appeared in the 1965 Census both as employees and as independent producers. See A. Corvalán, 'El Empleo en el Sector Agricola: realidad y Perspectivas', *Estudios de|planificación*, no. 52 (CEPLAN, Catholic University of Chile, Apr. 1976).

23. G. M. McBride, *Chile: Land and Society* (American Geographical Society, Research Series, New York, 1936).

24. W. Ringlien, 'Economic Effects of Chilean National Expropriation Policy on the Private Commercial Sector', PhD thesis, University of Maryland, 1971.

25. ODEPLAN, *Indices de producción agropecuaria, 1939-64* (Santiago, 1971).

26. Floto, 'Chile: The Secular Food Crisis'.

27. M. Zeitlin and R. Ratcliffe, 'Research Method for the Analysis of Dominant Classes: The Case of Landlords and Capitalists in Chile', *Latin American Research Review*, vol. X, no. 3 (Fall 1975), pp. 5-62.

28. Indeed, one could expect a violent reaction in this region where capitalist farming had developed out of a race war against the indigenous population in the nineteenth century and had been marked thereafter by continual encroachment on and usurpation of the lands of the Mapuche people.

29. For further details see Lehmann, 'Land Reform in Chile'.

30. The 1965 Agricultural Census gives 878,000, but as always in agricultural censuses, one must allow for double-counting because the same people appear in different places of work. See Corvalán, 'El Empleo en el Sector Agricola'.

31. Seventy per cent of the value of these bonds was indexed according to the official cost-of-living index, but they were repayable only after 25 to 30 years. The UP government tended to accord harsher terms than the Christian Democrats.

32. A survey conducted in the Central Valley in the late fifties found that only a quarter of *inquilinos*' income was accounted for by their own production (Ministerio de Agricultura, *Aspectos Económicos y Sociales del Inquilinaje en San Vicente de Tagua Tagua 1960*). Other studies find a higher proportion, reaching a maximum of 57.5 per cent: P. Ramirez, *Cambios en la Forma de Pago de la Mano de Obra Agricola* (ICIRA, Santiago, 1968) and A. Schejtman, 'Peasant Economies Within the Large Haciendas of Central Chile', unpublished BLitt thesis, University of Oxford, 1970. These differences are methodological and do not reflect changes over time.

33. In 1968 CORA was lending to *asentamientos* at 18 per cent while inflation was running at 21.9 per cent. Later the subsidy element in interest rates to *asentamientos* grew. INDAP lent to smallholders at even more favourable rates, carrying from 5 to 12 per cent in 1968.

34. Law No. 16,640.

35. S. Barraclough et al., *Diagnóstico de la Reforma Agraria Chilena* (Santiago, UNDP, November 1972) (known simply as *Diagnóstico*).

36. See Lehmann, 'Land Reform in Chile', p. 435.

37. The *Diagnóstico*, basing itself on the *Registro*, gives 6.3 per cent for the region, see Annex 6, Table 3.

38. A standard source is S. de Vylder, *Allende's Chile: The Political Economy of Unidad Popular* (Cambridge University Press, London, 1977).

39. World Bank, *Chile: An Economy in Transition* (Washington, DC, 1980), p. 542.

40. Floto, 'Chile: The Secular Food Crisis'.

41. World Bank, *Chile*, p. 437.

42. There were many parties and factions: Communists and Socialists were dominant, but there were also two left Christain groups, minority breakways from the Christian Democrats, the majority faction of the Radical Party, and offering critical support from without, the MIR – Left Revolutionary Movement.

43. Recipients of INDAP credit rose from 78,544 to 177,660 between 1970 and 1971. See also Table 6.4, especially the decline of the 'other' category.

44. *Diagnóstico*, Ch. III, Annex Table 1.

45. R. Cortázar and R. Downey, 'Effectos Redistributivos de la Reforma Agraria', *Estudios de Planificación*, no. 53 (CEPLAN, Santiago, 1976).

46. In 1972 Lehmann revisited five farms he had studied in 1969 and 1970 in Colchagua. The number of *asentados* had increased by an average of 113 per cent, mostly accounted for by the incorporation of people who had previously been working on the units as hired labourers, often on a permanent basis. See Lehmann, 'Land Reform in Chile', p. 384.

47. J. M. Caballero, 'Los Eventuales en las Co-operativas Costenas Peruanas: Un Modelo Analitico' in *Economia* (Lima, Aug. 1978), vol. I, no. 2, and C. Scott, 'Agrarian Reform and Agricultural Labour Markets' (School of Development Studies, University of East Anglia, Norwich, unpublished manuscript).

48. A. Jolly, O. Brevis and O. Lefevere, *Estudio Económico de los Asentamientos* (ICIRA, Santiago, 1968).

49. In the sample of 105 farms studied by Cortázar and Downey, 4 had escaped expropriation, 41 were expropriated under Frei and 60 under Allende. Only 2 were expropriated in terms considered 'less favourable' to the owner (with respect to reserves, period of repayment, amount of cash down payment and so on) under Frei, whereas under Allende this was the case for 20 farms. Cortázar and Downey, 'Effectos Redistributivos'.

50. D. Stanfield *et al.*, 'The Impact of Agrarian Reform on Chile's Large Farm Sector' (Development Economics Department, World Bank, June 1975), Background Paper for the Comparative Study of Land Reform in Latin America. This short paper contains much of the raw material for the Cortázar and Downey paper already quoted.

51. Between 1965 and 1970 credit to commercial farmers from the Development Corporation (CORFO) and the State Bank multiplied by 2.8 while credit for *campesinos* from the Agrarian Reform Corporation (CORA) and the Institute for Agricultural Development (INDAP) multiplied by 6.3. However, most of this latter increase was accounted for by CORA and thus by the beneficiaries of the Land Reform, until 1970 when INDAP's credit programme shot ahead. See World Bank, *Chile*, p. 447, and Barraclough *et al.*, *Diagnóstico*. INDAP lent 103.8 million escudos in 1965, 167 million in 1969, 135.4 million in 1970 and 204.1 million in 1971 (in constant 1971 escudos), while the number of recipients remained around the 50,000 mark from 1965 to 1970, rising to 106,000 in 1971.

52. Tractors per standard hectare: Reserves – 0.032; subdivided units – 0.023; intact estates – 0.016; and *asentamientos* – 0.011. Stanfield *et al.*, 'Impact of Agrarian Reform' p. 30. The total number of tractors in the sample rose from 103 to 161 during the period.

53. According to the *Diagnóstico*, there were 350 farms still in this situation in mid-1972 (Ch. III, 1.3).

54. By April 1972, 7,461 beneficiaries had been assigned land in definitive ownership, and they were organised in 153 co-operatives. *Diagnóstico*, p. III. 5.

55. The tables on which these and all the foregoing statements are based were obtained from the Oficina de Normalización Agraria, Sub-Programa de Programación y Control, Santiago.

56. ICIRA (Instituto de Capacitación e Investigación en Reforma Agraria), *Análisis de las Situación de los Asignatarios de Tierras a Diciembre de 1976* (Santiago, Nov. 1977). See also World Bank, *Chile*, pp. 475ff, and 181.

57. The World Bank report does not specify, so I have referred to the global average of 10 standard — and therefore high-quality — hectares.

58. VECTOR, *Informe de Coyuntura Económica, April 1980* (Santiago, May 1980).

59. S. Gomez, J. M. Arteaga and M. E. Cruz, *Reforma Agraria y Potenciales Migrantes* (FLACSO, Santiago, 1979).

60. Cf. Cortázar and Downey, 'Effectos Redistributivos'.

61. World Bank, *Chile*, pp. 437, 439.

62. See the account by S. Gómez, 'Que se Coman las Vacas', *Boletín del GEA*, no. 4 (July-Oct. 1979).

63. World Bank, *Chile*, p. 354. Total value of agricultural exports in 1977 was $204 million, still only 8 per cent of the total.

64. Stanfield, 'The Impact of Agrarian Reform'.

7 AGRARIAN REFORM AND RURAL DEVELOPMENT IN NICARAGUA, 1979-81*

Peter Peek

I Introduction

Since the Sandinist victory of July 1979, the new government of Nicaragua has adopted a wide range of social and economic policies to bring about profound changes in the structure of the economy. One of the most important measures adopted is an agrarian reform programme designed to restructure the highly inegalitarian agrarian economy inherited from the Somoza regime. This programme, initiated only recently, promises to be one of the most important agrarian reforms carried out in Latin America in the post-war period. So, far many large farms have been expropriated. The small-farm sector has also been subject to important reforms. Most small farms have been organised in co-operatives and provided with cheap credit, while land rental contracts have been regulated, effectively reducing the rental price. Agricultural credit is now exclusively provided by the state through nationalised banks. Equally important, the state now buys and markets all export crops and a large part of food crops. As a result of these measures, the state's role in the generation and extraction of agricultural surplus has assumed great significance, even though two-thirds of cultivable land still belong to the medium-sized and large-scale private farms.

This chapter seeks to place these policies in the context of the overall transformation of the Nicaraguan economy and analyse their impact on the rural economy. It also focuses on the particular problems and difficulties that were faced in implementing these policies.

Needless to say, such an analysis can only be tentative at this stage. The new regime is now just over two years old, and its agrarian policies

*I would like to thank Solon Barraclough, Dharam Ghai and Samir Radwan for their helpful comments on an earlier draft. Any remaining errors of misinterpretation are obviously my responsibility. Many data presented in this paper were obtained from the report of a recent IFAD (International Fund for Agricultural Development) mission headed by Solon Barraclough, and from several studies on employment made by PREALC (Programa Regional de Empleo para América Latina y el Caribe).

273

are still in the process of being implemented. Also, civil war prior to July 1979 severely damaged the economy, and many of the policies adopted since then were designed to repair this damage; it is often difficult to separate policies designed to reconstruct the economy from those designed to restructure it. In the course of the civil war factories were destroyed, cities were left in ruins and land remained uncultivated. It has been estimated that the value of this damage amounted to US$580 million.[1] Moreover, nearly 2 per cent of the country's population, 50,000 persons, were killed during this period. In addition to these complicating factors, analysis is also handicapped by the scarcity of data for the post-revolutionary period. Sometimes newspaper reports provided the only source of information, and the few data available are often inconsistent and frequently based on individual estimates rather than on official sources.

Despite these limitations, a study of this kind can serve a useful purpose. The Nicaraguan agrarian reform programme involves a process of transition from private farming to state-owned and co-operative farming. The characteristics of this transition process are of obvious theoretical and practical interest, particularly in view of its consequences for rural poverty and unemployment. Given that there are few quantitative studies analysing such transitions, even a preliminary analysis of this unfolding experience can be useful.[2]

It must be pointed out that this analysis limits itself to describing the agrarian reform measures and analysing their impact on the rural economy. No attempt is made to describe the political and economic process that led to agrarian reform and that determined its nature. Such description would be essential to a more complete study of Nicaragua's agrarian reform. It would also be extremely interesting, given the alliance of different socio-economic groups that provided support to the insurrections against the Somoza regime. However, this study will mainly be concerned with examining the agrarian policies and their socio-economic impact.

The following section describes the agrarian structure prior to 1979. It focuses on the distribution of land, the structure of the labour market and the levels of living of the rural population. Section III describes the agrarian reform measures adopted during the first two years after the revolution. Section IV describes how these measures have affected, so far, agricultural production as well as rural incomes and employment. The final section notes some of the problems that were encountered in implementing the agrarian reform, and reviews the emerging structure of the rural economy.

Before analysing the pre-revolutionary agrarian structure, some general observations will be necessary so as to provide an appropriate context. Nicaragua is relatively sparsely populated compared to most Central American counttries. It has large areas of arable land, mainly in the eastern part of the country, which remain uncultivated. Nicaragua produces a number of export crops – coffee, cotton, sugar and tobacco. Thus it has a varied resource base and it does not face the problems of the monoculture economy characteristic of many countries in the region. Another important aspect is the relatively low level of industrialisation. Only 9.5 per cent of the labour force is employed in manufacture and industry, a proportion much lower than that in the majority of Central American countries.

The rural areas of Nicaragua can be divided into three geographical regions with distinct agrarian structures. The central part, where the altitude goes up to 1,500 m in certain areas, is inhabited by around 30 per cent of the total population, and it covers around 20 per cent of the arable agricultural land.[3] Most of the households are engaged in small-scale farming producing food crops. Nearly all the coffee-producing farms, most of which are medium size, are also located in this area. The lowlands near the Pacific coast constitute 50 per cent of the cultivable land, and they are inhabited by 60 per cent of the population. Landlessness is far more significant in the lowlands than in the central highlands. Around one-third of the labour force in the Pacific zone works, on either a seasonal or permanent basis, on the large farms, many of which produce cotton. Compared to the central region, the percentage of rural households producing food on small farms is lower here. The third region is along the Atlantic coast; it is the least populous, containing only 10 per cent of the total population, although it is the largest in terms of surface area. The Atlantic region is economically the most backward; its population largely consists of indigenous communities engaged in subsistence food production.

II The Rural Economy before 1979

Agriculture was, and still is, the most important economic activity. If the processing of agricultural products is included, the agricultural sector accounted for nearly 35 per cent of the gross national product in 1971.[4] It also employed nearly 47 per cent of the economically active population, and it produced 73 per cent of total exports.[5]

The structure of agricultural production largely resembled the so-

Table 7.1: Distribution of Farms and Land by Farm Size, Nicaragua (1963-71) (in manzanas[a])

Farm Size	No. of Farms				No. of Manzanas			
	1963 Absolute	%	1971 Absolute	%	1963 Absolute	%	1971 Absolute	%
Less than 10	51 936	50.8	37 521	43.8	190 098	3.5	128 552	2.2
10-50	27 976	27.4	27 150	31.7	614 135	11.1	641 728	11.2
50-200	17 240	16.9	16 130	18.9	1 447 603	26.4	1 387 659	24.1
More than 200	5 049	4.9	4 834	5.6	3 209 326	58.5	3 578 603	62.5
Total	102 201	100.0	85 635	100.0	5 461 162	100.0	5 736 542	100.0

Note: a 1 manzana = 0.7 hectare.

Sources: For 1963, FAO, *Report of the 1960 World Census of Agriculture 1966-1970* (Rome, 1970), vol. I/B, pp. 180-1; for 1971 P.F. Warnken, *The Agricultural Development of Nicaragua: An Analysis of the Production Sector* (University of Missouri-Columbia, Columbia, 1975), pp. 47-8.

called 'capitalist agro-export model'.[6] The most fertile land was used for export crops and it was concentrated on medium-sized and large farms owned and controlled by a small group of landowners and commercial entrepreneurs. On the other hand, there were a large number of small-scale farms cultivating the least fertile land growing food. The households on these farms also provided wage labour to the larger farms in order to complement their farm revenues which were usually insufficient to cover their consumption needs. Table 7.1 shows that, in 1963, 85 per cent of cultivated land belonged to farms of more than 50 manzanas, which constituted only about 22 per cent of the total number of farms.[7] The bulk of the farm population lived on small farms; around 50 per cent of all farms, those with 10 manzanas or less, occupied only 3.5 per cent of cultivated land.

Table 7.2: Economically Active Population in Agriculture, by Major Occupation, Nicaragua (1978)

Major Occupation	No. of Workers (000)	%
Large farm owners or administrators	1.6	0.4
Medium-size farm owners or administrators[a]	38.7	8.9
Family-size farm owners or operators[a,b]	54.6	12.7
'Mini' farm owners or operators[a,b]	164.8	38.3
Permanent wage workers	32.3	7.5
Seasonal wage workers	138.0	32.1
Total	430.0	100.0

Notes: a. Includes unpaid family workers.

   b. Since only the major occupation is reported, these figures do not show the extent of wage work among family-size and 'mini' farm owners and the unpaid family workers.

Source: IFAD, *Informe de la Misión Especial de Programación a Nicaragua* (Rome, 1980), p. ix.

As regards the composition of the rural labour force, Table 7.2 indicates that the largest single group consisted of workers of small-farm households. The second largest group were the seasonal wage workers — all landless — employed in the harvests of agro-export products on the medium-sized and large farms. The fact that the number of small-farm workers was larger does not necessarily imply that the small farms absorbed more labour than other farms. Table 7.2 only reports the major occupation. It is generally known, however, that a substantial number of small-farm workers were also employed on the large farms during harvest time. Unfortunately, there are no data available to show precisely how much labour the small-farm population provided to the medium-sized and large farms.

Table 7.3: Crop Cultivation by Farm Size, Nicaragua (1971, percentages)

| Crop | Farm Size (manzanas[a]) | | | | |
	0-10	10-50	50-500	More than 500	Total
Cotton	1.6	7.2	54.7	36.5	100.0
Coffee	5.1	17.9	57.0	20.1	100.0
Sugar	1.0	4.0	20.0	75.0	100.0
Beans	23.7	34.5	38.7	3.1	100.0
Corn	18.7	43.1	28.0	10.2	100.0
Sorghum	16.6	23.0	31.5	28.9	100.0

Note: a. 1 manzana = 0.7 hectare.

Source: IFAD, *Informe de la Misión Especial*, p. 38.

Table 7.3 shows that food crops were predominantly cultivated on the smaller farms, and export crops mainly on the larger farms. Around 75 per cent of the export crops were cultivated on farms of more than 50 manzanas, and more than 50 per cent of the food crops came from farms of less than 50 manzanas.

Prior to the mid-nineteenth century, Nicaragua was ruled by an oligarchy of large landowners raising cattle and growing food crops.[8] The bulk of the rural population consisted of small-scale farmers renting land from the oligarchy through a variety of rental agreements. After independence was declared in 1821, close economic contacts were developed with other countries, such as the United States and Great Britain, resulting in an increased economic and political dependency. One of the reasons was that Nicaragua occupied a strategically important area ideal for constructing a coast-to-coast canal. Another important development was the expansion of coffee production, started in 1847, which profoundly changed Nicaragua's political and economic structure.[9] The production and commercialisation of coffee gradually took over from livestock production as the main economic activity. This was accompanied, among other things, by increased investments in different sectors of the economy, and by the establishment of the first liberal government of José Santos Zelaya in 1893.[10] During this period a political and economic struggle took place between the traditional landowners, who had major economic interests in the livestock sector, and the agro-export producers, which was accompanied by direct interventions, both militarily and financially, from the United States.[11] It was also during this period that a nationalist guerrilla movement led by Sandino was created whose main purpose was to free Nicaragua from foreign intervention. The movement was crushed and a regime supported by an alliance of agro-export producers and the landed elite came to power in 1934. Foreign investments expanded significantly since then, as did the political and economic power of the Somoza family.

The production of cotton, the other major export crop, began to develop in the early 1950s. This occurred largely at the expense of food production. As the profitability of cotton farming increased, landowners bought up small farms or cancelled rental contracts in order to cultivate more land with export crops. Thus the peasantry was gradually displaced from land in the zones favourable to cotton production. Table 7.4 shows that the production of coffee, sugar and cotton increased respectively by 148, 249 and 282 per cent between 1960 and 1979. During this period the production of the principal food crops expanded only by around 60 per cent. Table 7.5 provides further evidence of the displacement of the peasantry. It shows the area cultivated with food crops per rural inhabitant for each department. One observes that the *per capita* area mostly declined in the first six departments between 1950 and 1977. These were primarily departments in the Pacific zone where agro-export production strongly expanded during this period. A reverse trend is observed in the remaining departments; the *per capita* area cultivated with corn and beans mostly increased between 1950 and 1977. Agro-export production expanded only marginally in these areas.[12]

The displacement of the peasantry in the agro-export zones stimulated migration both to the central region and to the agricultural

Table 7.4: Crop Production, Nicaragua (1960-79) (in thousands of quintals[a])

Year	Coffee	Sugar	Cotton	Corn	Beans	Sorghum
1960-1	571.4	16,897	2,139.3	2,588	605	847
1965-6	697.8	20,108	7,071.5	3,723	1,075	1,075
1970-1	856.9	42,174	5,209.6	4,136	790	977
1975-6	1,068.2	56,603	7,189.6	4,176	962	1,366
1978-9	1,417.5	58,968	8,167.0	NA	NA	NA
Increase (1960-78)	148%	249%	282%	61%	59%	61%

Note: a. 1 quintal = 46 kg.

Source: IFAD, *Informe de la Misión Especial*, pp. 11 and 38.

Table 7.5: *Area Cultivated with Corn and Beans, per Rural Inhabitant, per Department, Nicaragua (1950-77)*

Department	Corn (manzana[a]/rural inhabitant)				Beans (manzana[a]/rural inhabitant)			
	1950	1963	1971	1977	1950	1963	1971	1977
Chinandega	.49	.23	.21	.21	.06	.02	.02	.02
Leon	.27	.25	.24	.16	.06	.02	.02	.03
Managua	.53	.21	.23	.16	.21	.05	.02	.04
Masaya	.18	.23	.20	.15	.07	.08	.10	.07
Carazo	.17	.14	.15	.17	.13	.08	.08	.09
Granada	.31	.16	.21	.15	.15	.09	.09	.19
Rivas	.28	.12	.19	.12	.09	.04	.03	.10
Nueva Segovia	.26	.30	.35	.26	.06	.07	.09	.11
Madriz	.23	.20	.22	.20	.12	.08	.16	.16
Estelí	.18	.25	.25	.20	.09	.10	.29	.17
Boaco	.23	.31	.37	.20	.04	.04	.10	.09
Chontales	.46	.51	.64	.31	.06	.07	.11	.10
Río San Juan	.16	.36	.48	.43	.06	.12	.16	.37
Zelaya	.08	.31	.40	.57	.03	.07	.07	.29
Country	.27	.28	.30	.33	.08	.07	.09	.13

Note: a. 1 manzana = 0.7 hectare.

Source: IFAD, *Informe de la Misión Especial*, p. 15.

frontier in the Atlantic zone where agro-export production did not expand because of poor-quality land. Part of the displaced population of the Pacific zone also migrated to the urban areas to find employment in the service sector. However, it appears that the majority of displaced population stayed in order to seek wage employment on the export crop farms.[13]

An important element of the 'capitalist agro-export' model was the subservient role assigned to the small-farm sector. Its main function was to facilitate the growth of export agriculture. This function had two main components.[14] First, the small-farm sector supplied food (Table 7.3). Second, the small farms supplied labour to the agro-export farms, particularly during harvest time. This labour was available at relatively low cost since the peasantry had a small but steady income from the land so that landowners could pay a wage below the subsistence level.

Table 7.6: Distribution of Bank Credit by Crop, Nicaragua (1976)

Crop Type	Area Cultivated (manzanas[a])		Area Financed (manzanas[a])		Amount Credit (million cordobas[b])	
Export	475.800	47.3%	285.558	75.2%	510.300	90.3%
Food	531.800	52.7%	94.500	24.8%	54.400	9.7%
Total	1,007.600	100.0%	380.058	100.0%	564.700	100.0%

Notes: a. 1 manzana = 0.7 hectare.
    b. 7 cordobas = US$1 (1976).

Source: IFAD, *Informe de la Misión Especial*, p. 42.

This subservient role of the small-farm sector was secured through several mechanisms. For example, the peasantry was denied access to additional land which would have increased their incomes thus undermining their subservient status. The state and the legal system functioned in such a way that small-scale farmers were unable to acquire more land to grow food crops. Also, the peasantry was denied access to credit. Table 7.6 shows that the food-producing sector, accounting for about 53 per cent of the cultivated area, obtained only 9.7 per cent of total agricultural bank credit in 1976, and most of this very probably went to the larger farms producing food.

This agrarian structure, characterised by a dominant agro-export

Figure 7.1: Seasonal Variations in Employment by Crop,
Nicaragua

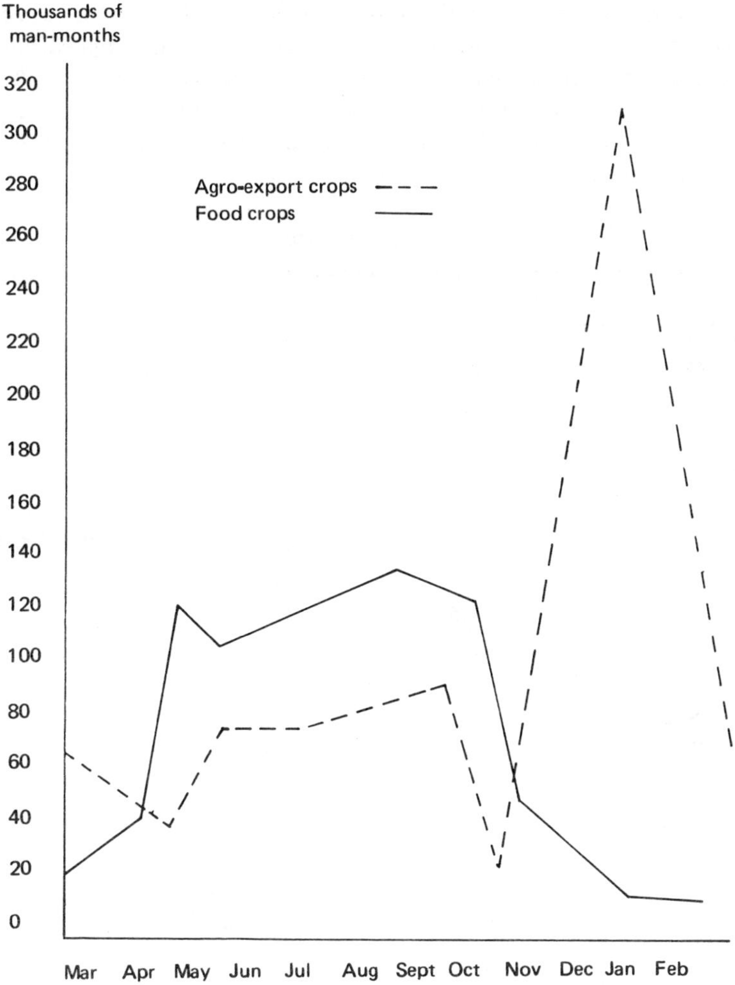

Thousands of
 man-months

Source: O. Nuñez, *El Somocismo: Desarrollo Contradicciones del Modelo
Capitalista Agroexportador en Nicaragua (1950-75)* (Centro de Estudios
sobre América, Havana, Cuba, 1980), p. 39.

sector and an impoverished peasantry, was also reflected in the labour
process. First, Table 7.2 already shows that around 78 per cent of the
rural labour force consisted of small farmers and landless workers.
Second, a large transfer of labour from small to large farms took place
each year. The incomes from small-scale farming were insufficient so

that the households had to sell labour to outside farms. This was made possible by the fact that harvests of the agro-export crops took place during the slack season on the small farms. Figure 7.1 shows that when small-farm employment was at its lowest level, large-farm employment peaked. However, despite the absorption of small-farm labour on the export farms, there was substantial unemployment during most of the year. Table 7.7 shows that the rural labour force was fully employed for only four months in a year.

Table 7.7: Underemployment in Agriculture, per month, Nicaragua (1975-8, thousands of man-years)

Month	Labour Demand	Economically Active Population	Under-Employment Rate
January	352	352	–
February	283	325	13
March	253	325	22
April	232	325	29
May	261	325	20
June	275	325	15
July	331	331	–
August	348	348	–
September	347	347	–
October	190	325	42
November	220	325	32
December	273	325	16
Average	280	332	16

Source: PREALC, *Empleo y Salarios en Nicaragua* (Santiago, Chile, 1980), p. 23.

This agrarian structure naturally produced a highly unequal income distribution. Available data for 1971 (Table 7.8) indicate that 63.1 per cent of rural incomes went to only 3.5 per cent of the economically active population, whereas rural workers, constituting 51 per cent of the economically active population, earned a meagre 7.5 per cent of rural incomes. This inequality was associated with a high incidence of poverty among the rural population. In 1972, the annual average income of the poorest 50 per cent of the rural population was estimated

Table 7.8: Rural Income Distribution by Occupation, Nicaragua (1971)

Occupational Category	Percentage Econ. Active Population	Percentage Income
Owners medium-sized and large farms	3.5	63.1
Own account workers	45.5	29.4
Wage workers	51.0	7.5
Total	100.0	100.0

Source: IFAD, *Informe de la Misión Especial*, p. 2.

at US$35 *per capita*.[15] Data of a 1976 survey in the central interior region show that 75 per cent of the household heads earned less than 1,000 cordobas, or US$143 annually (Table 7.9).[16] This area had a high percentage of small-scale farmers, and since it certainly cannot be considered as the poorest area in Nicaragua it is probable that the figures in Table 7.9 underestimate the incidence of poverty. It must be noted that this figure does not include the incomes of other working household members. However, they are generally so low that the inclusion of additional wage incomes is unlikely to alter the picture significantly.[17]

Table 7.9: Distribution of Annual Incomes of Household Heads, Interior Central Region, Nicaragua (1976)

Annual Income (1976 cordobas[a])	Percentage Household Heads
Less than 1 000	75
1 000 to 2 999	18
3 000 to 4 999	6
More than 5 000	1
Total	100

Note: a. 7 cordobas = $US1 (1976).

Source: V. H. Gillespie, *Analysis of Rural Social Survey Conducted in 1976 by INVIERNO* (USAID, Managua, 1978), quoted in PREALC, *Diagnóstico de las Estadísticas y Bibliografía Sobre el Empleo Rural en Centroamerica y Panama* (Santiago, Chile, 1979), Ch. 5.

*Table 7.10: Per Capita Daily Consumption of Calories,
Proteins and Fat, by Income Strata, Nicaragua (1970)*

Calories	Lowest 50%	Lower Medium (50-20%)	Upper Medium (20-5%)	Upper 5%
Animal	197.2	337.3	497.8	727.8
Vegetal	1 570.0	2 366.2	2 757.3	3 203.4
Proteins (grammes)				
Animal	12.6	22.7	33.8	49.1
Vegetal	34.0	49.8	56.5	62.8
Fat (grammes)				
Animal	13.4	22.6	33.3	49.4
Vegetal	18.3	31.6	44.6	65.3

Source: FAO-SIECA, *Perspectivas para el Desarrollo y la Integración de la
Agricultura en Centroamérica* (Guatemala, 1974).

Other indicators also suggest a high level of rural poverty. A survey
in the interior central region showed that only 15.6 per cent of the
households had access to electricity.[18] It also reported that less than 20
per cent of the households had a primary school nearby. Furthermore,
only 3.9 per cent of the households had running water in their homes
and 8.4 per cent had access to local health facilities. Finally, Table 7.10
gives data on the consumption of calories, proteins and fat by different
income strata. These show that the daily *per capita* consumption of the
lowest 50 per cent is very small and that it is a small proportion of what
is consumed by the upper 5 per cent. The *per capita* daily calorie
consumption reported for the lowest 50 per cent, 1,767 calories, is
only 79 per cent of the average minimum requirement for the population of Nicaragua, 2,244 calories, according to the World Bank.[19]
Given these socio-economic conditions, the massive support which the
population gave to the Sandinist movement was hardly surprising.

III The Agrarian Reform Programme 1979-81

Almost immediately after the victory over the Somoza regime agrarian
reform measures were introduced. The basic orientation of these policies
was defined by the political and economic situation at that time.[20]

Different socio-economic groups collaborated to overthrow the Somoza regime. These included most of the peasants, rural and urban workers as well as an important segment of the middle class such as skilled and professional workers, and even some businessmen and large landowners. The policies adopted since 1979 reflected, to an extent, the interests of this heterogeneous group. It is also important to point out that the agrarian reform measures formed part of a much larger programme; they were part of a programme designed to restructure the entire economy. Banks, both national and foreign, were nationalised, as were insurance companies. The same happened to the industrial units belonging to the Somoza group.

III.1 Reforms in the Agro-export Sector

In the first instance, all land belonging to the Somoza group was nationalised. Expropriation is perhaps not the proper word; most of the members of the Somoza group had left the country before July 1979 and the state simply took over ownership of their land.[21] Many of these landholdings were already occupied by the peasantry after the Sandinist victory was proclaimed. It has been estimated that a total of 1,132,352 manzanas, i.e. 23.2 per cent of the cultivable land, was nationalised.[22] Around 91 per cent of the confiscated land belonged to farms of over 500 manzanas (Table 7.11). As a result of this measure, nearly 43 per cent of the land belonging to this category of farms become state-owned. Thus large-scale private farming was substantially reduced, although it still remains important as shown in Table 7.12.[23]

The nationalised farms, 2,000 in total, are now state-owned.[24] A rather centralised system has been created to organise and supervise production on these farms. They have been consolidated into 827 productive units called Unidades de Producción (UPES) which in turn have been organised into 170 production complexes. These are administered by the Nicaraguan Institute of Land Reform (INRA) which is part of the Ministry of Agriculture. Production targets, employment, investments and wages are decided on by INRA, and its decisions are carried out by the administrator of each UPE, who is directly responsible to INRA.

Furthermore, it has been decided to create a *consejo de producción* (production council) in each UPE. The members of these councils are workers' representatives, the administrator and other representatives of the farm's managerial staff such as supervisors and engineers. It has not yet been agreed what the role and responsibilities of these councils

*Table 7.11: Distribution of Confiscated Land by Farm Size,
Nicaragua (1980,* in manzanas[a])

| Farm Size | Land Confiscated[b] | | Percentage Land |
	Absolute	%	
0-10	379	–	0.2
10-50	3 245	0.3	0.5
50-500	99 125	9.0	4.8
500 or more	999 996	90.7	42.9
Total	1 102 745	100 (average)	23.2

Notes: a. 1 manzana = 0.7 hectare.
   b. These figures do not include the departments of Rio San Juán and Zelaya.
Source: Computed from IFAD, *Informe de la Misión Especial*, pp. 78-9, and
Table 7.1.

will be. However, it appears that INRA will continue to decide on
production targets and investments, whereas the councils will decide
on issues related to employment, for example hiring of workers and
fixing wages. The few *consejos de producción* that already exist do not
appear to be functioning well.[25] For example, in some cases the workers
have accused their representatives of collaborating too closely with
INRA. However, it is still too early to judge whether or not these
councils would make genuine workers' participation feasible on the
state farms.

III.2 Reforms in the Small-farm Sector

Important agrarian reform measures were also introduced in the small-
farm sector. These served, in part, to increase food production. The
decline in food production during the previous years and the fact that
the planned increases in the incomes of the lowest-income groups
implied an increase in the demand for food necessitated a concentrated
effort to augment food production. The agrarian policies affecting the
small-farm sector also served to increase the standard of living of the
rural population which had provided ample support to the Sandinist
movement during the insurrection and which now constituted the
major socio-political base of the government.

   The agrarian reforms in the small-farm sector consisted of two parts.
First, they aimed to replace individual farming with co-operative
farming in the long run. Second, they aimed to increase, in the short

Table 7.12: *Distribution of Productive Resources by Land Tenure, Nicaragua (1978-80, percentages)*

	State Farming		Land Tenure[a] Private Large-scale Farming		Private Small-scale Farming	
	1978	1980	1978	1980	1978	1980
Farms	—	1.3	23.9	22.6	76.1	76.1
Land	—	21.5	86.0	64.5	14.0	14.5
Value agricultural production	—	14.2	76.7	65.5	23.3	20.3
Workers	—	13.3	53.2	39.9	46.8	46.8
Agricultural credit	—	25.6	90.2	53.6	9.8	20.8

Note: a. These figures do not include the departments of Rio San Juán and Zelaya.

Source: IFAD, *Informe de la Misión Especial*, p. 84.

run, the returns from small-scale farming by providing cheap credit and lowering the rental price of land.

Interestingly, the first co-operatives were not created on the small farms but on the large farms which belonged to the Somoza group. These had already been established in the zones which the Sandinists controlled prior to the final collapse of the Somoza regime. These Co-operatives Agricolas Sandinistas (CAS) consisted of between ten and thirty households which cultivated the land on a communal basis. The land and all other means of production were collectively owned. These co-operatives were created primarily for stimulating food production, which had declined rapidly during the revolution, but they also helped to ensure the safety of the existing installations and machinery on the farms in the face of attempted sabotage by the previous owners.

After the revolution, most of these co-operatives were transformed into UPES. This was because food production had started to increase after the food-producing areas in the interior central region came under the control of the Sandinists. Another reason was that most of the co-operatives had begun to grow food crops on land normally used for agro-export crops. The CAS farmers apparently were not sufficiently skilled to handle the complex technical operations of agro-export cultivation, neither were they capable of marketing these crops. Thus, continuing these co-operatives would have meant a reduction in badly needed foreign exchange. Certain government circles regretted that these CAS co-operatives were replaced since it was felt that workers' participation in their management was more significant than on the UPES.

After the final victory various types of co-operatives were set up. These included primarily Co-operativas de Credito y Servicio (CCS), but also some CAS-type co-operatives. The CCS consist of small farmers who cultivate the land individually but who obtain credit, sell products or buy inputs through the co-operatives. Table 7.13 shows that a large number of co-operatives were organised in a relatively short period. Only one year after the revolution more than 2,500 co-operatives were created. These had more than 73,000 members, accounting for 75 per cent of the small-scale farm households.

Policies were also introduced affecting the provision of credit to the small-farm sector. After the revolution, Nicaragua received a substantial amount of credit at low cost from other countries and international organisations. A significant part of this was used to issue short-term loans to small-scale farmers. Members of co-operatives, and CAS, paid interest of only 7 and 8 per cent, whereas the rate of

*Table 7.13: Number of Agricultural Co-operatives,
Nicaragua (30 June 1980)*

Type of Co-operative	No. of Co-operatives
Cooperativas Agricolas Sandinistas (CAS)	1 327
Cooperativas de Credito y Servicio (CCS)	1 185
Total	2 512

Source: IFAD, *Informe de la Misión Especial*, p. 97.

inflation was estimated to be at least 30 per cent.[26] Moreover, the terms
for paying back the loans were very flexible. Farmers who for some reason could not pay off the principal were treated with leniency and
there were many cases where debts were cancelled. Table 7.12 shows
that the percentage of total farm credit absorbed by the small-farm
sector increased substantially; it rose from 9.8 per cent in 1978 to
20.8 per cent in 1980 while the proportion of land area operated by
small farmers remained roughly constant (around 24 per cent).

Besides the provision of cheap credit, a number of decrees were
issued with a view to improving rental agreements.[27] Decree No. 263
stipulated that a tenant will not have to pay more than 100 cordobas
annually for a manzana of food crop land.[28] It has been estimated that
this measure lowered the rental price of land by more than 25 per
cent.[29] Furthermore, the peasantry were provided with greater tenure
security. According to Decree No. 230, landowners were obliged to
offer land for rent to the person who cultivated it the previous year
before offering it to anyone else. Another decree, No. 293, relating
to reclaimed land stipulated that peasants were to receive part of the
future revenues of land which they had made cultivable. This decree
was introduced in order to eliminate the practice of many landowners
of renting out unused land to the peasantry (who brought it under
cultivation) and subsequently cancelling their rental contract. It appears
that, so far, these rental controls have been strictly adhered to. One
of the main reasons is that compared to previous rental prices, the new
prices are much closer to the level at which the supply of rental land
equals demand.[30]

These policies were complemented by others. For example, the state
fixed the selling and buying price of food, and purchased around 48 per

cent of the food crop harvest in 1981 through its agency ENABAS.[31] The producers' prices were fixed at a level above the costs of production so as to encourage farmers to produce more. The state subsidised food consumption by fixing consumer prices below producers' prices.

Furthermore, a large-scale literacy campaign was organised covering all rural and urban areas. It started in April 1980 and lasted six months. More than 120,000 teachers, students and members of the Sandinist movement participated in this campaign, which has been estimated to have cost around US$20 million. The *alfabetizadores* who were sent to the rural areas lived there during the campaign and organised courses on a daily basis. The objective of the campaign was not only to increase literacy but also to inform the population about the objectives and policies of the regime. The latter objective was particularly important in isolated rural areas. The campaign also served another purpose. During the Somoza period there existed a paternal relationship between the poorer sections of the rural population on the one hand, and the landowners and the government on the other. It was felt that this could not be changed simply by the land reforms but that an extra effort was needed to make the population realise that, henceforth, they would be considered equal partners in the process of development. The campaign also provided an opportunity to promote a spirit of co-operative farming among the small-scale farmers who were traditionally accustomed to farming the land individually. It is very likely that the programme was able to reduce some of the reservations towards co-operative farming among the peasantry.

Also, a trade union was created to defend the interests of rural wage workers and small-scale farmers. The Asociación de Trabajadores del Campo (ATC) participates in the decision-making concerning minimum wages, together with representatives of private farm owners, UPES and the Ministry of Labour. It also represents the workers in the production councils of the UPES. The fact that membership of the ATC has grown very rapidly probably indicates its success. Within two years more than 130,000 members, around one-third of the rural labour force, were registered.[32]

III.3   The 1981 Decree on Land Reform

On the day of the second anniversary of the revolution, 19 July 1981, the government announced a land reform decree affecting both the large- and small-farm sectors. According to this decree the following categories of land will be expropriated:[33]

- landholdings of 500 manzanas or more in the Pacific zone and of 1,000 manzanas or more in the rest of the country which were not cultivated during the preceding two years;
- similar landholdings of which less than 75 per cent was cultivated;
- similar landholdings, intended for livestock, with less than one animal (Pacific zone) or two animals (rest of the country) per hectare;
- landholdings of more than 50 manzanas (Pacific zone) or 100 manzanas (rest of the country) that are not cultivated by the owner directly (sharecropping, etc.);
- landholdings that were abandoned and on which land, equipment, etc., are in a state of decay.

As can be seen from this list, the expropriation concerns uncultivated or insufficiently cultivated land as well as land not cultivated by the owner. It is not known yet how much land will be expropriated. According to a provisional estimate, the first two categories alone will increase the percentage of expropriated land from 23 (Table 7.12) to 36 per cent.[34] The important thing about this new decree is that much of the land previously belonging to large farms will be assigned to individual farmers, who will be organised in co-operatives. Article 9 of the decree stipulated that the expropriated lands will be allotted to those farmers (renters, sharecroppers, etc.) currently cultivating the land, to landless workers and to those who participated in the struggle against the Somoza regime.[35] A minor part of this land will be apportioned to existing or new state-owned farms.[36] This largely distinguishes the new decree from the earlier one issued in 1979 which transferred the ownership of confiscated land primarily to the state. Another distinction between the two decrees is that the 1979 measure concerned the land of only those who had already left the country, but the 1981 decree will affect the holdings of landowners who stayed in the country after 1979.

It is also important to note that the individual beneficiaries of the 1981 decree will only have a 'limited' ownership of the land. Article 11 mentions that those receiving land in ownership are not entitled to sell it.[37] Land can only be transferred without being divided, through inheritance. Also the ex-landowners will be compensated by the government for the expropriated land, except for the land which they have abandoned. Government bonds will be issued based on the land's fiscal value, which is generally much below its market value.

IV The Impact of Agrarian Reform

Before analysing the impact of the agrarian reforms it must be repeated that these were mainly designed to redistribute the agricultural surplus rather than land. The preceding sections showed that the agrarian reforms, apart from the 1981 decree, sought to revise the distributive mechanism of agricultural surplus without changing the distribution of land. This is demonstrated by Table 7.12 which indicates, first, that the proportion of private farm land is still around 80 per cent, and, secondly, that small farms maintained their land share at around 14 per cent. The justification of this policy has been described as follows:

> With a dependent and backward economy, it is not as important to control all means of production as it is to control all surpluses. This is because, with an economy of this type, the training of the people, including technicians, is really insufficient to achieve state control over the whole production ... we must seek to put to good use the productive potential of all our citizens, including those in the private sector who have experience. If a medium landowner produces and holds, to take an example, 500 manzanas of cotton, his product will allow us to gain foreign exchange. Besides the interest he has to pay to the State Bank for the money lent to him, he has to pay land taxes, national taxes, production taxes, other levies when he exports, income taxes and then − very impor-tant − higher wages to farm workers.[38]

Clearly, an analysis of the impact of the agrarian reform can only be indicative given the scarcity and quality of data, and the fact that only two years have passed since the process was initiated.[39] Another complicating factor is that the reforms were carried out simultaneously with policies to reconstruct the economy. For example, the unemploy-ment rate in agriculture dropped substantially between 1979 and 1981, from 32 per cent to around 20 per cent.[40] It is very difficult to deter-mine what proportion of this decline was due to the reforms and what part was due to the restoration of activities which had been disrupted by the war.

The impact of agrarian reform can first be considered within the context of the objectives set by the government. These, as mentioned earlier, were mainly concerned with raising production levels and redistributing agricultural incomes. As regards the second objective, it appears that a substantial redistribution has taken place, although the reforms have been introduced only recently. First, the incomes from small-scale farming, in all probability, increased substantially. Data on farm incomes after 1979 are not available, but the fact that there was a large increase in area under food crops (see Table 7.14) indicates that incomes on small farms went up. This is not surprising given the rise in credit and the more favourable conditions of renting land.

Table 7.14: Area cultivated with Different Crops, Nicaragua (1974-80, in thousand manzanas[a])

Crop	Average Area Cultivated (1974-8)	Planned Area Cultivated (1980)	Area Actually Cultivated (1980)
Cotton	281	170	148
Coffee	123	140	130
Sugar	57	56	62
Corn (first harvest)	184	180	349[b]
Corn (second harvest)	61	60	74[b]
Beans	80	100	110[b]
Sorghum	76	80	80
Total	862	786	953

Notes: a. 1 manzana = 0.7 hectare.
  b. These figures exclude the departments of Rio San Juán and Zelaya.
Source: Ministerio de Planificación, *Programa de Reactivación Economica en Beneficio del Pueblo* (Managua, 1980).

The standard of living of the rural population has also been significantly affected by the subsidies and assistance programmes set up after 1979. Government revenues had increased substantially, partly because a large proportion of private profits from agro-export crops were transferred to the state. For example, a progressive *ad valorem* tax on coffee exports went into effect in October 1979. Similar taxes were introduced later in 1979 on other export crops. In 1980 alone, the taxes on coffee yielded over US$20 million and those on other exports US$8 million.[41] Increased revenues permitted the government to set up a number of assistance programmes and subsidy schemes. For example, food products were subsidised up to a value of 50 per cent of the market price. This subsidy amounted to US$32 million, or US$13 *per capita* during 1980.[42] Public facilities such as transport were also heavily subsidised. Furthermore, the total government expenditure increased by 38 per cent in real terms between 1978 and 1980.[43] Expenditure on health and education will increase from 17.7 per cent of the GNP in 1978 to an estimated 28.9 per cent in 1981.[44] Also, the literacy campaign led to a sharp decline in the number of adults unable to read or write. In the 1981 Economic Plan, it is estimated that the percentage of illiterate persons of 15 years and older declined from 50.3 per cent in 1978 to 12.9 per cent in 1981.[45]

It also appears that the number of seasonal and permanent workers earning less than the minimum wage declined. Data show that around 30 per cent of rural workers earned less than the minimum wage in 1978.[46] A survey of cotton workers in 1981 showed that only 3 per cent earned less than the legally established wage.[47] This reduction is not only the result of large landowners no longer being able to coerce workers into employment at lower wages. It can also be explained by the fact that small-farm incomes had gone up, and that the peasantry were not obliged any more to seek as much work outside their farms.

On the other hand, real agricultural wages have declined since 1979. For example, the government increased the basic nominal wage for harvesting cotton from 24 to 29 cordobas per quintal between 1979 and 1980.[48] However, this increase, 21 per cent, is lower than the rate of inflation which was estimated to be 30-35 per cent.[49] Furthermore, the wage rate of permanent agricultural workers declined. On 1 June 1980 the daily minimum wage rate increased from 21.20 cordobas to 25.30 cordobas.[50] This increase of 19.3 per cent was also well below the rate of inflation. The IFAD report estimates that, on the whole, rural and also urban workers experienced a decline of around 19 per cent in wage incomes.[51] The decline in real agricultural wages does not necessarily imply that all rural wage workers experienced a decline in their standard of living. First, many seasonal workers came from farm households and the rise in farm earnings may have more than compensated for the fall in wage earnings.[52] Second, the harvest workers like other workers benefited from government subsidies such as the food subsidy described earlier. However, the decline in real wages was most probably much larger than the food subsidy.[53] Thus the landless workers very likely experienced a decline in their standard of living.

The decline in the agricultural wage rate was a direct result of government policy. The wages for seasonal work are fixed by the

Ministry of Labour in consultation with workers' and landowners' representatives. The wage policy adopted by the government largely follows the principle that wage increases should only be permitted within the capacity of the economy to meet the resulting increase in consumer demand so as to avoid higher prices.[54] Since redistribution of income from the rich to the poor results in a rise in demand for necessities, particularly food, it was decided to limit wage increases so as to control inflation.[55] The fact that prices of agricultural products increased by only 19 per cent during 1980, whereas the overall rate of inflation was more than 30 per cent, indicates that the government has been successful so far in limiting demand-pull inflation.[56] However, this seems to have been achieved at the expense of the landless workers who have benefited the least from post-revolutionary policies.

The agrarian reforms affected not only the absolute incomes of the rural poor but also their relative incomes. Although data are not available, upper incomes very likely declined since 1979, and for a number of reasons. First, rental and price controls reduced the returns from landed property. Second, a salary limit was fixed at 10,000 cordobas which reduced the ratio between the highest and the lowest wages from 73:1 to 8:1.[57]

However, the gap between the lowest incomes and the medium-level incomes increased, according to the IFAD report. It estimated that the salaries of the medium-level income groups have increased by 9.2 per cent in real terms since 1979.[58] This rise can largely be attributed, according to the report, by the scarcity of technical and professional personnel. The lack of such personnel was felt particularly in the public sector which has expanded rapidly since 1979.

Available data suggest that the redistribution of agricultural incomes was accompanied by an increase in food production. As mentioned earlier, the area cultivated with food crops rose substantially (Table 7.14), surpassing the target set by the Ministry of Planning.[59] In fact, the level of food production in 1980 was the highest ever reached. But notwithstanding this increase, the supply of food was insufficient to meet demand. Increased food production had led to a higher level of consumption among farm households, while the demand for food of the non-farm population had also risen as a result of the redistributionary policies. The result was that in 1981 imports of corn and beans were much larger than had been foreseen in the Economic Plan.[60]

The policies concerning agro-export production were less successful. As Table 7.14 shows, the total area cultivated with export crops was below the planned level. This was particularly disappointing given that the targeted levels were relatively low compared to those for food crops; this applies in particular to cotton. Using another criterion, that of foreign exchange earnings, the performance of the agro-export sector was also below expectations. As Table 7.15 shows, actual total earnings were 10 per cent below the planned level. The situation looks much worse comparing foreign exchange earnings with expenditures. The 1981 Economic Plan reports that the actual foreign exchange deficit on the trade balance amounted to US$300 million, compared to a planned deficit of US$250 million.[61] The increase in the deficit was mainly due to a large increase in food imports and also in intermediate inputs such as construction material. Preliminary estimates for 1981 indicate that the foreign exchange deficit will be even larger, mainly because of further increases in imports. Thus it appears that the agro-export sector, despite the increase in production, will not be able to meet the rising demand for foreign exchange.

To explain the relatively slow recovery of cotton production, it must be pointed out first that around 65 per cent of cotton is cultivated by individual large-scale cultivators, many of whom rent the land on an annual basis.

Table 7.15: Foreign Exchange Earnings of Selected Agro-Export Crops, Nicaragua (1980, US $ million)

Product	Planned	Actual
Coffee	162.0	164.8
Cotton	33.7	30.8
Sugar	35.7	25.9
Meat	71.5	55.7
Total	302.9	277.2

Source: Ministerio de Planificación, *Programa de Reactivación*, p. 64.

Cotton production became less profitable as a result of the agricultural policies adopted since 1979. Previously, cultivators were able to obtain very large profits growing cotton, but this situation has changed as prices and credit are fixed by the state. Moreover, they faced a substantial labour shortage during the harvests of 1980/1. Traditionally, small-scale farmers sold their labour to the agro-export farms during harvest time, but they felt much less of a need to do so after the revolution as their socio-economic condition had improved considerably. Thus, it is very probable that, because of these factors, cultivators decided to rent less land or to sow less cotton than in previous years.

Finally, casual observations also suggest that the land reform programme was accompanied by a drop in agricultural productivity on the large farms. Preliminary indications showed that in the summer of 1980 the unit costs of production on the state farms were running far above what was expected.[62] This, no doubt, was partly due to the fact that many skilled labourers had left the expropriated farms and increasing difficulties were experienced in hiring qualified personnel. Moreover, it has been reported that the average length of the working day on the state farms decreased by 30 per cent.[63]

One can justifiably argue that these problems — slow recovery of export production and lower productivity — were unavoidable in a period when important structural changes were taking place. It would have been surprising if they had not occurred given that material damage caused by the civil war was extensive, and that the rural policies introduced very substantial modifications in the agrarian structure. Nevertheless, they must be considered as causes for worry and need to be satisfactorily resolved in the near future.

V Problems and Prospects: Concluding Observations

Before summarising and interpreting the findings of the preceding sections, it is necessary to point out that this analysis only covers the first two initial years of a land reform programme that is still in progress. The final outcome of this on-going process will no doubt be an agrarian structure much different from the one observed today. Still it is useful to evaluate and interpret the land reforms already carried out, though it should be borne in mind that these observations are tentative and may need to be modified once further reforms are introduced.

First, the measures to control the agricultural surplus enabled a redistribution of rural incomes, but mainly to the small-scale farmers. Landless workers, comprising one-third of the rural labour force, very likely experienced a decline in their real incomes. Second, food production expanded substantially reaching historically unprecedented levels, while agro-export production, particularly that of cotton, did not attain the targeted level of the 1980 Economic Plan. This has been one of the reasons why the foreign exchange balance has worsened considerably.

The preceding sections also mentioned a variety of reasons for these achievements and failures. It would be incorrect, however, to consider these in isolation. For example, the fact that the food-producing sector performed much better than the agro-export sector is not coincidental. The performances of the two sectors are interrelated. This is largely because the food crop and agro-export sectors compete to a large extent for the same resources. This applies not only to agricultural credit but also to labour. Policies to increase incomes and employment in the small-farm sector will reduce the labour provided to the agro-export sector. To some extent, the two sectors also compete for the same land. A certain proportion of the land can be used profitably to cultivate both food and agro-export crops. Thus the relatively poor performance of the agro-export sector can be viewed, in part, as a consequence of the rise in food production. This competition between sectors for the same resources makes it difficult to identify policies that will benefit one sector without harming the other.

The latest decree on land reform indicates that the government is adopting, at least for the moment, a policy to favour the food-producing sector. This decree, as described earlier, is designed in part to introduce small-scale farming on the land belonging to large farms previously used for growing export crops and raising cattle. The choice of this strategy is based on both political and economic considerations. The small-scale farmers constitute a major socio-political base of the government, and this strategy essentially benefits them. By ensuring their access to land, credit, etc., food production will rise implying a genuine redistribution of incomes in favour of the small-farm sector. Such a choice makes sense also from other points of view. For example, relying too heavily on crop exports makes the country too dependent on international crop prices which are known to fluctuate widely. This dependency can influence negatively the development of the Nicaraguan economy. Also, the government realises that it does not have an adequate organisation and sufficient technically qualified personnel to manage a major expansion of agro-export production.

It would have been difficult, if not impossible, to achieve a comparable reduction in poverty among small-scale farmers through the alternative strategy of favouring agro-export production. The latter would lead to an expansion of the production of export crops on the medium-sized and large farms, likely at the expense of small-scale farming. That is to say, less credit and land would be available to small-scale farms. Thus, small-farm households would increasingly need employment outside their own farms. Wage employment would probably increase more rapidly than under the 'pro-food' strategy, first, because the area cultivated with export crops would increase, and second, because higher foreign exchange earnings would make it possible to develop industry and manufacturing more rapidly, thus increasing urban employment. However, it is unlikely that the increase in the number of rural and urban jobs would be sufficient to absorb workers unable to find work in the small-farm sector. First, the increase in employment on agro-export farms would mainly consist of seasonal work.[64] Second, it is uncertain whether sufficient urban employment opportunities would be created as a result of the development of industry and manufacturing. The so-called modern sectors are so small in size, producing only 15 per cent of GNP, that output would have to increase enormously in order to absorb any significant number of workers from rural areas.

But even if a 'pro-export' strategy generates sufficient wage employment for the small-farm population, more is required to ensure that this strategy will reduce poverty among this population. A rise in food availability is also needed to meet the higher demand for food resulting from higher employment. A 'pro-export' strategy may not permit a sufficiently large increase in food availability, as an increasing proportion of credit and land is absorbed by the agro-export sector. Given these arguments, it is not surprising that the government, determined to raise the standard of living among its largest support group, has adopted a 'pro-food' strategy.

However, the above arguments in favour of a 'pro-food' strategy do not imply that such a strategy is without its problems and that it can be carried out for an indefinite period. First, the development of small-scale farming may lead to a polarisation in rural areas and give rise, as is often argued, to an emerging class of prosperous farmers and the re-emergence of a group of marginalised farmers, though one has to remember that such a danger is inherent in a 'pro-export' strategy as well. The government has announced its intention to organise all farmers into co-operatives so as to prevent such a situation. Moreover, it has taken the initiative of providing only a limited type of land ownership which prohibits land sales under the 1981 decree. However, there is the danger that these policies may not be sufficient to counteract the spirit of individualistic farming which is still strong among a population which has not known anything else for centuries.

Also there is the question as to what extent food production should be favoured. Some have argued that it should be stimulated beyond the point where self-sufficiency is reached so as to export food.[65] Several advantages of this strategy can be identified. International food prices are more stable than international agro-export prices. Also, food production involves a much lower level of technology than agro-export production, thus implying less economic dependency. However, although these advantages cannot be disputed, it remains a fact that food exports generate far less foreign exchange. It has been estimated, for example, that one manzana of cotton generates five times more foreign exchange than one manzana of corn.[66]

Given that the supply of foreign exchange largely influences the rate at which the Nicaraguan economy will develop, this means that a policy of stimulating food exports will lead to limited economic expansion. This applies in particular to the non-agricultural sectors where increases in production are largely facilitated through imports of material, equipment, etc. Limited economic growth also means a reduced expansion of non-agricultural employment which will adversely affect the position of urban workers. Section IV showed that this group, like the rural workers, experienced a decline in their standard of living since 1979. If the expansion of the non-agricultural sectors is further limited, they are likely to worsen their economic position *vis-à-vis* the small-scale farmers. The same applies to the landless rural workers who largely depend on agro-export production for employment. Thus one can foresee a growing disparity between the standard of living of small-scale farmers, on the one hand, and urban and rural wage workers, on the other hand. This would be inconsistent with the government's objective of raising the standard of living of the poorest segments of the population. The rural and urban wage workers, like the small-scale farmers, are part of the political base of the current government and their continuing support is essential to the successful implementation of the structural reforms in the future. To ensure this, it makes sense to favour agro-export production only after self-sufficiency in food has been attained, rather than to adopt a strategy of continuing a 'pro-food' policy indefinitely with a view to exporting food.

Notes

1. United Nations, Economic and Social Council, *Nicaragua: Repercusiones Económicas de los Acontecimientos Políticos Recientes* (Santiago, Chile, 1979).

2.   Similar transitions have been analysed in D. Ghai *et al.*, *Agrarian Systems and Rural Development* (Macmillan, London, 1979).
3.   Oficina Ejecutiva de Encuestas y Censos, *Encuesta Demográfica Nacional* (Managua, 1978).
4.   FAO, *Nicaragua: Misión de Identificación y Formulación de Proyectos* (Programa de Cooperación Tecnica, Managua, 1979), p. 9.
5.   O. Núñez, *El somocismo: Desarrollo y Contradicciones del Modelo Capitalista Agroexportador en Nicaragua (1950-75)* (Centro de estudios sobre America, La Habana, 1980).
6.   Ibid.
7.   1 manzana = 0.7 hectare.
8.   UNRISD, 'Food Systems and Society: The Case of Nicaragua' (United Nations, Geneva, 1981), mimeo.
9.   J. Wheelock, *Imperialismo y Dictadura*, 3rd edn (Siglo Veintiuno Editores, Mexico City, 1979).
10.   H. Ortega Saaverda, '50 Années de Lutte Sandiniste', in A. Jacques, *Nicaragua: La Victoire d'un Peuple* (Edition L'Harmattan, Paris, 1979), pp. 22-104.
11.   Ibid., pp. 26-32.
12.   Indirect evidence on the displacement of the peasantry is also provided by Table 7.1. These data show that the number of small farms decreased between 1963 and 1971 whereas the number of farms of 50 manzanas or more increased during the same period.
13.   Núñez, *El Somocismo*.
14.   A. de Jainvry and C. Garramón, 'The Dynamics of Rural Poverty in Latin America', *Journal of Peasant Studies*, vol. 4, no. 3 (April 1977), pp. 206-16.
15.   FAO, *Nicaragua*, p. 20.
16.   V. H. Gillespie, *Analysis of a Rural Social Survey Conducted in 1976 by INVIERNO* (USAID, Managua, 1978).
17.   The survey also indicated that the minimum wage level was very low and that 65 per cent of the rural wage workers were reported to earn less (11.60 cordobas in 1975/6 or US$1.66 a day).
18.   Ministerio de Agricultura y Ganadería, *Estudio de Comunidades en la Región Interior Central* (Managua, 1977).
19.   World Bank, *World Development Report 1981* (Washington, DC, 1981), p. 176.
20.   For a detailed analysis see J. Lopez *et al.*, *La Caida del Somocismo y la Lucha Sandinista en Nicaragua* (EDUCA, Costa Rica, 1980).
21.   In fact Decree No. 3 issued on 20 July 1979 stipulated that all the belongings would be nationalised of those persons who left the country after August 1977.
22.   Documentation of CIERA quoted in IFAD, *Informe de la Misión Especial de Programación a Nicaragua* (Rome, 1980), p. 76.
23.   The credit records of the newly created state banks show that there are still private farms of over 1,000 hectares.
24.   On some of these farms, rentee farmers and farm workers who previously had no land were loaned small plots without charge to grow food products. This, however, involved only a very small proportion of the land on these farms.
25.   IFAD, *Informe de la Misión Especial*.
26.   *El Nuevo Diario* (Managua, 20 July 1980), p. 7A.
27.   Data for 1963 indicate that around 15 per cent of the farmers cultivated rented land. This percentage probably rose in the 1970s as a result of increased parcellisation and absentee landownership.
28.   *El Nuevo Diario*, p. 36.

29. UNRISD, 'Food Systems and Society', p. 20.

30. IFAD, *Informe de la Misión Especial*.

31. Ministerio de Planificación, *Programa Económico de Austeridad y Eficiencia* (Managua, 1981), p. 76.

32. *Barricada* (Managua, 22 Aug. 1981).

33. Ibid.

34. Estimate by CIERA, Managua.

35. *Barricada*.

36. Ibid.

37. Ibid.

38. Interview of Jaime Wheelock, Minister of Agriculture and Agrarian Reform, 'We are interested in controlling the economic surplus', in *CERES* (FAO, Rome), vol. 14, no. 5 (Sept.-Oct. 1981), pp. 47-51.

39. The analysis will be limited to examining the impact of the policies adopted in 1979 and 1980, excluding, for example, the 1981 decree on land reform which is too recent.

40. Estimates for 1981 varying from 18.0 to 21.8 per cent are found in Ministerio de Planificación, *Programa Económico*, p. 86, and PREALC, Empleo y Salarios en Nicaragua, 1980, p. 10.

41. World Bank, *Nicaragua: The Challenge of Reconstruction* (Washington, DC, 1981), p. 14.

42. D. C. Deere, 'Nicaraguan Agricultural Policy: 1979-81', *Cambridge Journal of Economics*, no. 5 (1981), pp. 195-200.

43. World Bank, *Nicaragua*, p. 15.

44. Ibid.

45. Ministerio de Planificación, *Programa Económico*, p. 103.

46. IFAD, *Informe de la Misión Especial*, p. 130.

47. P. Peek, *Seasonal Employment on Cotton Farms in Nicaragua: Results of a 1981 Survey* (ILO, Geneva, forthcoming, 1983).

48. Ministerio del Trabajo, *Decreto no. 198: Ley de Tabla Salarial Minima para las Labores de Corte de Algodon, Ciclo Agricola 1979-1980* (Managua, 1979).

49. Ministerio de Planificación, *Programa Económico*.

50. World Bank, *Nicaragua*, p. 105.

51. Presumably this calculation was based on the wage rates for both seasonal and permanent agricultural work, IFAD, *Informe de la Misión Especial*, p. 132.

52. The survey of cotton workers referred to earlier showed that one-third came from farm households, Peek, *Seasonal Employment*.

53. This is indicated at least by a simple hypothetical calculation. Given that the *per capita* food subsidy amounts to $15 annually, the average household receives a subsidy of $65 or 650 cordobas, assuming that average household size is five. On the other hand, the reduction in the real wage rate for cotton pickers was 9-14 per cent. Assuming a poor household with two persons doing six months of seasonal work, their household earnings will be 9,000 cordobas; 9-14 per cent of their revenue is much larger than the value of the food subsidy.

54. Ministerio del Trabajo, 'Algunas Consideraciones Teoricas para la Interpretación de la Problematica Salarial en Nicaragua' (Managua, 1980).

55. For example, in an Indian study it was estimated that 'one monetary unit of income transferred from the richest to the poorest classes reduced demand for 0.02 units of foodgrain but creates a new demand for 0.50 units, and imbalance in the ratio of 30 to 1'. From J. Mellor, 'Food Price Policy and Income Redistribution in Low-income Countries', *Economic Development and Cultural Change* (Chicago, Ill.), vol. 27, no. 1 (1978), p. 25, quoted in K. Griffin and J. James, 'Managing the Transition to Egalitarian Development', paper pre-

sented at the Sixth World Congress of the International Economic Association, Mexico, 4-8 August 1980, p. 2.
56. Deere, 'Nicaraguan Agricultural Policy', p. 200.
57. IFAD, *Informe de la Misión Especial*, p. 132.
58. Ibid.
59. Ministerio de Planificación, *Programa Económico*, p. 31.
60. Deere, 'Nicaraguan Agricultural Policy'.
61. Ibid.
62. Deere, 'Nicaraguan Agricultural Policy', p. 199.
63. Ibid.
64. For example, on cotton farms of over 50 manzanas six permanent workers only were employed per 100 manzanas. Thus if the area cultivated with cotton were to be doubled, only 8,400 additional permanent jobs would be created, i.e. less than 3 per cent of the rural labour force; see Peek, *Seasonal Employment*.
65. IFAD, *Informe de la Misión Especial*, pp. xix-xxiii.
66. Estimate by CIERA (Managua).

References

Barricada (1980, 1981) (Managua, 25 Apr. 1980 and 22 Aug. 1981)
CERES (1981) 'We are Interested in Controlling the Economic Surplus' (Rome, FAO), 14 (5) (Sept.-Oct.), 47-51
Deere, D. C. (1981) 'Nicaraguan Agricultural Policy: 1979-81', *Cambridge Journal of Economics* (London) (Sept.)
FAO (1970) *Report of the 1960 World Census of Agriculture 1966-70*, Rome, vol. I/B
—— (1979) *Nicaragua: Misión de Identificación y Formulación de Proyectos*, Programa de Cooperación Tecnica, Managua
FAO-SIECA (1974) *Perspectives para el Desarrollo y la Integración de la Agricultura en Centroamérica*, Guatemala City, Guatemala
Ghai, D. *et al.* (1979) *Agrarian Systems and Rural Development*, Macmillan, London
Gillespie, V. H. (1978) *Analysis of Rural Social Survey Conducted in 1976 by INVIERNO*, USAID, Managua
Griffin, K. and James, J. (1980) 'Managing the Transition to Egalitarian Development', paper presented for the Sixth World Congress of the International Economic Association, Mexico, 4-8 Aug.
Hintermeister, A. (1980) 'La Agricultura, Base para la Construcción de una Nueva Sociedad', *Cuadernos de Marcha* (Mexico City) (Jan.-Feb.), 17-33
—— (1981) 'El Empleo Agricola en una Estructura en Transformación: El Caso Nicaragua', PREALC paper presented at workshop on 'Conceptualization of Rural Employment in Latin America and Policy Recommendations', Mexico City
IFAD (1980) *Informe de la Misión Especial de Programación a Nicaragua*, Rome
de Jainvry, A. and Garramón, C. (1977) 'The Dynamics of Rural Poverty in Latin America', *Journal of Peasant Studies*, 4 (3) (Apr.)
Kaimowitz, D. and Thome, J. R. (1980) 'Nicaragua's Agrarian Reform: The First Year (1979-80)', Land Tenure Center, University of Wisconsin, Madison, mimeo.
La Prensa (1982) (Managua, 13 Feb.)
Latin America: Weekly Report (1980) (London, 5 Sept.)

306   *Agrarian Reform in Nicaragua*

Lopez, J. *et.al.*(1980) *La Caida del Somocismo y la Lucha Sandinista en Nicaragua.* EDUCA, Costa Rica

Mellor, J. (1978) 'Food Price Policy and Income Redistribution in Low-income Countries', *Economic Development and Cultural Change* (Chicago, Ill.), 27 (1)

Ministerio de Agricultura y Ganadería (1977) *Estudio de Communidades en la Región Interior Central*, Managua

Ministerio de Planificación (1980) *Programa de Reactivación Economica en Beneficio del Pueblo* Managua

—— (1981) *Programa Económico de Austeridad y Eficiencia*, Managua

Ministerio del Trabajo (1979) *Decreto No. 198: Ley de Tabla Salarial Minima para las Labores de Corte de Algodon, Ciclo Agricola 1979-80*, Managua

—— (1980) 'Algunas Consideraciones Teoricas para la Interpretación de la Problematica Salarial en Nicaragua', Managua

Nuevo Diario (1980) (Managua, 20 July and 13 Aug.)

Nuñez, O. (1980) *El Somocismo: Desarrollo y Contradicciones del Modelo Capitalista Agroexportador en Nicaragua (1950-75)*, Centro de Estudios sobre América, Havana, Cuba

Oficina Ejecutiva de Encuestas y Censos (1978) *Encuesta Demográfica Nacional*, Managua

Ortega Saaverda, H. (1979) '50 années de lutte Sandiniste' in A. Jacques, *Nicaragua: La Victoire d'un Peuple*, Edition L'Harmattan, Paris, pp. 22-104

Peek, P. (1983) *Seasonal Employment on Cotton Farms in Nicaragua (1981): A Survey Analysis*, ILO, Geneva, forthcoming

PREALC (1972) *Situación y Perspectivas del Empleo en Nicaragua* 2 vols., Santiago, Chile

—— (1979) *Diagnóstica de las Estadísticas y Bibliografía Sobre el Empleo Rural en Centroamerica y Panama*, Santiago, Chile

—— (1980) *Empleo y Salarios en Nicaragua*, Santiago, Chile

United Nations Economic and Social Council (1979) *Nicaragua: Repercusiones Económicas de los Acontecimientos Políticos Recientes*, Santiago, Chile

Regional Report of Mexico and Central America (1981) (Latin American Newsletters Ltd, London, 14 Aug.)

UNRISD (1981) 'Food Systems and Society: the Case of Nicaragua', United Nations, Geneva, mimeo.

Warnken, P.F. (1975) *The Agricultural Development of Nicaragua: An Analysis of the Production Sector*, University of Missouri, Columbia

Wheelock, J. (1979) *Imperialismo y Dictadura*, 3rd edn, Siglo Veintiuno Editores, Mexico City

World Bank (1981) *World Development Report 1981*, Washington, DC

—— (1981) *Nicaragua: The Challenge of Reconstruction*, Washington, DC

PART FOUR

IN SEARCH OF MODERNISM

8 THE AGRARIAN QUESTION IN IRAN

Homa Katouzian

I Introduction

This chapter is a study of recent agrarian reforms in Iran, their historical setting and genesis, and their consequences for agriculture and rural
society in relation to developments in the rest of Iran's economy. It
starts by providing a historical perspective — the evolution of traditional organisation, relations and techniques of agricultural production
and their broader socio-economic context — and then proceeds to
present an analysis of the origins, methods and implications of the
agrarian reform measures. This is followed by a discussion of the
economic position and performance of agriculture in the post-reform
period (1963-78), with special emphasis on the adverse effects of the
ill-conceived modernisation programmes, inspired by the phenomenal
growth of oil revenues, on the country's agrarian economy. The next
section then examines the social consequences of these economic events
for the rural population: the very large gap between rural and urban
incomes and welfare, rural unemployment, the high rate of rural-urban
migration and the consequent transfer of rural poverty to urban areas,
and the level of impoverishment of the peasantry. The fifth and final
section draws together some broad lessons of the Iranian experience
with agrarian reforms.

II Traditional Agriculture: The 'Aridisolatic' Society[1]

Iran never had feudalism in the classical sense of the term. Private
property in land was weak and tenuous, based as it was on various land
assignment systems associated with military/bureaucratic *privileges*
rather than with aristocratic *rights*. The state itself directly owned a
significant share of the agricultural land, and indirectly controlled the
'ownership' of the remainder. There existed neither a manorial system
(landlords characteristically made up an *urban* social class), nor serfdom. Peasant obligations were limited to the payment of crop shares
and taxes to various exploitative agents. Towns and cities were relatively

309

large and numerous, commerce was extensive and elaborate, money played a significant role in the urban sector, and — at least as early as a thousand years ago — there existed a widespread network for credit transfer (and, to a limited extent, credit creation) through the circulation of bills of exchange. Consequently, 'politiconomic'[2] power was historically concentrated in cities (*Shahr* = 'country'). Political power was both absolute and arbitrary (regardless of whether or not there existed a large centralised bureaucracy), social mobility was high, and there existed no aristocratic peerage. The whole of Iranian history is characterised by a succession of despotic states, separated from one another by periods of (internal or external) upheavals and socio-economic involutions.

The traditional Iranian village provided the socio-economic foundation for the functionally despotic state. It was normally made up of households with a traditional right of cultivation (the *Nasaq*-holders), households without such rights (the *Khusnishīns*), and a number of traders and moneylenders who supplied small amounts of credit in cash or in kind at high (implicit) interest rates by way of advance purchases of a part of the crop. In many villages, some *Nasaq*-holders and/or *Khushnishīns* (known as *Gāvbands*) lent one or two pairs of oxen to other cultivators against a share (normally one-fifth) of the crop. From the early twentieth century the development of *Ijāreh Karī* (a form of small-scale tenant farming) in some parts of the country had encouraged the use of wage labour which was usually supplied by the *Khushnishīn* community.

The traditional *system of production* was communal. The Iranian village 'commune' is described by various terms in different regions: *Buneh* and *Sahrā* are most commonly used. This does bear a similarity to the old Russian village commune, the *Mīr*, but was a looser and less comprehensive institution.[3] The *Buneh* owed its origins to the fact that, except in one or two small regions, water was the country's most scarce resource. This encouraged communal co-operation for the construction and upkeep of underground water channels (*Qanāt* or *Kārīz*) as well as the distribution of water among the cultivators; hence also the existence of the *Ābyār* (*Owyār*) or 'water assistant', and his deputies. Thus the climatic aridity did not turn Persia into a 'hydraulic society' *à la* Wittfogel.[4] Typically, the village community itself, as one unit, organised the supply and distribution of water.

Once the *Buneh* came into existence, however, it developed other socio-economic functions, such as decisions concerning crop rotation, fallow fields, etc. It also often tried to ensure an average equality of fertility for the holdings of all the cultivators and consequently peasant holdings were usually 'open' and 'scattered'. In general, the more arid the location, the stronger was the *Buneh*, and the greater was the likelihood of 'scattered' holdings. The landlord (who could be either the state itself, or a land assignee, or the board of trustees of a charitable endowment) was generally an outsider and, in any case, was not a member of the *Buneh*; only his local agent (the Mubā-shir) provided the necessary link between the two.

The traditional method of distribution of output was based on the 'five input rule', the inputs being land, water, seeds, oxen and labour. The landlord would take the shares of land and water, the peasant would take the share of labour, and the two shares of seeds and oxen would go to their respective suppliers. In practice, the mode of distribution diverged from this rule, though not to the extent of making it irrelevant, primarily because there existed a variety of land and other taxes which evolved through history and varied across regions.

Therefore, both the internal socio-economic structure and relations, and the external (geographical as well as 'politiconomic') conditions, made the Iranian village an independent social unit, with few (if any) links with other — usually distant — villages, and little or no interaction with the urban outsiders who only came at harvest time to take 'their shares' of the village output. As mentioned earlier, Iranian agriculture and peasantry were not dependent on the state for the provision and regulation of water supplies, or anything else. On the contrary, it was the state that depended on isolated and scattered villages for the agricultural surplus which it either directly requisitioned or assigned to landlords and tax-farmers.

Thus, the model of the 'aridisolatic' society that evolved historically may be briefly described as follows. The inter-related phenomena of the aridity of land and the paucity of population led to the development of isolated and internally self-sufficient units of agricultural life and labour which made up the Iranian village. As a rule, none of these autonomous villages could singly produce a sufficiently large surplus to afford a feudal power base, but all of them taken together produced a collective surplus large enough to form the basis of an economically viable state. The martial force necessary for the survival of a despotic state was originally provided by invading *nomadic* tribes, and thereafter both by the existing as well as incoming nomads who succeeded in setting up various urban states at different stages of history. The size of the direct and indirect collective agricultural surplus was large enough to enable these despotic states to spend on transport, communications, military and bureaucratic organisations, etc., which both maintained their hold on the land and prevented the emergence of a feudal aristocracy in the village. Such were the origins, the logic and the history of the functionally despotic (as opposed to merely 'absolutist') state which has dominated Iran for millennia.[5]

III Agrarian Reforms: A Review

The revolution of 1905-9 that overthrew the despotic Qajar state marked a watershed in the development of Iran's agricultural economy. The first National Assembly (*Majlis*), formed in 1906, abolished the traditional land assignment system and institutionalised the existing ownership of land, turning the landlords into a class of large private landowners. It was therefore not surprising that Rezā Khān's rise to power (1921-6) was generally resisted by landowners, but helped by rootless 'modernist' army officers, bureaucrats and intellectuals who made up the actual and potential clients of the state. For landowners (as well as urban traders) saw in his efforts to concentrate all political power in his own hands and centralise the bureaucratic and military networks a clear threat to their own economic and political power. Events proved them right, because in spite of Rezā Shah's modern administrative reforms (1926-41) concerning land registration, etc., the large landowners (as a class) lost a great deal through land tax, the state monopoly of trade in the main agricultural products (particularly corn), loss of political influence, and the confiscation of some of their estates by the Shah himself.

Rezā Shah's abdication and the succession of the young crown prince in 1941 made it possible for large landowners to regain much of their economic and political strength. After the first few months of Dr Musaddiq's government (1951-3) the landlords joined the opposition because they were afraid of the consequences of the deadlock in the Anglo-Iranian oil dispute and the radicalisation of the popular movement which backed Mussaddiq. They therefore joined forces with the Shah and other disaffected groups in a campaign which eventually resulted in the *coup d'état* of August 1953.

After the 1953 *coup*, the landlords succeeded in effectively dismantling the law – first enacted by Qavām's government in 1947 and later extended by Musaddiq's – which obliged them to pay 10 per cent of their annual share of the village output back to the peasants and a further 10 per cent into a public fund for rural development.

But the inflow of substantial oil revenues and American aid increased the Shah's power at their expense, and there was a brief confrontation in 1960 when a mild and meek 'land reform' Bill (proposing to put an upper limit on the size of private holdings) was submitted to the National Assembly. It was eventually passed but soon became a dead letter. In 1960, an acute balance-of-payments deficit, rising inflation and internal political opposition began to alter the existing balance of power, and in 1961 Dr Ali Amini at the head of a group of reformist politicians formed a government.[6]

The new government seriously intended to carry out a comprehensive programme of land distribution. The programme intended to divide up the land among the *Nasaq*-holding peasantry (roughly about 65 per cent of the peasant population), who would compensate the landlords by annual instalments. They had not and, in the circumstances, probably could not have thought through the whole of their scheme, but by making the receipt of land conditional to membership of rural co-operatives, Arsanjānī, the Minister of Agriculture, may have intended to preserve something of the old communal system of production. There was only one alternative programme put forward as a serious rival to Arsanjānī's: Khalīl Malekī's (the Socialist leader) proposal for the democratisation (*Mellī kardan*) of land and water resources. This did not mean nationalisation; it suggested the collective transfer of the title of ownership from large landowners to groups of peasants, which would dispossess the landlords by one stroke and without complicated legal wrangles, avoid the colossal administrative task of defining every single peasant holding in every village, prevent the emergence of scattered and small individual holdings which (by further fragmentation through inheritance) could be reduced to plots so small as to have to be sold to a few large holders resulting in concentration of ownership, and prevent the dispossession of the *Khushnishīn* community, and its social and economic consequences. The most important implication of the scheme was that it would preserve the communal mode of village life and labour, and the 'egalitarian' modes of holding and distribution while getting rid of the large landlords. The idea, however, was too advanced and (seemingly) too radical for it to be acceptable to the dominant political forces in the country at the time.[7]

The Land Reform Law, drafted by the Amini government, began to be applied in January 1962. It had envisaged the distribution of land among the *Nasaq*-holding peasantry in various 'stages' of the reform. The state would value the land and pay the landlord, and the peasant households would reimburse the state in ten annual instalments.

There were a number of significant exemptions. Orchards and 'mechanised' farms were excluded, but the most important exemption clause was that which authorised the landowners to retain one whole ('six-*dang*') village or its equivalent scattered among a number of villages. In June 1962, the Shah dismissed Amini, and the law underwent a series of amendments in favour of the affected landowners. For example, the law had declared that 'for the purposes of the provision of this law, a household is regarded as one person'. This definition was confirmed by the 36th Session of the Land Reform Council. But the 43rd Session of the LRC (7 June 1963) reversed this decision by declaring that 'women landowners in the same way as men and irrespective of their position in the household [i.e. even if their husband is also a landowner] are entitled to the upper legal limit of ownership'. This was followed by a further decree of the LRC (issued on 25 August 1963) which also entitled the dependent children of the landowners to the upper limit of ownership. Thus, for example, a landowning family with four dependent children would be legally entitled to retain six whole villages.[8] There were a number of other such amendments which violated the spirit of the original reform programme, but the most obvious violation of its letter is manifested in the Supplementary Articles of the Law, known as the Second Stage. At the same time as the Supplementary Articles were decreed, the Shah held a referendum to launch what he described as the White Revolution. The programme included six statements of principle, namely land reform, the 'enfranchisement' of women, the nationalisation of forests, the establishment of a conscripted Literacy Corps, the entitlement of industrial workers to a share of the firm's profit, and the sale of a few state monopolies to help finance the land reform programme.

According to the Supplementary Articles on the Second Stage — which began to be applied in 1966 — one of the following five possibilities was open to the landowners: they could (i) lease their land to peasant cultivators; (ii) divide it with them according to the ownership of the traditional five inputs; (iii) sell it privately to the peasants; (iv) establish private corporations with peasants, the corporation shares being distributed according to the ownership of the five inputs; or (v) buy the peasants' rights and employ them as wage labourers.[9]

Table 8.1 shows the results of the application of both stages of land reform after their completion. According to this table, 24.6 per cent of all peasant households (or 22.4 per cent of the rural population) received land through the operation of the First Stage, while a further 48.6 per cent of the peasant households (or 47.9 per cent of the rural population) were affected by the provisions of the Second Stage for tenancy reform.

Table 8.1: Results of the Land Reform

	Villages and Hamlets	Peasant Households	Rural Population
1. Country-wide total	76 682	3 200 000	17 600 000
2. First Stage: distribution			
(i) Total number	16 593	787 000	3 942 000
(ii) Per cent of country-wide total	21.6	24.6	22.6
3. Second Stage: tenancy reform			
(i) Total number	NA	1 556 480	8 432 000
(ii) Per cent of country-wide total	NA	48.6	47.9
4. Total (2 + 3)			
(i) Number	NA	2 343 480	12 374 000
(ii) Per cent of country-wide total	NA	73.2	69.3

Sources: Bank Markazi Iran, *Annual Report*, 1972/3, Appendix Table 71; M.A.H. Katouzian, 'Land Reform in Iran: A Case Study in the Political Economy of Social Engineering', *Journal of Peasant Studies* (Jan. 1974), Table 1, p. 230.

In sum, less than 70 per cent of the rural population were affected by the operation of both stages, and only 22.6 per cent received land through the original distribution programme. However, the official figures are likely to be over-estimates, and the total number of villages and hamlets affected by the Second Stage has not been made publicly available. Apart from that, it is not clear how many of the 16,593 villages, apparently affected during the First Stage, were 'whole villages' and how many of them were hamlets and smaller villages.[10]

However, long before the 'completion' of land reform, the growth of oil revenues via rapid annual expansions of the quantity of exported crude oil had initiated a series of fundamental changes in the social and economic structure of the country. In particular, it was rapidly increasing the financial and — therefore — political power of the state, replacing rural property as a source of sustenance for the state and its clientele, and reducing the dependence of the urban-industrial sector on the domestic agricultural surplus for food, raw materials and foreign exchange. Historically, the collective agricultural surplus had been the main source of the financial autonomy and the despotic power of the Persian state. This had been a collective rent received by the state and its urban clientele (landlords, tax officials, etc.) from outside the urban sector for use inside it. That was the basic logic of the aridisolatic society. The growth and, later, explosion of the oil revenues reshaped this historical framework into what could be described as the petrolic society: oil revenues became the new source of collective economic rent for the state.[11] The significant differences between the old and the new frameworks are, first, that (unlike the collective agricultural surplus) increase in oil revenues are (in part) in the nature of windfall gains – and, in particular, 'production' of oil engages a negligible percentage of the country's total labour force; second, that the revenues are received wholly by the state, there being no question of a direct 'assignment' of a part of the oil resources to some of its clientele; and third, that (even before the oil revenue explosion of 1973) the revenues became so large that the state could widen its circle of clientele and – apart from that – extend its direct interventions to the Iranian village.[12] Historically, the state had (directly or otherwise) taken the village surplus, but left the village community undisturbed; but the petrolic state entered the village in order to destroy its traditional framework and to control peasant life and labour directly.

The psychology was that a large traditional agriculture was a 'shameful' evidence of backwardness; the sociology was that the state was trying to integrate the historical 'outsiders', i.e. the peasantry, for the first time in Iranian history; and the economics were that the state was no longer dependent on the agricultural surplus as a source of food and exports, because the oil revenues could take care of these. In effect, (a) the state had no real interest in the development of the country's large agricultural sector or a growing prosperity for its vast rural population, and (b) it had every interest in creating a small 'modern' agriculture and turning the majority of the peasant population into urban wage labourers. It was the Shah himself who, early in 1973, even before the oil revenue explosion, boasted that by 1980, there would be no more than 2 million people (that is 300,000 workers) on the land. At that time the rural population was about 18 million, and the country-wide average annual rate of population growth was 2.9 per cent!

In accordance with this vision, a new programme to develop farm corporations was launched. The farm corporations were apparently intended to prevent fragmentation of peasant holdings into very small units, to increase the scale of agricultural production, and to encourage the use of modern technical inputs. The risk of fragmentation presumably arose as a consequence of the original land distribution programme. But a revitalisation of communal cultivation would have ensured a reasonably large scale of production in spite of the small size of individual holdings. And the application of relevant modern inputs and equipment would have been possible through the introduction of an appropriate co-operative network as well as the extension of financial and technical services without the direct intervention of state officials in the daily life and labour of the village and the rural society.

Such an outlook, however, was wholly inconceivable to the protagonists of the farm corporation project. Indeed, the project quickly put its emphasis on the establishment of farm corporations beyond a single village — the age-old Iranian unit of agricultural production — and tried to set up larger corporations to encompass a number of villages taken together. If there could be any justification for the formation of single-village corporations, there could be no justification at all for 'breaking the boundaries of the Iranian village'. There were obvious economic, sociological, technological, transport and other costs involved in the project, but its benefits were far from clear.

In the areas where farm corporations were established, the ownership of shares was theoretically determined by the relative size of peasant holdings. This converted the ownership of land into the ownership of paper shares. And it was only a matter of time before small shareholders would begin to sell their shares to large ones. Hence, absenteeism of bigger farmers could now replace the traditional absenteeism of landlords. The large shareholder could now afford to live in a nearby town and engage in other activities, while the army of landless peasants — the *Khushnishīn* village community — would provide labour for cultivation. The significant disincentives created by the new system for the small peasant shareholders are probably too clear to elaborate. But, above all, the corporation made it much easier for state officials to control every aspect of social and economic activity in incorporated villages.

The establishment of the farm corporation units was supposed to be voluntary. In practice, however, villages in the most favourable locations were officially selected for the purpose, and the 'volunteers' were theoretically given a chance to reject the proposal through the ballot. Even in theory the majority decision of a single village against incorporation was not sufficient, as long as the majority vote of all the affected villages, taken together, went in its favour. In practice, once the official decision had been taken in Tehran, the peasants were already deemed to have 'volunteered' to accept it. There were the usual constitutional window dressings concerning the annual general meeting of the shareholders, the 'elected' board of directors, etc. In fact, state officials were sent directly from the relevant ministries to manage the corporations, and to implement all the major decisions which were, nevertheless, taken not by themselves but by ministers and higher officials in the state departments.

In 1968 the strategy for concentration and 'modernisation' took yet another turn with the policy of setting up large agricultural factories described as 'agro-business' or 'agri-business'. This policy intended to create large-scale capitalistic farms conceived as a curious blend of Latin American latifundias and Californian agricultural firms. The agri-business companies were to be set up by a combination of foreign and domestic (private as well as public) capital, although in practice the Iranian government owned more than 60 per cent of the shares of all of them taken together. And they were to operate in regions where the fertility of the soil and availability of water resources – officially described as 'the poles of land and water resources' – made it most attractive land for extensive farming.

The policy had originated in a World Bank pilot project for the demonstration of modern farming methods and technology which the Bank had made a condition for its significant financial contribution, in 1959, to the construction of Dez Dam, in the south-western (oil) province of Khūzistān. The pilot project had encompassed 58 villages covering about 50,000 acres of farmland, but neither the landlords nor, after land distribution, peasants showed much enthusiasm for it. Meanwhile, the management of the project had passed on to the provincial water and power authority, and when in 1968 the Agri-business Law was passed, the same authority was charged by the government with the responsibility of requisitioning the project area and leasing it out to various agri-business firms.

The villages were levelled, and the peasantry were moved to five newly created service towns and provided with accommodation in houses made of cinder blocks (which are entirely unsuitable for the intolerably hot and humid environment) with a total floor space of 40 square metres per 10 residents. Both peasant holders and *Khush-nishīns* lost the opportunity of small-scale livestock farming which they used to carry out in and around their former dwelling places. The peasant holders were paid administrative prices for the loss of their holdings, which were further reduced by the full amount of their debt obligations to the state as well as the cost of their new habitats. They lost their communal relations, became (mainly part-time) wage labourers for the agri-business firms, had to walk for miles to and from their place of work, and had to purchase their food requirements from the market. In all, there were 40,000 peasants affected in that particular area. These agri-business ventures were followed by a few more in other parts of the country, notably in the western province of Kirmānshahān, and the north-eastern province of Khurāsān.[13]

IV Agrarian Reforms and Agricultural Growth, 1963-78

IV.1 Basic Features of Agricultural Growth

The most important factor which in this period influenced state policies and thus economic growth and structural change in Iran was the growth and, later, explosion of oil revenues. Oil was *the* independent variable of the development strategy which led, among other things, to a rapid change in the country's economic structure, the plight of the agricultural sector, a continuous rise in the rate of rural-urban migration, the creation of acute imbalances between and within various economic sectors, a higher capital-output ratio, and an increasing rate of inflation.

In particular, the development strategy and other state policies turned the social and economic scale against agriculture and the peasantry in every respect. A peasantry-based development strategy was quickly replaced by a strategy for the creation of a small 'modern' agricultural sector and the eventual dissolution of peasant farming; the five-year economic plans allocated inadequate funds to agriculture, both in relation to its requirements and relatively to public expenditures on industry and services; most of the planned allocations and other extensions of funds to agriculture were channelled to the farm corporations and agri-businesses; concentrration of public expenditure in the urban sector rapidly widened the already large income and welfare gaps between the two sectors of the economy; the highly capital-intensive technology of modern industry made impossible the absorption of (peasant) immigrant labour and condemned them to a subterranean existence in the urban peripheries; and the easy availability of foreign exchange encouraged a revaluation of the Iranian currency, and thus facilitated rapidly increasing food and raw materials imports, both of which went against agriculture.

The oil revenues exploded late in 1973: the previous decade (1962-72) covered the period of the Third and Fourth (Development) Plans, and the following quinquennium was the Fifth Plan period. Tables 8.2 and 8.3 show that in 1972 GDP was two-and-a-half times what it had been in 1962, and that in 1977 it was more than twice what it had been in 1973. Unfortunately, figures in Tables 8.2 and 8.3 are not comparable, because those in Table 8.2 are based on 1959/60 prices, whereas those in Table 8.3 are based on 1974/5 prices.

Table 8.2: Gross Domestic Product by Sector, 1962-72 (000 million rials at constant, 1959/60, prices)

	1962/3	1967/8	1971/2
1. Agriculture	88.8	111.1	122.3
2. Manufacturing and mining	41.5	72.5	118.0
3. Construction, water and power	16.3	33.8	50.4
4. Services	119.8	187.0	317.9
5. Oil	67.3	127.4	221.9
6. Total GDP at factor cost	333.7	531.8	830.5

Source: Appendix Table 1.

Table 8.3: Gross Domestic Product by Sector, 1972-8 (000 million rials at constant, 1974-5, prices)

	1972/3	1975/6	1977/8
1. Agriculture	271.0	324.0	339.0
2. Manufacturing and mining	224.8	360.9	468.2
3. Construction, water and power	108.6	171.6	216.1
4. Services	629.4	1 029.1	1 281.3
5. Oil	449.2	1 224.9	1 422.0
6. GDP at factor cost	1 683.0	3 110.5	3 726.6

Source: Appendix Table 2.

Table 8.4 shows the annual growth rates of GDP and its various sectoral components for the three successive plan periods. It can be observed that both GDP and its various components *other than agriculture* grew at rapid as well as increasing rates throughout the period 1962-77. For reasons that were briefly described above, the growth of the oil sector should be regarded as the autonomous or independent variable which influenced the growth of GDP and the other sectors *as well as* the observed differences in sectoral growth. Agriculture grew at the rate of 4.5 per cent during the Third Plan period,

2.6 per cent during the Fourth Plan period, and 4.9 per cent during the Fifth Plan period. However, these (relatively moderate) growth rates, based as they are on official figures, are very likely to be over-estimates. For example, official figures indicate that the annual rate of growth of agriculture for the whole of the decade 1962-72 was 3.6 per cent (see Table 8.7 below), whereas independent observers and research workers have generally put it at or below 3 per cent.[14] Given the population growth rate of 2.9 per cent, this implies that the rate of growth of *per capita* agricultural output was zero or negative.

Table 8.4: Annual Growth Rates of GDP by Sector, Selected Periods, per cent

	Third Plan Period (1962-7)	Fourth Plan Period (1967-72)	Fifth Plan Period (1972-7)
1. Agriculture	4.5	2.6	4.9
2. Manufacturing and mining	11.2	11.7	14.9
3. Construction, water and power	14.1	9.6	15.2
4. Services	9.4	13.0	14.4
5. Oil	12.7	14.0	20.6
6. Total GDP at factor cost	10.0	11.0	15.3

Sources: Appendix Tables 1 and 2. Growth rates have been calculated by using semi-logarithmic regression equation $Y = Ae^{\beta t}$.

The relative decline of agriculture is clearly demonstrated by the data in Table 8.5: between 1962 and 1977, its percentage share of GDP is observed to have steadily decreased from 26.6 per cent to 9.1 per cent. It is worth noting that in 1977/8 agriculture and manufacturing *together* contributed less than 22 per cent of aggregate output, while services accounted for over 34 per cent of it. The share of manufacturing includes both traditional and modern and both rural and urban manufacturing output. In 1977/8, the share of oil and services together was more than 72 per cent of the national output.

Agricultural activities in Iran divide into two main sectors: arable farming and animal husbandry. These two sectors together produce a variety of food, cash crops and industrial raw materials. Arable

Table 8.5: The Distribution of Gross Domestic Product by Sector, 1962-78 (per cent)

	1962/3	1967/8	1972/3	1977/8
1. Agriculture	26.6	20.9	16.1	9.1
2. Manufacturing and mining	12.4	13.6	13.4	12.6
3. Construction, water and power	4.9	6.3	6.4	5.8
4. Services	35.9	35.2	37.4	34.4
5. Oil	20.2	24.0	26.7	38.1
6. Total GDP	100.0	100.0	100.0	100.0

Source: Tables 8.2 and 8.3.

farming is dominated by the production of wheat (which covers about 60 per cent of total cultivated land) and barley (which covers another 15 per cent of the area). In the remaining 25 per cent of arable land a large variety of crops, including rice, cotton, sugar-beet, oil-seeds, pulses, fruits, vegetables and tobacco, are cultivated. Cotton and rice (which is more of a cash crop in the case of Iran) top this list with a percentage share (respectively) of 3.5 and 3.3 per cent in total land. Livestock farming consists mainly of sheep, goats and poultry production. The supply of mutton has been the dominant contribution of this sector, although wool, leather, dairy products and poultry are also among the outputs. Apart from arable farming and livestock production, there are some forestry, and relatively insignificant fishing, activities. In sum, wheat and barley are the dominant staple food products, cotton, rice, fruits, sugar-beet, oil-seeds and tobacco are the important cash crops, and mutton and dairy products are the most important livestock products.

Iranian agriculture has always been partly settled and partly nomadic. At the beginning of the nineteenth century, the settled and nomadic rural populations were roughly equal, but by the end of that century there had been both an absolute and a relative increase in settled agricultural population: they were, respectively, about 6 and 2.5 million in the first decade of the twentieth century. A consequence of the centralist policies of Pahlavi rule (1926-79) was a further absolute decline in the nomadic population. Although reliable statistics are not available, by 1963 there were probably no more than 1.5 million nomads in the country. The state policy of enforced settlement led to armed resistance, symbolised by the revolt (in the mid-1960s) of the Qashqā'ī nomads in the southern province of Fars. But once the Qash-qā'ī revolt had been ruthlessly crushed, there could be no effective resistance against the settlement policy. However, the organisation of some settled agriculture is also tribal, and tribal agricultural activities can be both nomadic and non-nomadic. According to a 1976 estimate, the total tribal agricultural population was about 5 million, or 28 per cent of the entire agricultural population (see Etemad Moghadam, 1978).

Throughout the 1960s and 1970s, demand for agricultural products — and especially food — grew very fast. This was, to a limited extent, a consequence of the rapid increase in population (at an average annual rate of 2.9 per cent); but, much more significantly, it was due to the high rate of growth of *per capita* (urban) incomes as a result of the fast growth of oil revenues.

Food output was generally unresponsive to the pressure of demand. The changing whims of the State concerning land tenure and agricultural organisation were bound to be a cause of permanent insecurity for the peasantry. Modern large-scale production units were ill-designed and badly managed, while they tended to monopolise the best natural resources and received most of the financial and technical assistance from the state. The fast growth of population, towns and urban incomes led to a high rate of migration of skills from the rural sector to towns and cities. The growth of oil revenues made it possible to rely on food and raw materials imports which, on the one hand, made agriculture economically dispensable and, on the other hand, robbed agriculture of economic incentives which may have resulted from changes in the domestic terms of trade. The policy of confrontation with the nomadic population and the state monopolisation of scarce pasture lands spelled disaster for the supply of livestock — and especially mutton — products. Yet the state officials continuously blamed 'the weather' and 'the peasants' lack of response' for the plight of agriculture.

Table 8.6 provides a summary view of the composition and distribution of various types of agricultural activity in the initial and terminal years of the Third through to Fifth Plans. It appears that, throughout the fifteen-year period, the shares of crop farming and livestock production in total agricultural output have remained constant, respectively, around 65 and 32 per cent. But since the values are given in current prices, the near constancy of these shares conceals the fact that the price of livestock products has increased much faster than that of crop output. This point may be verified from the figure in Table 8.7

for the period up to the year 1971/2 (beyond which date the relevant data measured at constant (1959/60) prices are not available). Here we observe that the share of crop farming grew from 67 to 72 per cent, while that of livestock production declined from 32 to 26 per cent. And there is no reason to believe that this trend was reversed in the latter periods.

Table 8.6: The Composition and Distribution of Agricultural Output, Current Prices, 1962-77

	1962/3 000 Million Rials	1962/3 Per-centage Share	1967/8 000 Million Rials	1967/8 Per-centage Share	1972/3 000 Million Rials	1972/3 Per-centage Share	1977/8 000 Million Rials	1977/8 Per-centage Share
1. Crops	64.0	66.1	87.6	68.2	135.6	67.2	311.7	64.0
2. Livestock	31.7	32.7	39.2	30.5	62.1	30.8	158.7	32.7
3. Forestry and fishing	1.2	1.2	1.6	1.3	4.1	2.0	14.6	3.3
4. Total	96.9	100.0	128.4	100.0	101.8	100.0	485.0	100.0

Sources: Bank Markazi Iran, *Annual Report* and quarterly *Bulletin*, various dates; *National Income of Iran, 1959-72*.

The above tendencies are reflected in the total and sectoral growth rates of agriculture over the period 1962-72: the overall growth rate appears to have been 3.6, that of crop production 4.5, and that of livestock production 1.5 per cent. These official figures probably reflect the relative differences in the growth performance of the two main agricultural sectors accurately. But the actual growth rates themselves are likely to have been less, because growing consciousness of the poor state of Iranian agriculture prompted the government to show 'better figures'.

As regards the performance of crop farming and livestock production, the following observations may be made. First, the higher growth of crop production in the period 1962-7 (as opposed to its much lower growth rate in the next, 1967-72, period) tends to support the view that the original land distribution programme did not have a harmful effect on crop production, whereas the subsequent 'modernisation' schemes did (see Table 8.7). Second, the better performance of crop output as compared to livestock production, over the whole of the decade 1962-72, may be explained (a) by considerable land reclamations

Table 8.7: Composition, Distribution and Growth Rates of Agricultural Output, 1962-71 (constant 1959/60 prices)

	1962/3		1967/8		Growth Rate 1962-7	1971/2		Growth Rate 1971-2	1962-71 Annual (compound) Growth Rates 1962-71
	000 Million Rials	Percentage Share	000 Million Rials	Percentage Share		000 Million Rials	Percentage Share		
1. Crops	59.4	66.9	80.8	72.7	6.4	88.0	72.1	2.2	4.5
2. Livestock	28.2	31.7	29.1	26.1	0.6	32.2	26.4	2.6	1.5
3. Forestry and fishing	1.2	1.4	1.2	1.2	0.0	1.9	1.5	12.2	5.3
4. Total	88.8	100.0	111.1	100.0	4.5	122.1	100.0	2.6	3.6

Sources: Based on Bank Markazi Iran, *National Income of Iran, 1959-72*, and *Annual Report*, various dates.

over this period, and (b) by the phenomenal growth of the application of modern inputs (see Tables 8.7 and 8.8). Third, the near-zero growth rate of livestock output during the period 1962-7 is likely to be the result of the blow to mutton and dairy production, in part as a result of the policy of enforced settlement of the nomadic population, and in part because many peasants were deprived of the use of natural pastures.

Table 8.8: The Application of Technical Inputs to Iranian Agriculture (1963-72)

Year	Nitrogenous Fertilisers (hundred metric tons)	Potash Fertilisers (hundred metric tons)	Phosphate Fertilisers (hundred metric tons)	Tractors (000 units)
1963	73.0	18.0	86.0	7.5[a]
1964	77.0	23.0	93.0	n.a.
1965	127.0	17.0	141.0	n.a.
1966	155.0	20.0	150.0	16.0
1967	330.0	13.0	280.0	17.5
1968	460.0	19.0	269.0	20.0
1969	490.0	20.0	300.0	20.0
1970	652.0	42.7	293.0	21.0
1971	1 072.9	47.3	693.0	21.5
1972	1 238.0	86.5	758.2	23.0
1973	1 940.8	240.0	1 333.4	106.6
1974	2 485.7	250.0	1 731.2	500.0

Note: a. 1962.

Sources: Food and Agriculture Organization, *Production Yearbook*, various issues.

The differential rates of growth of crop and livestock outputs, however, do not by themselves explain the greater increase of livestock prices: the supply of all these products continuously lagged behind demand, and there was a firm commitment by the state to relieving the subsequent excess demand for them all by a liberal import policy. The main reason must therefore be sought in the greater homogeneity in use between domestically produced and imported arable farm products — and especially corn — than that between home-produced and foreign livestock products. For reasons of taste, cultural habits, the general superiority of fresh over frozen meat as well as the greater difficulties of obtaining and distributing foreign supplies in the right quantities and at the right times, prices of domestically produced red meat went on soaring in spite of unrestricted imports.

Rice production and prices provide a reasonably strong test for the above argument. Iranian rice is (in general) distinctly superior to the varieties produced elsewhere and — partly as a result of that — it has always been a relatively 'high income-elasticity' grain product. Apart from that, rice is a complement for mutton in the consumption of the higher-income groups. The high water-intensity of rice production makes it very difficult to expand its output by significant land reclamation (though this has been partially achieved by the further deforestation of the rice-producing regions), but it is possible to increase output through increases in the yield per hectare. This indeed is what happened over the 15-year period 1963-78. Nevertheless the rate of growth of production of rice did not keep pace with that of demand for it. Therefore, imported rice of inferior quality had to make up the deficit, with the result that home-produced rice of the higher grades sold at about twice the price of imported rice in the shops.

An examination of the data on output, although not very reliable, helps to clarify the picture. For the purpose of Table 8.9, wheat and barley have been selected both because of their predominance in terms of the area under cultivation, and as representatives of staple food products. Rice is a 'high income-elasticity' crop, while red meat and milk are 'high income-elasticity' livestock products.

In the period 1962-72, the quantity of wheat output grew at an annual rate of 5.4 per cent, faster than all other products in the sample. But this was largely because of a rapid expansion of the sown area; the rate of increase of yield per hectare was only 0.6 per cent. Output of barley grew at the rate of 4.3 per cent, but this is entirely due to its fast growth in the first five years of the decade; in any case, yield per hectare declined at the rate of 2.2 per cent. However, rice output grew at a rate of 4.2 per cent, and its yield per hectare increased at the rate of 2.8 per cent per year (see Table 8.9).

A glance at Appendix Table 3 reveals the much greater *variations* in the quantities of output and yields per hectare for the staple food products as compared to rice. Indeed, if the data for 1967 and 1968 (which happened to record exceptionally good climatic conditions for production) are excluded, the resulting annual growth rates for wheat and barley production will be much lower. In other words, a

Table 8.9: Quantities of Output and their Growth, Selected Agricultural Products, Selected Years

	1962/3	1967/8	1971/2	1962-72
1. Wheat				
(a) Quantity of output (000 tons)	2 700	3 800	3 700	–
(b) Annual growth of total output (%)	–	5.8	– 0.9	5.4
(c) Output per hectare (kg.)	720	1 050	700	–
(d) Annual growth of output per hectare (%)	–	1.4	– 0.1	0.6
2. Barley				
(a) Quantity of output (000 tons)	765	1 035	900	–
(b) Annual growth of total output (%)	–	8.3	– 3.5	4.3
(c) Output per hectare (kg.)	830	830	721	–
(d) Annual growth of output per hectare (%)	–	5.1	– 6.6	–2.2
3. Rice				
(a) Quantity of output (000 tons)	700	960	1 050	–
(b) Annual growth of total output (%)	–	4.8	2.6	4.2
(c) Output per hectare (kg.)	2 000	2 549	2 727	–
(d) Annual growth of output per hectare (%)	–	3.8	1.9	2.8
4. Meat				
(a) Quantity of output (000 tons)	243.2	265.5	303.0	–
(b) Annual growth of output (%)	–	1.9	3.2	2.4
5. Milk				
(a) Quantity of output (000 tons)	1 699	1 732	1 900	–
(b) Annual growth of output (%)	–	0.1	1.8	1.7

Source: Appendix Table 3. Growth rates have been calculated by using a semi-logarithmic regression equation.

simple comparison of annual growth rates conceals the fact that growth of rice production was more sustained than that of the other crops.

An interesting implication of the above detail is that, even within the crop farming sector itself, the pattern of output and productivity changes was uneven. In particular, the output of rice, cotton and sugar-beet grew more steadily, and their yields per hectare increased at significantly higher rates than those of wheat and barley which together cover about three-quarters of the cultivated area. Given the fact that both the international and the domestic prices of cash crops have been rising faster than the staple food products, it follows (a) that the rate of growth of total farm output reckoned in value (rather than quantity) terms must have been very disproportionately influenced by the much greater price *and* productivity increases of cash crops;[15] (b) given the fact that cash crops cover no more than 25 per cent of the cultivated area, this would mean that the bulk of the peasantry who produce staple food products experienced a much slower growth of their incomes than would appear from the overall performance of Iranian agriculture; and (c) given the regional concentration of cash crop production (mainly in the northern provinces of Gīlān and Mazandarān), the unevenness of the regional growth of agricultural incomes — due to this one factor alone — must have been considerable.

Table 8.10 helps to clarify some of the above points. In Table 8.10, output of wheat, barley and rice have been evaluated by their import prices (see column 4). These import prices are in fact representative of the domestic market prices in spite of some obvious objections to their use as proxy data. For example, the import price of rice refers to (the final) rice *grain*, whereas the quantities of home-produced rice refer to *paddy*; but given the fact that the market price of the home product is nearly twice the price of imported rice, the import price would not exaggerate the value of Iranian paddy rice. The use of import prices as proxy data would, in any case, not pose a serious problem, because here we are interested in a *comparison* of the output values of the sample crops.

Table 8.10: Farm Products, Selected Crops, 1974/5

	(1) Percentage Share in Cultivated Area	(2) Total Yield (million kg.)	(3) Import Price (rials per kg.)	(4) Estimated Value [(2)x(3)] (million rials)	(5) Percentage Share in the Total Value of Farm Output
1. Wheat	61.0	4 700	14.6	68 620	34.8
2. Barley	14.5	863	12.4	10 701	5.4
3. Rice	3.8	1 313	46.6	61 186	31.0
4. Total [(1)+(2)+(3)]	79.3	—	—	140 507	71.2
5. Other	(20.7)	—	—	(56 677)	(28.8)
6. Total arable output	100.0	—	—	197 184	100.0

Sources: Appendix Tables 3 and 4, Bank Markazi Iran, *Annual Report*, 1974, 1975.

Column 5 of Table 8.10 shows that (in 1974/5) the share of wheat production in the value of total farm output was under 35 per cent, rice production 31 per cent, barley production over 5 per cent, while the remaining 29 per cent was accounted for by all other (mainly cash) crops. Thus, about 75 per cent of the cultivated area is used for the production of wheat and barley, which yields only 40 per cent of the value of total arable output and 25 per cent of the value of total agricultural production. Although for a variety of historical, climatic and technical reasons yields per hectare in Iranian agriculture are generally lower than in comparable countries such as Egypt, this is particularly pronounced in the case of wheat and barley which cover three-quarters of the cultivated land.

It seems clear that as part of a serious effort to raise peasant incomes, it is the production of wheat and barley — rather than that of 'high income-elasticity' food and export crops — which needs to be raised. The output and productivity of these crops were restricted by the shortage of water and technical inputs in the peasant sector and this reflected both the poverty of the peasantry and their exclusion by the state from the support programmes. On the other hand, the so-called modern systems which enjoyed state support produced no better results. Between 1962 and 1974, the number of tractors used in Iranian agriculture increased from 75,000 to half a million, the use of various chemical fertilisers grew in much higher proportions, the expenditure on high dam construction took a large share of public investment projects, and yet the yields per hectare of wheat and barley production remained constant (see Table 8.8 above).

IV.2 Systems of Production and Growth Performance

From the discussions above, it is clear (a) that the 'traditional' sector of Iranian agriculture had inadequate access to funds, technical assistance and productive resources provided by the state; (b) that it was under the threat of appropriation in favour of 'modernisation'; and (c) that it suffered from unhelpful interference by state officials. In contrast, farm corporations and agri-businesses, which covered only a small proportion of the cultivated area, had a disproportionately large share of fertile land, water resources, financial credit and technical inputs. Did this 'modernising' adventure pay off? As far as the overall performance of agriculture and, especially, the consumption and welfare of Iranian peasantry is concerned, the answer is plainly in the negative. But it would still be interesting to discover whether or not the 'modern' systems turned out to be significantly more efficient than the traditional system, for the answer to the latter question is important for future policy purposes.

Table 8.11 gives the percentage share of the agricultural and non-agricultural sectors in the development expenditures of the three plan periods. On an average, only 10 per cent of the state plan funds were spent on agriculture. Furthermore, more than 50 per cent of the funds destined for agriculture were allocated to the 'moderns', which accounted for no more than 2.8 per cent of the total agricultural land, and the rest to the 'traditional' sector, which *includes* independent (i.e. non-corporate) capitalist farms (see Table 8.12).

Table 8.11: Percentage Distribution of Development Expenditures, 1963-76

	Third Plan (1963-7)	Fourth Plan (1967-72)	Fifth Plan (excluding 1977)	Average 1963-76
1. Total agriculture	12.5	8.1	10.7	10.4
(i) Traditional	(6.7)	(3.7)	(4.9)	(5.1)
(ii) Modern	(5.8)	(4.4)	(5.8)	(5.3)
2. Non-agricultural	87.5	91.9	89.3	89.6
3. Total	100.0	100.0	100.0	100.0

Source: Based on Government of Iran, Plan and Budget Organisation.

Table 8.12 shows the distribution of landholdings among various systems of agricultural production. In 1976, the 'traditional' or non-corporate sector held 97.2 per cent of the total area, while the 'modern' or corporate sector held the remaining 2.8 per cent, 1.7 per cent of which was held by farm corporations, and 1.1 per cent by agri-businesses. However, the distribution of holdings *within* the traditional, non-corporate sector itself may not be entirely accurate: direct data on the holdings of independent capitalist farms are not available, but indirect data (e.g. the size of holdings and the use of wage labour in production) indicate that this should be about 2.7 million hectares or 15.2 per cent of total landholdings. Thus, peasant holdings accounted for about 82 per cent of the cultivated area. Finally, although data on the distribution of population and labour force among these various systems of production are not available, the capital-intensive technology of the corporate sector and their insignificant share of total landholdings indicate that their share of the agricultural labour force must likewise have been insignificant.

Table 8.12: The Distribution of Landholdings by System of Production, 1976

	Million Hectares	Percentage Share in Total
1. Traditional (non-corporate)	17.2	97.2
(i) Peasant	(14.5)	(82.0)
(ii) Independent capitalist	(2.7)	(15.2)
2. Modern (corporate)	0.5	2.8
(i) Farm corporations	(0.3)	(1.7)
(ii) Agri-businesses	(0.2)	(1.1)
3. Total	17.7	100.0

Sources: Based on the Statistical Centre of Iran, *Agricultural Census 1971* (compilation date 1973, in Persian), and Ministry of Corporatives and Rural Affairs, various reports, 1971-6 (in Persian).

Table 8.13: State Loans and Grants to the Peasant Sector and the Farm Corporations, 1968-75

	Total 1968-75 (million rials)
1. The peasant sector: loans	108 895
2. Farm corporations:	7 841
(i) Free grants	(6 322)
(ii) Loans	(1 519)

Source: Based on F. Etemad Moghadam, 'The Effect of Farm Size and Management System on Agricultural Production', unpublished D Phil thesis, University of Oxford, 1978, Part I, Tables 1 and 16.

Table 8.13 shows the amount and distribution of state loans and non-returnable grants paid (over the period 1968-75) to the peasant and farm corporation sectors. It should be noted (a) that 6,322 million rials or 81 per cent of total payments to farm corporations were in the form of non-returnable grants, and (b) that the remaining 1,519 million rials (or 19 per cent of the total) consisted of 5- to 15-year loans carrying *administrative* interest rates ranging from 2 to 8 per cent but an *effective* interest rate of 1 per cent.[16] On the other hand, the credit extended to the peasant sector through the co-operative societies consisted entirely of short-term (mainly annual) loans carrying an effective interest rate of 6 per cent. Moreover, the total amount of loans to the peasants, shown in Table 8.13, is almost certainly an overestimate; it was a widespread practice among co-operative societies to record outstanding loans (which the peasants were unable to repay) as new credit.

The farm corporations, however, had a much smaller share of landholdings in the earlier years than they did in 1975. For this reason, figures in Table 8.14, showing the overall payments per hectare of land between 1968 and 1975 and their annual averages, provide a better comparison. The figures are eloquent and require little comment.

Table 8.14: State Loans and Grants to the Peasant Sector and Farm Corporations (rials), 1968-75

	Per Hectare 1968-75	Per Hectare per Annum
1. The peasant sector: loans	7 810	939.0
2. Farm corporations: grants	95 544	11 943.0
3. Farm corporations: loans	26 839	3 355.0
4. Total farm corporations [(2)+(3)]	122 383	15 298.0

Source: As in Table 8.13.

A main state agency for the supply of low-interest, long- and medium-term credit was the Agricultural Development Bank. Theoretically, the Bank would lend only to producers the size of whose farms was no less than 500 hectares; but in practice, producers with smaller holdings could also have access to its loans and credits. Assuming that the Bank was prepared to lend to all farm sizes above 100 hectares, it has been estimated that 88 per cent of holdings — i.e. virtually the whole of the traditional sector — would have been excluded from the Bank's lendings.[17] This being the case, peasant producers could hardly have access to ordinary commercial banks, whose interest charges and other conditions for giving credit were in any case non-preferential.

The methods of payments and receipts were themselves very different in their socio-economic implications; agri-businesses, farm

Table 8.15: The Distribution of Capital by Various Agricultural Systems

	(1) No. of Units	(2) Total Holdings (1 000 hectares)	(3) Total Capital (million rials)	(4) Capital Per Hectare (rials)
1. Co-operative societies	2 925	14 500.0	8 385	578.3
2. Farm corporations	93	300.0	1 515	5 050.0
3. Agri-businesses[a]	4	63.5	1 700	26 771.6
4. Capitalist farms	NA	2 700.0	NA	NA

Note: a. A sample of four agri-business firms in the Dez Dam area (1970-3) covering about one-third of the total area held by all the agri-business companies. Figures for total capital exclude the construction costs of the Dez Dam.

Sources: Based on Bank Markazi Iran, *Annual Report*, 1977/8, table on p. 102, and Etemad Moghadam, 'Effect of Farm Size'.

corporations and large capitalist farms received state grants and 'loans' directly through their own managements, but peasants had to depend on the favours of the bureaucratic heads of their co-operative societies, who in turn had to depend on the heads of the co-operative unions, who in their turn had to depend on the higher officials in the Agricultural Co-operative Bank and the Ministry of Co-operatives and Rural Affairs. Apart from all this, embezzlement and corruption were more easily possible through the co-operative network than through other systems, and the allocation of credit to peasants was generally influenced by the relative financial and social standing of different groups of peasantry.

A consequence of lack of access to long-term credit is low capital investment. Tables 8.15 and 8.16 show the pattern of distribution of capital stock among different systems of agricultural production. From column 4 of Table 8.15 it can be seen that, in 1977/8, capital per hectare of land held by agri-business firms was 46 times, and that of farm corporations nearly 9 times, that in the peasant sector. Table 8.16 extends the comparison to the membership of the peasant co-operative societies and farm corporations. The membership of farm corporations was only 1.2 per cent of the total membership of both systems taken together, whereas it claimed 15 per cent of their total capital; the corresponding figures for the peasant sector being, respectively, 98.8 of total membership and 85 per cent.

Table 8.16: The Distribution of Membership and Capital: Peasant Co-operatives and Farm Corporations, 1977/8

	Membership (000)	Total Capital (million rials)	Percentage Distribution of Total Membership	Percentage Share in Capital
1. Co-operatives	2 983.0	8 352	98.8	85.0
2. Farm corporations	35.4	1 515	1.2	15.0
3. Total	3 018.4	9 867	100.0	100.0

Source: Based on Bank Markazi Iran, *Annual Report*, 1977/8, table on p. 102.

The basic differences between the three systems were in the quality of land, the availability of water, the scale of production, the choice of technique, the mode of ownership, employment and distribution, and the managerial organisation. The agri-business system was well endowed in land and water resources, used modern capital-intensive techniques, produced on a very large scale, divorced ownership from management and labour, and had a bureaucratic management system. The same characteristics applied to the farm corporations system, though less strongly. In particular, the scale was smaller, technology relatively mixed, much of the labour was supplied by smaller shareholders, and bureaucratic managers were to a limited extent bound by the corporate spirit. Finally, the traditional sector was dominated by small-scale production, dependent on the use of traditional techniques and inputs as well as family labour, and in general suffered from very serious water constraints.

Unfortunately, no rigorous analysis of the comparative performance of farm corporations and peasant producers is feasible. But there are strong indications from both casual observation and partial comparisons that the performance of farm corporations has not been better than peasant producers, *in spite of* their various advantages in terms of finances, technical inputs, etc., over the latter. For example, the Garm-sār Corporation, which was officially held as a case (and probably the only case) of success, has performed no better, and probably worse, than the neighbouring traditional village of Risān; while a study of the Shams-ĀBĀD Farm Corporation in Khūzistān has shown the output to be below the 1960 level (when the Corporation did not exist).[18]

A rigorous and reliable study of the performance of agri-business farms and the traditional system is, however, available. The study employs rigorous statistical techniques for the analysis of output and productivity data, a lot of which have been obtained by sample surveys carried out in the field.[19]

Five villages and four agri-business farms were selected for the comparative study. The four agri-business farms are all from the Dez Dam area in the fertile Khuzistān province, their command area ranging from over 10,000 to over 20,000 hectares of cultivable land. The five villages are from heterogeneous regions in terms of climate and crop pattern. Four of these villages are from relatively advanced areas, while the fifth is located in a backward area, and was under the threat of official expropriation at the time of study. The size of holdings varies greatly both between and within the five villages (with a minimum of 0.1 and a maximum of 330 hectares), but the average range is between 1 and 15 hectares.

The main statistical methods used were analysis of variance and analysis of co-variance; the variables consisted of both total productivity and partial productivity indices. Three different concepts of total productivity have been applied: gross output per unit of total output; net output per unit of total input (net of the depreciation allowance); and net value added (i.e. net output minus current inputs) per unit of total input − net of depreciation as well as current inputs − all of them expressed in value terms. The partial productivity indices refer to output per unit of labour input, output per hectare, and output per unit value of water input.

The results were nothing short of staggering. First, within the village system itself, the medium-sized peasant holdings proved to be significantly better performers than both the small peasant holdings and the larger independent capitalist farms. It was further found that as the capitalist farms themselves became larger in size, there was a tendency for a decline in total productivity. Apart from that, the villages entirely based on peasant holdings displayed higher total productivity ratios than those which included independent capitalist farming units.

Second, a comparison of peasant production with the agri-business system also showed a significant superiority of peasant production. Different sets of total productivity indices were calculated on the basis of different assumptions. In particular, a set of total productivity indices for peasant production was calculated by inputing the (lower) land and water unit costs of agri-businesses to them. In this case, the total productivity indices for peasant production in all the villages turned out to be significantly higher than those for the agri-business firms. Another set of total productivity indices for peasant production, this time calculated by applying their own (higher) land and water costs, turned out to be significantly higher than those for the agri-businesses in most cases.

Third, with respect to partial productivity indices, (a) land productivity indices for four of the five villages were higher than those for the agri-businesses; (b) labour productivity indices for three of the five villages were higher than those for the agri-businesses; and (c) water productivity indices did not produce a clear pattern.

Finally, specific comparisons of independent capitalist farms (within the village system) with the agri-business firms displayed the same pattern of higher productivity performance for the former. It may be argued that the agri-business firms were still 'infants' and that they had yet to reap the full benefit of their dynamic potentials. This, however, amounts to speculation since the possible sources of such dynamic benefits are difficult to envisage.

V Income Distribution, Poverty and Unemployment

V.1 The Distribution of Income and Welfare

The consequences of state policies towards agriculture since 1963 for the distribution of rural incomes as well as for employment and levels of living of the rural population have been, as might be expected, far from desirable. The difference in the general level of welfare (including private consumption and state services) between the urban sector and the rural society rapidly increased and, especially after 1973, became more pronounced. Apart from that, there was an increase in the degree of inequality between the *Nasaq*-holding peasants who had received land through the land reform and those (including the *Khushnishīns*) who had not; the degree of inequality among the *Nasaq*-holders themselves also increased; already significant regional differences in rural incomes and welfare tended to get worse; and poverty and unemployment among the rural immigrants into towns – itself being a conse-quence of agricultural depression and peasant destitution – became the single most pressing social problem about which, however, absolutely nothing was done. In this section we hope to provide some arguments and evidence for the above observations.

Up to 1972 separate figures for urban and rural consumption were made publicly available. Since such data are not available for the latter

Table 8.17: The Sectoral Distribution of Private Consumption and Population, Selected Years (per cent)

	1963 Consumption	1963 Population	1968 Consumption	1968 Population	1973 Consumption	1973 Population	1978 Consumption	1978 Population
1. Urban	53.9	35.0	59.9	38.7	70.6[a]	42.4	75.5[a]	46.1
2. Rural	46.1	65.0	40.1	61.3	29.4[b]	57.6	24.5[b]	53.9
3. Total private consumption	100.0	100.0	100.0	100.0	100.0	100.0	100.0	100.0

Note: a. The difference between total private consumption and value added in agriculture, as percentage of total private consumption.
b. Total value added in agriculture as percentage of total private consumption.

Sources: Based on Bank Markazi Iran, *Annual Reports*, various dates, and official population statistics.

years, we have obtained separate estimates for urban and rural consumption by simply assuming that rural consumption has been equal to
total value added in agriculture and, therefore, that urban consumption
is equal to the difference between total private consumption and value
added in agriculture. The data presented in Table 8.17, which show the
percentage distribution of private consumption and population between
urban and rural sectors of the economy, should therefore be taken as
indicative only of general trends.

Keeping in mind the limitations of the data, we can make the
following observations. First, there is clear evidence that rural-urban
consumption inequality, already considerable in the initial period
(1963), had significantly increased by 1978. Second, it should be
emphasised that a large proportion of the growth of urban population
was due to peasant migration into towns, and the consumption of
immigrant peasants was probably not much more than previously;
hence, the use of official figures for the urban sector's share in total
population tends to understate the growth of consumption inequality
between the two sectors. Third, the 'consumption data' for the rural
sector in 1973 and 1978 are none other than the total value added in
agriculture which are very likely to over-estimate actual rural consumption. Finally, the welfare benefits of high and rapidly rising state
consumption expenditure (not shown in the table) were almost exclusively enjoyed by the urban sector.[20]

Table 8.18: The Sectoral Distribution of Private per Capita *Consumption (1963-78), Selected Years, US dollars*[a]

	1963	1968	1973	1978
1. Urban *per capita* consumption	209.8	288.2	401.6	1 442.2
2. Rural *per capita* consumption	97.3	123.1	161.8	382.8
3. Country-wide *per capita* consumption	137.1	186.9	316.0	892.2

Note: a. 1963: 80 rls = $1; 1968: 75 rls = $1; 1973: 73 rls = $1;
1978: 70 rls = $1.

Source: As in Table 8.17.

Some of the above observations can be more readily grasped from
the distribution of private consumption *per capita* as shown in Table
8.18. In 1963, urban and rural consumption *per capita* were, respectively, US$210 and US$97; in 1978, US$1,442 and US$383. It may be

further observed from Table 8.19 that between 1963 and 1978 the ratio of rural to national *per capita* consumption fell from 0.7 to 0.4. These estimates are all based on current prices and, in particular, they do not take into account the deterioration of the terms of trade against agriculture over the period.[21]

Table 8.19: Sectoral per Capita *Consumption Relative to Country-wide Private* per Capita *Consumption, 1963-78, Selected Years*

	1963	1968	1973	1978
1. Urban	1.50	1.50	1.60	1.60
2. Rural	0.71	0.66	0.51	0.40
3. Country-wide	1.00	1.00	1.00	1.00

Source: See Table 8.18.

Table 8.20 shows the share of various expenditure categories in the consumption of urban and rural areas. It can be seen that food and clothing (i.e. bare subsistence) take up nearly three-quarters of the rural, as opposed to 50 per cent of the urban, private expenditure, even though it is likely that the figure for the share of rural subsistence in its total expenditure excludes a part of the consumption of their own (unpurchased) products. In other words, the share of mere subsistence in rural expenditure is likely to be even higher than three-quarters of the total. It goes without saying, however, that such percentage shares reveal nothing about the different *levels* of consumption, or the differences in qualities of consumption between town and village: for example, in rural areas, housing — for the most part — consists of no more than hovels made of mud and straw, meat and dairy consumption is rare, 'health' facilities are virtually non-existent, and so on.

So far we have been discussing consumption inequalities between 'urban and rural areas' and their trends. A no less important issue, however, is the question of changes in the distribution of land, income and expenditure within the rural sector itself: i.e. both within 'the village' and between different groups of villages in various regions of the country. First, let us make some observations on the distribution of ownership and holdings in agriculture. In 1960, between 35 and 40 per cent of the peasant population were *Khushnishīns* (i.e. those without a right of cultivation), and the rest were peasant (*Nasaq-*

Table 8.20: Percentage Distribution of Sectoral Household Consumption Expenditure by Category, 1972

	Rural Areas	Urban Areas
1. Food and clothing	73.6	49.9
2. Housing	9.9	24.0
3. Health and education	4.5	8.3
4. Travel	2.3	5.3
5. Others	9.7	12.5
6. Total	100.0	100.0

Source: Based on H. Azimi, 'Aspects of Poverty and Income Distribution in Iran', unpublished D Phil thesis, University of Oxford, 1979, Ch. 3, Table 18.

holding) cultivators who owned one-third of total agricultural land. Thus, by aiming to distribute the remaining two-thirds among *Nasaq*-holders, the original land reform had intended to turn about 60 per cent of the rural population into peasant proprietors owning all agricultural land in the country. As we have already seen, about 23 per cent benefited from the operations of the First Stage of the reform, while others were left to use the provisions of the Second Stage for purchasing land *directly* from the landlords, entering partnership with them, selling their own rights to them, etc.[22] Consequently, when the land reform officially ended, about 82 per cent of agricultural land belonged to peasant producers, the remaining 18 per cent being owned by independent capitalist farmers (who were either former landlords or others who had purchased their estates from them) and the 'modern' sectors (see Table 8.12 above).

Direct data for the proportion of peasant proprietors after the reform are not available. What is certain is that (a) *Khushnishīns* were not affected; (b) a certain proportion of peasants had owned their land before the reform; (c) 23 per cent of them received land through the First Stage; and (d) 18 per cent of total agricultural land still remained outside peasant ownership after the reform. Thus, while some *Nasaq*-holders improved their position, the *Khushnishīns* now became legally (and finally) landless.

More significantly, it has been shown[23] that the distribution of holdings was highly unequal before the land reform (1960), and that it became even more unequal after the reform (1972). Thus, the Gini

Table 8.21: The Distribution of Holdings, by Size and Population, among Nasaq-holders, 1960

Size of Holdings (hectares)	Holdings (%)	Nasaq-holders' Population (%)
Below 5	18.7	60.1
5-9	21.2	18.9
10-19	26.9	13.9
20-49	19.5	5.8
50 and above	13.7	1.3
Total	100.0	100.0

Source: Based on Azimi, 'Aspects of Poverty', Appendix Table 6/I-1, p. A-107.

Table 8.22: The Distribution of Holdings by Population among Nasaq-holders, 1960

Nasaq-holders	Agricultural Land (%)
1. The lowest 30%	3
2. The middle 30%	16
3. The higher 30%	41
4. The top 10%	40
5. Total	100

Source: Based on Azimi, 'Aspects of Poverty', Ch. 7, Table 5, p. 286.

concentration ratio for the distribution of land among various holdings significantly increased from 0.6355 in 1960 to 0.6923 in 1972. However, these ratios, and the growing inequality which they reflect, refer to the distribution of total agricultural land among *all* categories of holdings. The case of *Nasaq*-holders as a group was somewhat different. First, they collectively increased their ownership from 33 to about 82 per cent of total agricultural land. Second, the distribution of land being inherently in favour of smaller farm sizes led to a gain by the lower- and, especially, middle-sized groups relative to larger ones. Thus, Table 8.23 shows that although, as a result of land distribution, all sizes improved their share of the total, it was the lower and middle sizes (and especially those between 2 to 50 hectares) that gained most.

Table 8.23: The Distribution of Landownership among Nasaq-holders, 1960-72

Size (hectares)	1960 (%)	1972 (%)
Less than 1	52	81
1 to less than 2	41	80
2 to less than 5	33	82
5 to less than 10	29	81
10 to less than 50	25	77
50 to less than 100	44	80
100 and above	67	80
Total	33	80

Source: Azimi, 'Aspects of Poverty', Ch. 8, Table 2, p. 337.

To sum up, land reform increased peasants' share of land from about 33 to over 80 per cent and probably reduced the inequality of land distribution among the peasants, but it also created a large class of landless agricultural labourers (the erstwhile *Khushnishīns*). Over 15 per cent of agricultural land was left to the landlords who turned it — partly in words and partly in deeds — into relatively large independent capitalist farms; and the remainder was converted into 'modern' agri-business and farm corporation farming.

We may now turn our attention to the distribution of income and expenditure within the rural sector. A comprehensive study of this subject should include an analysis of income distribution among various regions of the country as well as among different strata of the rural population. Unfortunately, lack of data and the notorious problems of classification and disaggregation make it virtually impossible to carry out such a thorough quantitative analysis. The available data on the distribution of expenditure for the whole of the rural sector indicate that, between 1967 and 1972, the share of the poorest households in total expenditure increased slightly at the expense of the richest, the shares of all other expenditure groups remaining virtually unchanged (see Table 8.24). However, the developments since 1972 lead one to believe that the distribution of rural expenditure, income and welfare may have worsened in subsequent years.

Table 8.24: Percentage Distribution of Expenditure among Various Rural Income Groups

Income Groups	1967	1972
The poorest	5.5	6.6
The poorer	11.4	11.3
The poor	15.1	15.6
The ordinary	21.6	21.8
The richer	26.2	26.2
The richest 5 per cent	20.2	18.5
Total	100.0	100.0

Source: Based on Azimi, 'Aspects of Poverty', Ch. 8, Table 5, p. 344.

Yet, as is well known, the distribution of expenditure is not necessarily indicative of relative income distribution, and can frequently result in a distorted view of the latter. For lower-income groups, expenditure typically exceeds income (i.e. there is borrowing and dis-saving), while for higher-income groups, income exceeds expenditure (i.e. there is saving).[24] It is thus difficult to draw any unambiguous conclusion concerning changes in rural income distribution. We can only say that it is unlikely to have improved.

Finally, it must be emphasised that there are extreme inter-regional variations in rural incomes and welfare in Iran, the poorest regions being Kurdistān, Balūchistan, Kuh Kīlūyeh, parts of Khūzistān and the Fars province, and the fact that such extreme variations cannot be adequately quantified should not diminish the significance as well as urgency of this problem. In fact, this has a direct bearing on the question of poverty in Iran to which we will now turn our attention.

V.2 Poverty and Unemployment

One consequence of the general agricultural and rural crisis was the flight of labour from the land; according to official estimates, about 1.7 per cent of the rural population has been migrating annually from rural to urban areas in recent years. Given a country-wide population growth rate of 2.9 per cent, the rate of growth of rural population (net of migration) was about 1.2 per cent per year.

We can cite no data on rural unemployment, but it seems plausible to suppose that rural-urban migration essentially reflected deteriorating employment conditions in rural areas and operated as a mechanism for transferring poverty from rural to urban areas. In 1978, the country's total labour force was about 10 million, 3.5 million of whom were in the rural sector. The official estimate of the general rate of unemployment was 9 per cent.[25] One study referring to an earlier period provides some interesting insights. Table 8.25 provides official estimates of open unemployment for 1966. It may be observed that although the rate of open agricultural unemployment at the time was low, it was significant for those engaged in activities 'not elsewhere classified'. The new entrants were excluded by definition.

Table 8.25: Sectoral Rates of Open Unemployment, 1966

Economic Sectors	Unemployment Rate (%)
Agriculture	3.6
Construction	3.9
Transport and communications	1.7
Commerce and banks	1.5
Manufacturing and mining	1.0
Services	0.9
Public utilities	0.7
Activities not elsewhere classified	38.0

Source: G. Irvin, *Roads and Redistributions, Social Costs and Benefits of Labour-intensive Road Construction in Iran* (ILO, Geneva, 1975), Appendix Table 1, p. 103.

Disguised unemployment is notoriously difficult to measure. The above-mentioned study made some estimates on the basis of a measurement of part-time as well as seasonal work. Thus for the year 1969, the rate of underemployment (other than open unemployment) in Iranian agriculture was estimated at about 24.7 per cent among the employed agricultural labour force (see Irvin (1975), Table 4, p. 37).

These observations relate to the period when agri-businesses and farm corporations were still in their infancy, and the oil revenues had not rocketed in consequence of the fourfold oil price increase of 1973. However, the picture is unlikely to have changed substantially. The oil boom may have led to an acceleration in the rate of rural-to-urban migration, thus reducing the rate of open unemployment in rural areas.

But rural underemployment is unlikely to have been wiped out. As we observed earlier, the urban rate of unemployment continued to be high. The most important cause of urban unemployment was the high intensity of both state and private investment (in manufacturing as well as service activities) in modern, technologically advanced, capital equipment which, at one and the same time, led to a low rate of unskilled labour absorption, an acute shortage of skilled labour, a rapid increase in urban wage differentials (with significant inflationary consequences) as well as low capacity utilisation. Immigrant peasants were incapable of skilled urban work at any level, and generally sought employment in construction activities where pure physical effort could be gainfully used. Yet the chronic shortage of some imports, and the use of capital-intensive technology even in road construction (which can employ immigrant peasants as well as rural unemployed workers), reduced the labour absorption capacity of this sector.[26]

Therefore, unemployment, poverty and destitution in the urban sector should be regarded as a consequence of the crisis in agriculture; peasant immigrants whose very existence on the edge of the cities went completely unrecognised should be considered not as a section of the urban poor, but as the impoverished sections of the peasantry who had left the village but had not been admitted into the towns. They had come looking for a better life, and what they found was probably much worse than what they had left behind. Yet they could not, in the main, return because they had uprooted themselves, sold their effects, spent their cash, lost face, and still hoped that things might improve for them where they had arrived.

The assertion that the high rate of rural-urban migration reflected in part the poor state of the agrarian economy is strongly supported by the evidence on rural poverty. If migration were in fact inspired by growing employment opportunities in towns, one would not expect to find widespread poverty in rural areas. This is precisely what one finds, however. A recent comprehensive study of poverty in Iran by Hossein Azimi is available. On the reasonable assumption that under-nourishment is the most extreme form of poverty, the study's findings provide important insights into the nature and extent of poverty in the country. Having identified an index of Minimum Calorie Requirement, Azimi defines three 'poverty lines': Line A for calorie intakes of between 90 and 99 per cent, Line B for calorie intakes of between 75 and 90 per cent, and Line C for intakes of 75 per cent or less, of the Minimum Calorie Requirement. Applying these criteria to a country-wide sample of expenditure groups, Azimi has obtained some revealing results on which Tables 8.26 and 8.27 have been based. The figures in these tables speak for themselves; in 1972/3, the calorie intake of 16 million people (i.e. 52 per cent of the population) was less than the minimum requirement, 4 million of whom (i.e. 13 per cent of the population) were severely undernourished (see Line B in Tables 8.29 and 8.30). It is also interesting to note that (a) undernourishment in the rural sector seems to be relatively less than in towns, and (b) this difference is particularly pronounced in the case of severe (i.e. Line B) undernourishment. We shall briefly discuss the possible causes and implications of these sectoral differences later.

Table 8.26: Undernourishment in Iran, 1972/3, Country-wide Figures (millions of people)

	Line A	Line B	Total (A + B)
1. Urban areas	5.4	3.3	8.7
2. Rural areas	6.6	0.7	7.3
3. Total (1 + 2)	12.0	4.0	16.0

Source: Based on Azimi, 'Aspects of Poverty', Ch. 4, various tables.

Table 8.27: Undernourishment in Iran, 1972/3, Country-wide Figures (per cent of population)

	Line A	Line B	Total
1. Urban population	39.0	25.0	64.0
2. Rural population	38.0	4.0	42.0
3. Total population	39.0	13.0	52.0

Sources: Table 8.29 and official Iranian population data.

These results, based as they are on country-wide data, do not take into account the large degrees of regional variations in a country like Iran. Therefore, the author conducted similar analyses for each of the 23 Iranian provinces separately, and then arrived at aggregate figures for the whole country. Tables 8.28 and 8.29 are based on these results. It appears that the total undernourished population is 13.5 million (or 44 per cent of the population), which is less than the

16.0 million obtained from the country-wide analysis. However, these latter tables show that (a) there are about a million people (i.e. 3 per cent of the population) who are in Line C — that is, dangerously undernourished — most of whom are in the rural areas; (b) there are 6.0 million people (or 20 per cent of the population) in Line B, as compared with 4.0 million (13 per cent) shown by the country-wide analysis. Altogether, the latter results show that *21 per cent of the population (most of them in towns) are undernourished, 20 per cent (also mostly in towns) are severely undernourished, and 3 per cent (mostly in villages) are dangerously undernourished.*

Table 8.28: Undernourishment in Iran, 1972/3, Aggregated Provincial Figures (millions of people)

	Line A	Line B	Line C	Total (A + B + C)
1. Urban areas	4.51	3.7	0.30	8.5
2. Rural areas	2.01	2.3	0.70	5.0
3. Total (1 + 2)	6.5	6.0	1.0	13.5

Source: Based on Azimi, 'Aspects of Poverty', Ch. 4, various tables.

Table 8.29: Undernourishment in Iran, 1972/3, Aggregated Provincial Figures (per cent of population)

	Line A	Line B	Line C	Total
1. Urban population	34.0	28.0	2.0	64.0
2. Rural population	12.0	13.0	4.0	29.0
3. Total population	21.0	20.0	3.0	44.0

Sources: Table 8.31 and official Iranian population data.

The reasons for the differences between urban and rural results may be manifold and we have no scope here to engage in a thorough discussion of this subject. Briefly, the cultivating peasants can (at least partially) feed themselves from their own produce, their other subsistence requirements — housing, clothing, bedding material, etc. — are normally supplied by their own household labour (thus enabling them to spend a larger proportion of their incomes on food), while the unemployed landless peasants are more likely to fall into Line C than unemployed workers in towns. On the other hand, most of those in Lines A and B in the urban centres must themselves be recent immigrants from the rural areas, living in the urban fringe areas, being (at least) partially unemployed, and having lost the opportunity, which they previously had, of supplying some of their own food.

Apart from that, Azimi's individual results for each and every province bring out the regional dimension of the problem of inequality. In the rural areas of Kurdistan, for example, virtually every expenditure group, that is, almost the entire Kurdish peasantry, was found to suffer from some form of undernourishment; and Khūzistan, Kirman, Bakhtīyārī, etc. – i.e. the main areas of ethnic/tribal concentrations – were significantly worse off than other areas.

VI Summary and Conclusions

Agrarian reforms in developing countries tend to be motivated by aspirations for transforming traditional societies into industrial economies. Initially, and in line with these general aspirations, the land reforms in Iran had intended to reduce the size of large holdings, distribute land and institute rural co-operatives among the peasants who had traditionally cultivated the land without always owning it. The promoters of the Land Reform Law of 1962 had hoped that these institutional changes would help increase agricultural productivity and living standards, increase demand for manufacturing products, provide resources for urban-industrial development and create a more stable political atmosphere.

Such hopes were, however, quickly frustrated when the Shah – being both worried about the immediate reaction of large landowners and conservative religious leaders and, more importantly, afraid of the consequences of the success of the reformist government for his own power and position – decided to take over the idea and present it as a part of a wider 'modernising programme' which he described as the 'White Revolution'. The immediate consequence of this move for the land reform programme was to dilute the terms of the Land Reform Law (later described as the First Stage) in favour of large landowners, and turn its Second Stage into a tenancy reform programme as opposed to a land redistribution programme (1963-6).

In the meantime, the growth of oil revenues as a result of large and steady increases in both the price and the quantity of crude oil exports began to alter the country's economic and political situation and, in consequence, affect the position of Iranian agriculture more fundamentally than the land reform. The sustained and rapid rise in the oil revenues increased the spending power of the state, made property ownership (for the landowning class) less attractive, resulted in the cumulative (but uneven) growth of urban incomes and encouraged the notion that a peasant-based agriculture was dispensable, indeed anachronistic. Therefore, and in line with purely bureaucratic and technological blueprints for 'modernising' the whole of the economy, it was decided to impose a system of 'farm corporations', that is, joint-stock agricultural companies sometimes consisting of several villages, and then to promote agri-business complexes – i.e. extensive farms based on foreign and domestic capital and wage labour – in the most fertile regions of the country, officially described as 'the poles of land and water resources'.

The parallel strategy of rapid urban 'industrialisation' generally discriminated against agricultural development. Of the funds allocated to the agricultural sector, direct state investment was channelled primarily into agri-business companies, while most of the state credit and loans were granted to farm corporations. The peasants, constituting 55 per cent of the country's population and operating 82 per cent of the cultivable area, were left to obtain what credit they could through the complex bureaucratic network of co-operative units or, more often, from the local usurers and merchants.

The performance of agri-business farms was extremely poor; their unit costs of production were characteristically higher than those in the peasant sector. The reasons for this poor showing were several; the units were too large, their management structure was ill-suited to the local conditions, their imported technology was ill-adapted, and their work-force was typically made up of expropriated former peasant holders used as 'migrant labour'. Farm corporations performed relatively better than agri-business farms because of their smaller size and more relevant forms of ownership, management, technology and employment. Yet considering the moral and material support which they received from the state, their performance was also disappointing. The destruction of the social and economic boundaries of the traditional village, the immiserisation of the poorer peasants, and the direct bureaucratisation of decision-making which they involved were among the major reasons for their inefficient functioning.

The available evidence clearly indicates that, in spite of the anti-peasant state policies, the performance of peasant agriculture was better than agri-businesses and farm corporations, but it grew at a slow rate and remained economically undeveloped and socially underprivileged. The extension of the central bureaucracy to the villages, and the official contempt for the peasantry which was amply reflected in the attitude and behaviour of local (civilian as well as military) officials, created social instability, economic insecurity and psychological resentment among the peasants.

An important consequence of this bureaucratisation — which was symbolised by the so-called co-operative units — was the decline or disappearance of the *Buneh*, the traditional production commune, with its indispensable functions of putting small and fragmented holdings under the same crop, maintaining and extending the artificial irrigation networks (the *qanāts*), and distributing water among the cultivators. The Land Reform Law of 1962 had already contained the seeds of such an undesirable development, partly because of its exclusion of the *Khushnishīn* village community but mainly as a result of its creation of individual peasant proprietors rather than communal village ownership, which would have been well in line with the historical (including social, economic, technological and geographical) developments, and perfectly consistent with peasant aspirations.

In the meantime, the growth and, from 1973/4, explosion of the oil revenues was raising urban living standards, though at a very different rate for different social classes. This, given the deteriorating conditions of a section of the peasantry, created a strong incentive for peasant migration into towns. However, the ultra-modern technology applied to modern manufacturing and service production, even road construction, left little employment opportunity for purely manual labour, migrant peasants had no capital of their own and they were completely alien to the urban culture. The rapid rise of urban food and dwelling prices resulted in hunger and homelessness of the immigrant peasants. Thus, both in villages and in towns the scales were drastically turned against the Iranian peasantry, resulting in extensive undernourishment in a country with a *per capita* income of US$1,700 per annum. This was the context in which the revolution of 1979 took place.

The future prospects of Iranian agriculture are difficult to predict at this juncture. It is to be hoped that the lessons of past experience will not be ignored, and it will be recognised that meaningful social and economic progress cannot be achieved without fostering an efficient agricultural sector and a thriving rural community.

Appendix Tables

Appendix Table 1: Gross Domestic Product by Sector, 1962-72 (000 million rials, at constant 1959/60 prices)

	1962/3	1963/4	1964/5	1965/6	1966/7	1967/8	1968/9	1969/70	1970/1	1971/2
1. Agriculture	88.8	90.3	92.2	99.5	103.0	111.1	119.6	122.1	128.8	122.3
2. Manufacturing and mining	41.5	45.2	47.2	53.7	62.9	72.5	82.8	90.9	100.9	118.0
3. Construction, water and power	16.3	20.5	22.0	27.9	29.1	33.8	37.0	39.9	43.5	50.4
4. Services	119.8	126.0	140.9	159.5	172.6	187.0	212.5	236.4	269.2	317.9
5. Oil	67.3	73.9	83.6	93.9	108.5	127.4	145.4	167.8	195.0	221.9
6. Total GDP at factor cost	333.7	355.9	385.9	434.5	476.1	531.8	597.3	657.1	737.4	830.5

Sources: Based on Bank Markazi Iran, *National Income of Iran, 1959-72*, and Bank Markazi Iran, *Annual Report*, various dates.

Appendix Table 2: Gross Domestic Product by Sector, 1972-78 (000 million rials, at constant 1974/5 prices)

	1972/3	1973/4	1974/5	1975/6	1976/7	1977/8
1. Agriculture	271.0	286.5	303.3	324.0	341.7	339.0
2. Manufacturing and mining	224.8	264.4	312.9	360.9	423.0	468.2
3. Construction, water and power	108.6	123.3	123.9	171.6	207.4	216.1
4. Services	629.4	749.6	889.1	1 029.1	1 173.2	1 281.3
5. Oil[a]	449.2	863.7	1 441.6	1 224.9	1 481.6	1 422.0
6. Total GDP at factor cost[a]	1 683.0	2 287.5	3 070.8	3 110.5	3 626.9	3 726.6

Note: a. Figures include 'compensation for the valuation of terms of trade'.

Source: Based on Bank Markazi Iran, *Annual Report*, 1977/8.

Appendix Table 3: Quantities of Output, Selected Agricultural Products 1962-72

	1962/3	1963/4	1964/5	1965/6	1966/7	1967/8	1968/9	1969/70	1970/1	1971/2
1. Wheat										
(a) Quantity of output (000 tons)	2 700	3 000	2 600	3 000	3 190	3 800	4 400	4 100	4 260	3 700
(b) Output per hectare (kg.)	720	750	700	750	760	1 050	900	820	710	700
2. Barley										
(a) Quantity of output (000 tons)	765	740	718	935	1 080	1 035	1 160	1 140	1 083	900
(b) Output per hectare (kg.)	830	750	830	830	1 360	830	950	640	650	721
3. Rice										
(a) Quantity of output (000 tons)	700	860	800	845	873	960	980	1 020	1 060	1 050
(b) Output per hectare (kg.)	2 000	2 457	2 285	2 414	2 500	2 549	2 613	2 684	2 753	2 727
4. Red meat										
Quantity of output (000 tons)	243.2	242.9	247.2	253.3	261.0	265.5	270.7	277.5	286.0	303.0
5. Milk										
Quantity of output (000 tons)	1 699	1 632	1 600	1 600	1 603	1 732	1 806	1 846	1 800	1 900

Sources: Based on the Ministry of Agriculture, *Agricultural Statistics of Iran*, Bank Markazi Iran, *Annual Report*, and FAO, *Production Yearbook*. various dates.

Appendix Table 4: The Value of Selected Agricultural Imports and Outputs (1971-5)

	1971/2	1972/3	1973/4	1974/5
1. *Import value* (000 rials per ton)				
(a) Wheat	5 800	6 200	6 100	14 600
(b) Barley	3 800	4 900	11 700	12 400
(c) Rice	14 600	17 000	18 100	46 600
(d) Meat	57 900	62 400	92 300	112 600
2. *Output value* (based on import value) (million rials)				
(a) Wheat	21 460	28 185	28 060	68 620
(b) Barley	3 420	4 944	10 799	10 701
(c) Rice	15 330	20 400	24 145	61 816
(d) Meat	22 002	27 269	41 535	52 697

Source: Based on Appendix Table 3 and its sources.

Notes

1. The term 'aridisolatic' is made up of the words 'aridity' and 'isolation', and the text below will show its relevance to the historical development of Iranian society.

2. 'Politiconomy' is the etymologically consistent term for which 'political economy' is in use. 'Politiconomic' is the adjective of politiconomy. See, further, Katouzian (1980).

3. Purely linguistically, the *Mīr* means 'the world' (in Russian), whereas the *Buneh* only refers to a person's base as well as his roots. Apart from that, each Russian village had a unique *Mīr*, whereas Persian villages usually formed themselves into more than one *Buneh*, in consequence of which the composition of the *Buneh* varied over time (a household being able to leave one and join another).

4. Cf. Wittfogel (1957).

5. The above model of the aridisolatic society is closer to Marx's own concept of the Asiatic mode of production than its later developments, but in any case it is intended as a general model of the basic features of Iranian history and not a universal model which would be necessarily applicable to all Asiatic (and perhaps some other) societies. It differs from Wittfogel's well-known model of oriental despotism both in this regard and – more fundamentally – in that it does *not* explain the emergence of the functional state in terms of the need for a centrally organised supply (or control) of irrigation water. See, further, Katouzian (1981), Chs. 2, 4 and 15, and 'The aridisolatic society: A Model of Long-term Social and Economic Development in Iran', mimeo, University of Kent, Canterbury.

6. See, further, Katouzian (1981), Chs. 3-11.

7. Ibid., Ch. 11. For more details about the genesis and application of the reform see, further, Lambton (1970), Keddie (1972), Katouzian (1974, 1981).

8. See Katouzian (1974) for details and references.

9. See, further, Lambton (1970) and Katouzian (1974).

10. It should be pointed out (a) that about one-third of the land had already been owned by *Nasaq*-holders before the reform, and (b) that the mere number of villages distributed through the reform does not reflect the size of total agricultural land received by the beneficiaries. By 1972, over four-fifths of the agricultural land was owned, one way or another, by the peasant sector. See, further, section V below.

11. For a detailed discussion of this point, see Katouzian (1979).

12. See, further, Katouzian (1981), Chs. 11 and 15.

13. For more details, see Katouzian (1978, 1981) and Etemad Moghadam (1978).

14. See, for example, J. A. C. Brown (1977) and Price (1975).

15. In other words, at *current* prices, the rate of growth of the value of cash-crop products and, hence, the income of their producers would be much higher than is reflected by changes in output quantities alone. See further below.

16. The rest was allowed to be reinvested by the farms.

17. See Etemad Moghadam (1978), Ch. 2, pp. 60-3.

18. Ibid., pp. 78-82.

19. Ibid., Chs. II-VI.

20. For some evidence, see Farhad Mehran, 'Distribution of Benefits from Public Consumption Expenditures among Households in Iran', WEP Working Paper, WEP 2-23/WP57 (Geneva, 1977).

21. In a study of income distribution and inequality in Iran, Mehran observes

that 'while the average consumption per head in the urban area was about twice that of the rural area in 1959/60, it has grown to more than three times by 1971/72'. See Farhad Mehran, 'Income Distribution in Iran: The Statistics of Inequality', WEP Working Paper, WEP 2-23/WP 30 (Geneva, 1975).

22. See section III above.

23. See Azimi (1979), Ch. 8, p. 335.

24. Mehran's study, for example, showed that more than one-third of the rural households had a negative saving rate. See Mehran, 'Distribution of Benefits'.

25. The 9 per cent rate of unemployment is implicit in the Central Bank labour force statistics (*Annual Reports*, 1977/8).

26. See Irvin (1975), and this author's review of his book in *The International Journal of Middle-Eastern Studies* (1978).

References

Azimi, H. (1979) 'Aspects of Poverty and Income Distribution in Iran'. unpublished DPhil thesis, University of Oxford

Brown, J. A. C. (1977) 'Notes on Wheat and Red Meat Reports Prepared by Plan and Budget Organisation', Iran Planning Institute

Etemad Moghadam, F. (1978) 'The Effect of Farm Size and Management System on Agricultural Production', unpublished D Phil thesis, University of Oxford

Irvin, G. (1975) *Roads and Redistributions, Social Costs and Benefits of Labour-intensive Road Construction in Iran*, International Labour Office, Geneva

Katouzian, M. A. H. (1974) 'Land Reform in Iran: A Case Study in the Political Economy of Social Engineering', *Journal of Peasant Studies* (Jan.)

—— (1978) 'Oil versus Agriculture: A Case of Dual Resource Depletion in Iran', *Journal of Peasant Studies* (Apr.)

—— (1979) 'A Model of the Political Economy of Oil-exporting Countries', *Peuples Mediterranéen* (Nov.)

—— (1980) *Ideology and Method in Economics*, Macmillan, London

—— (1981) *The Political Economy of Modern Iran*, Macmillan, London, and New York University Press, New York

Keddie, N. (1972) 'Stratification, Social Control and Capitalism in Iranian Villages', in R. Antoun and I. Havik (eds), *Rural Politics and Social Change in the Middle East*, Indiana University Press, Bloomington, 1972, pp. 364-401

Lambton, A. K. S. (1954) *Landlord and Peasant in Persia*, Oxford University Press, Oxford

—— (1970) *The Persian Land Reform*, Oxford University Press, Oxford

Mehran, F. (1975) 'Taxes and Incomes: Distribution of Tax Burdens in Iran', World Employment Programme Working Paper, restricted

—— (1977) 'Distribution of Benefits from Public Consumption Expenditures among Households in Iran', World Employment Programme Working Paper, restricted

Nowshirvani, V. F. (1977) 'Production and Use of Agricultural Machinery and Implements in Iran', International Labour Office, Geneva, mimeo.

Price, O. T. W. (1975) *Towards a Comprehensive Iranian Agricultural Policy*, IBRD, Tehran, mimeo.

Rabbani, M. (1971) 'A Cost-benefit Analysis of the Dez Multi-purpose Project', *Tahqiqat-i Eqtesadi, VIII*, pp. 132-65

Safi-Nejad, J. (1974) *Buneh*, Tehran (in Persian)

Wittfogel, K. (1957) *Oriental Despotism*, Yale University Press, New Haven, Conn.

NOTES ON CONTRIBUTORS

Alula Abata is at the Institute of Development Research, Addis Ababa University, Ethiopia.

Leonardo Castillo is a Lecturer in Spanish and Latin American History at Cambridgeshire College of Arts and Technology, England.

Ajit Kumar Ghose is a Research Economist at the Rural Policies Branch, World Employment Programme, ILO, Geneva, Switzerland.

Homa Katouzian is in the Department of Economics, University of Kent, Canterbury, England.

Cristóbal Kay is in the Department of Economics, University of Glasgow, Scotland.

Fassil G. Kiros is at the Institute of Development Research, Addis Ababa University, Ethiopia.

David Lehmann is in the Faculty of Politics and Economics, University of Cambridge, England.

Peter Peek is a Senior Research Economist in the Rural Employment Policies Branch of the World Employment Programme, ILO, Geneva, Switzerland.

K.N. Raj is at the Centre for Development Studies, Trivandrum, India.

M. Tharakan is at the Centre for Development Studies, Trivandrum, India.

INDEX

359

362 *Index*